Concert Lighting

Techniques, Art and Business

THIRD EDITION

Dr. James L. Moody, Ed.D.

With contributing writer

Paul Dexter

ELSEVIER

AMSTERDAM • BOSTON • HEIDELBERG • LONDON • NEW YORK • OXFORD
PARIS • SAN DIEGO • SAN FRANCISCO • SINGAPORE • SYDNEY • TOKYO
Focal Press is an imprint of Elsevier

Focal Press

Focal Press is an imprint of Elsevier
30 Corporate Drive, Suite 400, Burlington, MA 01803, USA
Linacre House, Jordan Hill, Oxford OX2 8DP, UK

Library of Congress Cataloging-in-Publication Data
Application submitted

British Library Cataloguing-in-Publication Data
A catalogue record for this book is available from the British Library.

ISBN: 978-0-240-80689-1

For information on all Focal Press publications
visit our website at www.elsevierdirect.com

10 11 12 13   5 4 3 2 1

Printed in China

Working together to grow
libraries in developing countries

www.elsevier.com | www.bookaid.org | www.sabre.org

ELSEVIER   BOOK AID
International   Sabre Foundation

Dedication

I dedicate this edition to my students at the Theatre Academy at Los Angeles City College and all the people at colleges and seminars who have sat through my rambling and musing. To Paul Dexter, who gave such valuable suggestions, new material, and dedication to the continuation of this series. Finally, as before, to Trudie, my life partner, who puts up with all my time away. All my love.

The author would also like to thank the following people for their generous time and support in the preparation of this book: Jeff Ravitz, Richard Pilbrow, Andi Watson, Ronnie James Dio, Gil Moore, Kevin Cronin, Chip Monck, Joe Tawil, Bob Gordon, Susan Spolter-Vine, Wendy Dio, Steve Summers, Suzanne Dexter, Kenjie Ohashi, Toshie Oki, Peter Johanson, Lass Jendermark, Tobbe Berg, Ossian Ekman, Calle Brattberg, Mike Wood, Bruce Jordahl, Richard Bellview, Lewis Lee, Nick Archdale, Robin Alvis, Mark Hunt, Marty Wickman, Thomas Thompson, Nick Jackson, Larry McNeny, Kelly Sticksel, Gary Carnes, Jake Berry, and especially Paul Dexter.

Jim Moody
Studio City, CA

Contents

Foreword to the Third Edition

I figure that James and I may be only 10 years apart in age. We haven't seen each other in, I'd imagine, 35 years. I seem to remember we were a bit cautious of each other. He had his stable of acts, as did I, back then, and the only act we both worked on was Procol Harum—I, once, at the Hollywood Bowl, and he on tour.

I was flattered when he asked me to write a foreword for the third edition and asked for a *curriculum vitae/bio/résumé* so I could catch up. I received 51 pages from James—this was not the book, this was only his CV. My immediate reaction was to cancel all my plans to write … anything.

OK, reality. Here's a scholar who has received numerous accolades, a vet, a sponsor of numerous awards, a member of the Coast Guard, and a real professional at any type of performance, tour, one-offs, video, majors such as the Super Bowl *et al.*

I do, in fact, remember seeing a Dolly Parton presentation, by James at the Universal Amphitheater way back when. His palette was soft and understated, and Dolly looked beautiful. A lighting designer has but four brushes: color, angle, intensity, and, most critical of all, timing.

James had captured all four. I never went for beauty as he did. If you choose a harsh/cutting tune, then you are not meant to be beautiful. The lighting designer is then left with the fifth and sixth arts: camera and direction, which James has also well mastered.

I extend my congratulations to James for his accomplishments. Ladies and gentlemen, you now have the tools … apply them.

Fondly,
E.H.B. "Chip" Monck
Victoria, Australia

Foreword to the Second Edition

With the virtual explosion of new lighting technology, which brings innovation and change to this field almost weekly, this second edition of *Concert Lighting* by Jim Moody comes at a perfect time. Concert tours may be the fastest growing facet of the lighting industry, both in the United States and especially abroad, and Jim knows the field as well as anyone.

This book builds on the foundation that the first edition established 8 years ago and covers much more than concert and tour lighting. As I travel, I hear often of the value of the first edition, so I am pleased to tell these colleagues and friends that *the* expert has updated and expanded the "bible." Whether you're into bus and truck, part of a permanent staff in a tour venue or theatre, or a television designer, this is the guide you need.

As the late Lee Watson intimated in his foreword to the first edition, Jim Moody merges art and craft in this book. He shows you with practical examples and illustrations how to do it, but he also helps you get inside the designer's head to see *why* designers make decisions on placement, color, and movement. He builds on the history of the art form to help you understand the evolution of lighting control and equipment. This benefits you as you confront older equipment on tour and provides you with a sense of appreciation of new technology.

I have favorite sections, especially those relating to designer's perspectives on lighting and international touring. Given the cross-over work of lighting designers, Jim has wisely included sections on television, live theatre, and film, and we gain from his wealth of experience in each. I had the pleasure of collaborating with Jim a few years ago and can hear his "voice" in this book. He writes as he speaks—a practical, sensible approach to the artistry of the business. Jim is a great "teacher" and this book is a great tool.

Dick Durst
Dean of Fine and Performing Arts, University of
Nebraska–Lincoln
President, International Association of Designers,
Theater Architects, and Technicians (OISTAT)
April 1997

Foreword to the First Edition

Jim Moody is the ideal person to write about concert business and lighting. Not only is he one of the entrepreneurs in this new lighting industry, but he also writes with great clarity, organization, and readable simplicity

His specialized knowledge of lighting for the concert field embraces both the technical and the artistic, as well as pragmatic organizational know-how. With a solid background in theatre lighting, degrees from Southern Illinois University and UCLA, he moved into concert lighting. He has been a leading advocate for fusion of the innovations and techniques of concert lighting into other production forms.

He was honored in 1980 with the first Lighting Designer of the Year Award presented by *Performance* magazine, the leading publication on concert touring. The award was based on a readers' poll of concert industry contemporaries. Jim has also received awards in television as well as theatre.

Those who are specialists in other lighting design areas—drama, musicals, opera, dance, television, amusement parks, landscape, and architectural lighting— and educators in these fields will profit from reading this volume carefully. These days, almost all working lighting designers cross over into all areas of design use in light.

Jim Moody is in a unique position to contribute to our knowledge by sharing his experience and insight into concert lighting. We are fortunate that he has done so in this volume.

Lee Watson
Purdue University
West Lafayette, Indiana

Preface: Are We Legit Yet?

46 years! Amazing. Concert lighting has gone from being the new kid on the block to mature citizen. Walk around any lighting conference and the question as to whether or not this medium might survive is answered by row upon row of manufacturers from around the world showing things that didn't exist even a year ago, mostly spurred on by concert touring money. However, sales these days are also to theatres, theme parks, cruise ships, display and architectural clients ... the list goes on to include anyone who uses lighting for entertainment.

While sales of the second edition of this book went on and on way past the popularity of most books, the need for an update was long overdue. The reader will find several new and many expanded chapters. Most of the new chapters are on topics that didn't even exist when the last edition was printed: LEDs, digital lights, media servers, and piles of software. One of the sections I still enjoy the most is getting leading lighting designers to talk about a project and their take on their work and the business. This time around, a chapter where artists discuss the interaction between designer and artist expands our view of how the whole production comes together. So much has transpired in the past 13 years that every chapter has been expanded to fill in the gaps.

Two editorial issues need to be addressed. I have always been conflicted when writing the phrase as "rock & roll" over any of the following: Rock & roll, rock 'n' roll, rock and roll, and other variations of the phrase. As I compared the second edition to my current writing I found I was skipping back and forth, so I decided to check with the two authorities: *Rolling Stone* magazine, which uses *rock & roll*, and the Rock & Roll Hall of Fame. Both use the ampersand, so there you have it. I have adopted the usage of rock & roll from here on out.

The second issue has to do with what we call a *theatrical lighting fixture*. Starting with Vari*Lite years ago and culminating with the recently released Essential Skills for Entertainment Technicians (ESET) Body of Knowledge, as funded by the Entertainment Services and Technology Association (ESTA), the term *luminaire* is now the accepted term for "light sources contained in a housing" and is not just a term to refer to moving luminaires. Therefore, this edition will try, except inside quotes, to change to this 21st-century definition.

One other point I would like to make. This is not a textbook as much as it is a history lesson and a discussion by experts and practitioners of their craft who tell stories of their experiences. Sure, you will also find explanations of the equipment and high-tech gadgets that the industry has grown to love. I am a big believer that we learn more from history and listening to people who have survived the grueling road life to tell the reader what was tough, hard work but amazingly fun.

Finally, we are sadly at the juncture where we are saying goodbye to some of the early pioneers who believed in and stuck with this medium. Some have died way too young and without reaching their potential. Others have reached well beyond normal retirement age and are still going strong, although some have decided to slow down, if just a little. To the next generation, my own son included, I say we have only opened the gates ... you make it flood.

I

BACKGROUND AND ORGANIZATION

THE BIRTH OF ROCK AND THE RISE OF THE CONCERT LIGHTING FIELD

I t is difficult to pinpoint the actual beginning of concert lighting as we think of it today. Certainly the *Grand Tour* could be seen as having been the byproduct of opera in the mid-nineteenth century. The term was often given to a star's travels through Europe, presenting solo programs in the European cultural capitals. Later, the Grand Tour came to the Americas. Through the years it also came to include the popular figures of show business, encompassing not only opera but also the stars of dance halls, vaudeville, and the circus. In the late nineteenth century, despite their isolated locations, even small Nevada gold rush towns had opera houses to show the world how "cultured" they had become.

The swing bands of the 1920s and 1930s brought a big change to popular music and, some believe, sounded the first notes that would ultimately be recognized as *rock & roll*. Led by such greats as Duke Ellington, Count Basie, and Paul Whiteman, these bands emphasized instrumental solos—riffing, or playing a short phrase over and over, now considered a key ingredient of rock & roll.

Another milestone was the entrance of the *pop idol*. Although Benny Goodman is widely credited for igniting the first "teen hysteria" in 1938 at a Carnegie Hall concert, it would later be a teenager from Hoboken, New Jersey, Francis Albert Sinatra, who would endure a legion of young teenage girls screaming during his performances.

Enter the baby boom of the 1940s. Postwar American prosperity saw many cultural changes. An avalanche of consumer products became available to the average family, and the money to buy these products was also available in the boom of the postwar years. There was also a change in who had purchasing power. Before World War II, the head of the household made the purchase decisions for the family, but by the early 1950s manufacturers were witnessing the growing financial power of the teen market. Teenagers now had allowances, and it was estimated in a 1951 survey that they had $4.5 billion dollars to spend annually. It was estimated that $45 million of that was spent on 45 rpm single records. In the early 1960s, a large number of these war babies, who were 16 to 19 years of age, married, and their buying power increased even more. Producers and manufacturers were eager to figure out what this emerging class wanted to purchase.

Teens' listening tastes were having a decided impact on the music business. Disc jockeys could play a tremendous part in record sales, but it still was unclear what teens wanted. Stations relegated blues and country music to times when few adults listened. These listeners became known as the "late people." To be a disc jockey, you simply needed a sponsor, no experience required. One of these original late-night programs was called *King Biscuit Time* and aired on KFFA in Helena, Arkansas, and a disc jockey calling himself Howlin' Wolf had a show on KWEM in West Memphis, Arkansas. Both played blues and some country artists. Because this kind of music was not mainstream, the shows aired between 5 a.m. and 6 a.m.[1]

---

[1]Ed Ward, Geoffrey Stokes, and Ken Tucker, *Rock of Ages: The Rolling Stone History of Rock & Roll* (New York: Rolling Stone Press, 1986), pp. 68–70.

It may have been more than luck that these programs aired on what were called 50,000-watt "clear-channel" stations. With that kind of power and the phenomenon of AM signal skip due to the ozone, on a clear night, when the conditions in the ionosphere were right, you could hear stations that had "skipped" thousands of miles. Teens in Arizona might hear their first Mississippi Delta Blues from a station in Tennessee, or a group of teens in Wisconsin could tune in to hear young country singers like Carl Perkins or Buddy Holly.

Blacks owned no radio stations, white disc jockeys opened the doors to what was almost a secret society of black music and culture. Admittedly, it was a forced secret because of racism and segregation, but it was a distinct musical style that appealed to the white teens with money to spend.

In Los Angeles, one of the most popular programs was called *Huntin' with Hunter* on KGFJ featuring Hunter Hancock. His style was to mix blues, jazz, and spirituals with rhythm and blues. He was good friends with Johnny Otis, who owned a club called the Barrel House in a black section of Los Angeles called Watts. Teens began to flock to the club, white as well as black. Hancock's machine-gun delivery, growling, use of hip slang, and general carrying on like a madman caught on with white teens. Before long Hancock's program was transcribed to stations across the country. Hancock emphasized that he was bringing listeners the latest and greatest Negro performers. Although everyone assumed he was black, Hancock kept out of public view because he was white.

Wolfman Jack later adopted Hancock's style and for many years broadcast from a 50,000-watt station in Mexico without being seen in public. He had a tremendous influence on teens' taste in music in the 1960s. In his early days, many people also thought Wolfman Jack was black. He created a mystique that was widely held until he portrayed himself in the classic George Lucas film, *American Graffiti*.

The melding of country music and such regional sounds as rhythm and blues had been building until in 1951 a disc jockey named Alan Freed started the *Moondog Show* on WJW radio in Cleveland, Ohio. The name came from a tune by Todd Rhodes called "Blues for Moondog" that contained a wailing saxophone solo that Freed adopted as his theme song.

He'd leave the microphone open and howl like a coyote. It was demented and near anarchy, but it was what teens had been waiting to hear. In 1954, the name was changed to *The Rock and Roll Show*. *Rolling Stone* magazine wrote that the term was perfect because: "It was a way of distinguishing the new rhythm and blues from just plain blues and the old corny Mills Brothers style. After all, rock & roll didn't fit into any of the old categories."[2]

The benchmark of modern concert touring was set in the mid-1950s by the independent record companies in an effort to exploit the fledgling rock & roll recording artists. The tours were not a very radical departure from what swing bands and orchestras had been doing in the 1930s, 1940s, and 1950s—that is, playing dances in every town that had a community hall or theatre. After all, this was the way most musicians made their living—playing live dances. But now the pop singer was not just a member of the band; the singer, rather than the bandleader, came to front the band. Another change was that, rather than getting work through a booking agent separate from their manager, pop singers were promoted by an independent record producer, who also controlled the record, or sometimes an independent entrepreneur like Alan Freed. That way the record company made money not only from ticket sales but also, more important, by stimulating record sales. Many artists signed away their publishing rights for a small, one-time fee or were lied to by the record companies concerning a record's earnings.

CONCERT LIGHTING BEGINS IN THE UNITED STATES

Lighting did not attain a prominent position until after sound reinforcement made its first inroads in about 1960. The inadequate sound system in most buildings could not handle the demands of the recording artists, who had come to expect studio-quality sound (not to mention the new electronic

[2]Ed Ward, Geoffrey Stokes, and Ken Tucker, *Rock of Ages, the Rolling Stone History of Rock & Roll* (New York: Rolling Stone Press, 1986), pp. 96–97.

effects necessary to make their performances sound like the record). After the artists got used to absorbing the expense of carrying sound equipment from city to city, lighting soon followed.

One of the first artists to carry their own lighting equipment was Harry Belafonte in the mid-1960s. He had emerged on the record scene in 1957 from his native Jamaica and was truly ahead of his time. Chip Monck got his start with Harry. Generally, the middle-of-the-road (MOR) and country/western artists were the last to see the value of building a production around their music; however, folk acts such as the Chad Mitchell Trio, the Kingston Trio, and Peter, Paul, and Mary took notice of the added value that special lighting gave to the show and started hiring companies like McManus Enterprises in Philadelphia to provide lighting for their college dates.

THE SAN FRANCISCO LIGHT SHOW

What became known as 1960s *acid rock* was spearheaded by such bands as Big Brother and the Holding Company, Jefferson Airplane, Warlock, Grace Slick and the Great Society, and Quicksilver Messenger Service. All were based in San Francisco. Actually it was a nonmusical group, The Family Dog, at the close of 1965, who unwittingly created the first *light show*. Bill Graham, who was their manager, said that people would show up and ask if they could hang sheets on the walls, and he'd ask, "What are you talking about?" They would reply, "My screens; I'm a liquid projectionist." Light shows were not planned, and they were not even paid for. They were just part of what came together spontaneously. "Happenings" could include films, dance, music, mime, painting, and just about anything else people wanted to do, all going on at once.

Bill Graham was brought in to produce the now famous Trips Festival at the Longshoremen's Hall in January of 1966. Later that year, he rented the Fillmore Theatre at Fillmore and Geary from Charles Sullivan, who was black, to put on the second Mime Troupe benefit, but this production was billed as a "dance concert." After that, Graham could see live rock & roll music as being the main attraction. He split from the Mime Troupe and started promoting musical groups on his own. Graham arranged concerts that featured individual bands as the main attraction without ancillary features. Films, however, were shown during the set changes to keep audiences occupied.

When he started formally promoting concerts, Bill Graham continued to welcome the light shows and eventually started paying for in-house light shows that he could control. They were a visual explosion of color and design. The shows were based on liquid light projections, strobe lights, blacklights, and effect lighting to create a visual mood into which the band as well as audience was immersed. The liquid light projector was nothing fancier than the opaque projector your grade school teacher used to show photos and charts from books. For these light shows, though, the book was replaced by a pan holding oil or water into which paints were pushed, splashed, and injected. The pan was vibrated or tilted to add even more movement to the ever-changing patterns that this mixture created. These images could be projected onto dancers, walls, screens, the audience, even the performers.

Graham was not interested specifically in advancing lighting or film or anything in particular, but he was a bulldog when it came to his beliefs about how the audience should be treated. His background and training in New York theatre gave Graham a belief that with music alone he could create an art form. He wanted the audience to have a great experience, and he felt as though he was the only person looking out for the audience. The bands often didn't care if there even was an audience. If a good experience for the audience meant better lighting and sound, he encouraged and supported it. Graham provided the opportunity for many early lighting designers to push their limits and experiment. He was known for treating the people around him as family. That included yelling at them when he felt it was necessary. Graham would conduct regular meetings before a show for everyone involved, including the ushers.

Lighting was not important in those early concerts because people came more to "make the scene" than to listen to any one band. But, as individual band recognition grew and people came to see specific bands play, Graham encouraged expansion of production values as part of providing a better experience for the audience. Lighting had moved beyond the mask of the liquid light show.

OTHER CITIES AND VENUES

This is not to say that concert lighting did not exist outside of San Francisco. There were other venues at about the same time that received national attention. The Electric Factory in Philadelphia, circa 1967, was one. A very young Bill McManus fell into the role of local lighting guru by accident. A receptionist mistakenly sent a call meant for McManus's boss, MacAvoy, who was often called Mac, to a 19-year-old working in the shop of a theatrical lighting rental company. Bill "Mac" McManus saw an opportunity and met with a man who said he wanted to open "one of them psychedelic things." He had an old factory on 22nd and Arch and asked if Mac could look at it.

Before the Electric Factory opened, the largest live concert venue in town was the Latin Casino, which held about 3500 persons. The Electric Factory was to hold 5000, an unheard of size for its day. Field houses at universities held 5000 to 6000, and only a handful of artists dared perform in them. The Electric Factory was so successful that it was turning over the house (audience) two and three times a night.

Because of its experiences at the Fillmore, the Grateful Dead was one of the early bands to encourage light shows and film projection at its other gigs. A "family" quickly developed around the band. "People just started doing things, we just played and things happened. If it felt good we'd say do it next time," said Jerry Garcia in Graham's autobiography, *Bill Graham Presents*. Josh White noted, "Kip Cohen always said that the reason the light show worked so well was that musicians didn't realize that people had eyes as well as ears ... Jefferson Airplane would do twenty minute songs in the darkness in their street clothes with their backs to the audience."[3]

FILLMORE EAST

Bob See recalls that while he was a student at New York University in the School of the Arts theatre program, he just happened to walk through an open door with a group of friends into an abandoned movie house near campus. There were Chip Monck and some of the people with Joshua Light Shoe on the stage trying to put things together:

> We just kind of got involved. Because this was a real event, a real happening. From the standpoint that they were trying things that no one had ever tried before. It was the era of the '60s. So we started working there, Chris Langheart, Bob Gaddard, and John Chester. And as the place evolved we sort of took on jobs. I became the technical director and did lighting stuff with Chip Monck (Bob See, personal interview, 1996).

The Fillmore East opened in 1968 in the old Commodore Theatre. It was at the last minute that the venue took that name. It was to have been called the Village Theatre, but because of a legal threat the name was changed to the Fillmore East after the handbills had been printed. The first show was on March 8, 1968, featuring Big Brother and the Holding Company with Janis Joplin, Tim Buckley, and Albert King, recalls Josh White. Across the street at the Anderson Theatre, Gladys Knight performed using Pablo's Lights.[4]

FLEDGLING LIGHTING IN ENGLAND

While all this was happening in the United States, England was experiencing a blues explosion. Bands devoured any record they could get of Muddy Waters, Lightning Hopkins, B.B. King, and many others. There was only a club scene, no large venues like those starting to appear in the United States. The Marque Club in London and the Red Car Jazz Club, which was really a rock & roll club, were the places to play. The Round House, another club,

---

[3]Bill Graham and Robert Greenfield, *Bill Graham Presents* (New York: Doubleday, 1992), p. 259.

[4]Bill Graham and Robert Greenfield, *Bill Graham Presents* (New York: Doubleday, 1992), p. 232.

catered to visiting American bands like The Doors during the early 1960s. Ian Peacock said, "I remember it was more of a place to hang out than to hear the bands" (personal interview, 1996). Because these clubs were more sophisticated in their equipment, would-be designers came to see and hear what they were doing.

Most often the British bands found themselves playing in community halls and school assembly rooms where there was no stage lighting. Follow spots were even rarer. The British bands were hearing and sometimes seeing what the Americans were doing in a few spots like the Fillmores, but the British lighting industry did not have the equipment available to meet the needs of this new media. Consequently, many early systems were designed around American lights. It was up to people like Michael Tait to invent what they needed.

Tait was one of the early people to build lighting systems around the specific needs of a band. He started with YES in 1968. Tait had worked at a nightclub and somehow ended up driving the band somewhere. Tait says he just fell into the job. He started by doing anything the band needed. At first that included setting up their small public address sound system. In those days bands used only vocal microphones. No one thought of using microphones on the drums or guitars. Each band had its own sound system, so the crew completely changed that as well as the band gear after each set.

Tait remembers that the second show he worked was on a bill with The Who, Small Faces, Arthur Brown and the Mind Benders, and YES at the bottom of the bill. The show was in Newcastle, England, in the town hall. There might have been some white light above the stage, but that was all. Tim Murch, who was with Light and Sound Design for many years, remembers seeing Black Sabbath in a South Hampton Guild Hall with house lights around 1970.

Things developed slowly in England because the rather small audiences for most bands did not bring in enough money to support the band, let alone buy lighting. A few bands, such as The Who and Pink Floyd, were interested in production, but there still wasn't a lot of money to spend on such things, nor had the equipment been built to spend it on.

Other early British innovators were Jonathan Smeeton and Graham Flemming. Michael Tait remembered, and was echoed by Ian Peacock, that people just showed up and someone said, "OK, who's doing sound, who's doing lights tonight?" But soon, as in America, people gravitated to what they did best. No formal training, just desire and drive. "We were taking light bulbs and putting them in coffee cans," said Tait. He mentioned getting 12 automobile fog lamps, attaching them to a piece of pipe, and connecting them to wire-wound potentiometers to make mini-dimmers. "They worked for a few months, but they kept melting. But they had the desired effect, you got a narrow beam of light across the front of the stage," he said. "In the beginning, I didn't realize you need backlight and sidelight. I had them out front, and it looked flat and horrible. Then one day I couldn't find a place to attach my lights out front and hung them on the side. That's when I realized that side and backlight was what it was all about."

Tait's first go at lighting was at the Marque Club, which had a red light and a blue light above the stage with switches on the wall. YES's music had a staccato effect that lent itself to the flashing of these lights in time with the music. People went wild. But Tait still didn't consider lighting to be his skill: "YES first toured America as the opening act for Jethro Tull. I forgot where the first gig was but somebody said, well who's going to call the follow spots? Well, we never had any, so I was the one messing with lights, so it had to be me." He went on to say, "But I found I could do it. I found that I could cue four lights and that I knew what I wanted to do. It was just something that I just naturally could do." (All quotes are from a personal interview with Michael Tait in 1996.)

MOVING TO LARGER VENUES

Although these early converted night clubs, auditoriums, and factories got concert lighting off the ground, their success was short lived. The Fillmore was open from December 10, 1965, to the end of June 1968. The Fillmore East in the old Carousel Ballroom lasted from July 1968 until September

1971. Winterland in San Francisco became a Bill Graham venue before the Fillmore was closed so he could promote bands that demanded more money. It held 5400 people and had been built as a skating arena. It continued on for 7 more years.

Don Law, a promoter in Boston, was using the old Boston Garden to put on concerts at about the same time as the Electric Factory opened in Philadelphia. Both had about 5000 seats. When the Spectrum (a basketball arena) opened in Philadelphia in 1969, it drew the big acts away from the Electric Factory. All across the country, more sports arenas were being used to attract the artists that demanded 5000 to 7000 seats. All this was significant to concert lighting because these new buildings were not full-time clubs or venues designed for music. Most of them were sports arenas. They were hostile environments with bad acoustics, no permanent stages, no theatrical lighting, no intercoms, and no music sound systems (only public address). Portable lighting and sound companies became necessary. Local and regional lighting and sound companies started to appear, and the large fees they could charge meant they could invest further in their equipment. Banks were not generally willing to approve loans for these ventures so most capital to buy equipment came directly from cash received from the previous gig. Some of the early companies included McManus Enterprises in Philadelphia, Sundance Lighting in Los Angeles, TFA in Boston, and See-Factor in New York. Today, only one of these companies is still on the national scene, See-Factor. TFA was sold and eventually was merged into PRG, and Sundance sold their lighting equipment and became a design and consulting firm which today is known as Visual Terrain, Inc.

THE END OF BILL GRAHAM'S FILLMORES

Bob See recalls Bill Graham calling all the staff together—it was very much a family organization. Graham announced he was closing the Fillmore East even though every show was selling out. He said he just had too many irons in the fire. That was in 1971. The Fillmore East held only 2700 people. Acts were playing Madison Square Garden and making

four times what they could at the Fillmore; greed had set in. It all came down to dollars. Actually, all this didn't just happen by accident. The Monterey Festival took place in 1967 and Woodstock in 1969, and promoters realized that as many as 50,000 people would come to hear rock & roll music. Bill Graham closed the Fillmore East and went back to San Francisco to close the Fillmore. That last week in New York, Bob See remembers doing 28 shows in 27 days. The era of the concert hall as an ongoing venue was dying.

Bill Graham went on to promote and/or act as production manager for several national tours with the Rolling Stones, George Harrison, and Crosby, Stills, Nash, and Young, as well as Bob Dylan and the Band. To all these productions he brought his sense of theatricality, production values, and total dedication to making the event an experience for the public. Graham is quoted as follows in his autobiography:

> The greatest compliment I was ever given was at the Fillmore. ...Two guys looked straight ahead and one said, "Oh, shit, man, I forgot. Who's playing here tonight?" Without batting an eyelash, the other guy said, "I don't know man. What's the difference? It's the Fillmore."[5]

Bill died in a helicopter crash on October 25, 1991, returning from a Huey Lewis and the News show at Concord Pavilion. His efforts to promote great shows and his undaunted belief that the public deserved the best show possible gave encouragement to many who worked directly with him or for his shows. Graham wanted bands to give their audience a total experience. He encouraged people to try new things and opened the door for many of those who started this special form of lighting. Bands wanted bigger and more, more, more. To attract a paying audience to large venues meant you had to give them a show, and the shows had to now go on the road.

MELDING FORMS

Concerts moved from light shows to embracing more traditional music hall lighting simply because certain performers began to emerge from these groups and

[5]Bill Graham and Robert Greenfield, *Bill Graham Presents* (New York: Doubleday, 1992), p. 181.

gain star status. With this came audience recognition. I believe that the artist's ego was responsible for a move toward the more conventional musical comedy lighting techniques. Artists wanted to be in the spotlight, the traditional symbol of being a star.

As with any melding of the old with the new, and especially in light of the youth revolt of the 1960s, record artists started altering the rules. There was no comedian or animal act to separate the musical presentations as had been common in music halls or vaudeville. This was to be a show of solid music, lasting several hours. Lighting began to take on a more important role.

The greatest differences in concert lighting from traditional theatrical lighting are the use of vivid colors, heavy use of backlight, and absolute use of follow spots instead of balcony rail, torms, or front-of-house washes. The greatest advances have been in portable lighting structures, the PAR-64, the moving luminaire, and computerized lighting consoles.

LIGHTING BECOMES BIG BUSINESS

In the past 20 odd years, touring has turned into the *business* of touring. In 1995, Jere Harris formed the Production Resource Group (PRG) from an expanded Broadway scenic construction company called Scenic Technology. PRG rapidly began acquiring touring and rental sound, lighting, and video companies around the country through a global strategy to provide total production services for not only touring but also theatre tours, trade shows, themed entertainment, corporate presentations, and television and system installation. The strategy also included leaving the acquired companies largely intact under their former names. The facilities of these firms were then used to warehouse the growing services of PRG while retaining the good will of their clientele. The company has more than 20 offices and facilities in the United States, Japan, and England as of this writing.[6] PRG was also the first company in the industry to venture into the sale of stock in a private offering.

---

[6]PRG website, http://www.prg.com.

MERGERS AND BUYOUTS

Since the creation of PRG, several companies have tried to duplicate its success. The Obie Company in Torrance, California, merged with Westsun from Winnipeg, Canada, in 1999.[7] In 2008, Ed & Ted's Excellent Lighting joined forces with Q1 Production Technologies. Ed & Ted's is a Los Angeles-based company, while Q1 operates out of Vancouver and Winnipeg.[8] Ed & Ted's merge with Canadian IQ is now called Epic Productions.

Mergers between like entities are not the only form of growth witnessed over the past 20 years. Many of the existing lighting and sound companies added offices either overseas or in optimum rental points in the country such as Orlando and Los Angeles.

Another change in business for the touring concert industry was promoters buying or establishing their own lighting and sound companies. Also, promoters have bought out other promoters; for example, Bill Graham Presents was sold to Clear Channel Communications of Beverly Hills, California, in 2000. Clear Channel owned radio stations but had begun to operate as a promoter and provided complete equipment packages as part of their tour, promotion, and production services. The promotion and production departments were eventually spun off in 2005 into a separate company named Live Nation. It was stated that the company wanted to concentrate on their core business of outdoor billboards and radio station ownership. The split seems to have worked out for both entities. In 2005, Live Nation promoted or produced 28,500 events, including music concerts and theatrical shows, with total attendance exceeding 61,000,000. By September of 2005 they owned or operated 117 venues (75 in the United States and 42 overseas) and held the rights to book in 33 more venues. These figures were the total of all their enterprises, not just for concerts. Premier Global in Atlanta, Georgia, is another corporation adopting an all-inclusive approach. These companies differ in business style from the promotions in the 1980s of Bill Graham and Concerts West, neither of which insisted on providing production services.

---

[7]"Westsun Merges with the Obie Company," *Live Design*, April 1, 1999, http://livedesignonline.com/mag/lighting.
[8]"Q1 and Ed & Ted's Announce Merger," *Projection Lights & Staging News*, October 2008, pp. 1 and 8.

What this new twist means to lighting designers and independent rental companies in particular is that control has been taken away from them. A lighting designer or band may have a long history with a particular supplier and their crews, but this relationship is being challenged. Generally, it is part of contracts today that the promoter, not the artist or their designers, will contract for the lighting, sound, and video services. Although many companies can provide excellent equipment and crews, the loyalty factor is lost. The next tour by a band may have a different supplier and crew, and each tour has to start over with a new learning curve.

To the artist's management, such a deal seems like a win/win situation because the artist is paid a fee and bonus for a block of time, and the management is relieved of all responsibility of contracting for personnel or services. It is up to the promoter to book the dates, promote the show, provide the lights and sound, and basically ensure that everything is in place so the artist can simply walk in and perform. The tour promotion company can make block deals with larger suppliers for multiple tours and can negotiate reduced rates for the larger number of services. These deals may prove to be detrimental to the smaller suppliers. Some have already indicated that they have shifted their focus to corporate, themed, and special events and more local and regional accounts because they are losing the national tours.

One other area of mergers should be discussed. Manufacturers of lighting equipment have also seen a large number of mergers and acquisitions over the past 10 years. The largest to date was Philips Electronics buying ColorKinetics for a reported $791 million dollars.[9] ColorKinetics was an early leader in light-emitting diode (LED) lighting for entertainment. Philips apparently decided it was less expensive to buy the company than to spend the initial R&D money to ramp up. What the acquisition meant for ColorKinetics was the backing of a large company ($346 billion dollars in sales in 2006[10]) with the deep financial resources to allow them to continue to grow. Philips also acquired Genlyte, a company that had purchased the manufacturing and sales divisions

of Vari*Lite, Inc., in 2002.[11] Philips subsequently acquired Strand Lighting in 2006 as part of the Genlyte acquisition. The Strand name goes back to 1914 when it was opened to assist London theatrical productions. Years later Strand purchased the American company Century and became known as Century-Strand, Inc. Later the name was shortened to Strand Lighting, Inc. Vari*Lite was the first company to market computerized moving luminaires in the touring field (see Chapter 11). The interesting part about the acquisition of Genlyte is that Vari*Lite and Strand Lighting are minor components of Genlyte. Mike Wood has worked for several lighting firms including a 9-year stint as VP of Engineering and Chief Technology Officer for High End Systems, Inc. He also holds several lighting industry patents. Wood said their part of the sale probably flew below the radar but that Vari*Lite will look great at the Philips' stockholders meeting! Also significant was Barco's purchase of High End Systems, Inc., also a moving light manufacturer, for a reported $55 million dollars. In a statement released by the company: "The acquisition provides Barco … with an additional patent portfolio in the digital lighting market, along with increased distribution channels and products offerings worldwide."

General Electric (GE) apparently also believes in the future of LED technology, having recently purchased the remaining 49% interest in GELcore, LLC (Cleveland, Ohio) for $100 million dollars. In a press release, GE said their support for GELcore was their best way of penetrating the $12 billion dollar global lighting market.[12] This area of lighting is moving so quickly (see Chapter 16) that by the time this book is published there will no doubt be even greater innovations available due to these large investments in R&D.

Continued consolidation in the lighting rental and touring market is inevitable due to the economics of mergers and acquisitiovs. To survive, smaller companies will need to continue to diversify into nontraditional entertainment applications; however, this is not necessarily a bad thing. Such expansion will provide more work for young designers who will

[9]Philips website, http://www.Philips/Colorkinetics.com.
[10]Philips website, http://www.philips.com.

[11]"Philips Completes Acquisition of Genlyte," press release, Philips, Eindhoven, the Netherlands, March 2008.
[12]"GE Invests $100M in LED Lighting," *Projection Lights & Staging News*, October 2008, pp. 1 and 18.

have the opportunity to be tested on smaller projects. Consolidation often means that the smaller fish are forced to leave the waters of a shrinking lake to find refuge in a pond. That pond was once the concert touring business. Now is the time to find a new, hidden pond and grow in more protected waters.

CASINOS

Casinos have emerged as a lucrative and plentiful venue option for rock & roll, pop acts, and country bands to perform, particularly within the burgeoning American Indian casino community. These casinos are a relatively new venue for concert tours but have experienced extraordinary growth since 1988, when the rules for operation and regulation of Indian gaming were approved by the tribes and the U.S. government.

The first Indian casino had its problems. The Seminole tribe opened a high-stakes casino outside of Hollywood, Florida, in 1979. Soon after it opened, the state demanded that it be shut down. In their filing, the state of Florida challenged the right of the tribe to operate a gambling establishment contrary to state law. After the case spent 3 years in court, the casino was allowed to open again. It was ruled that the tribes are outside state jurisdiction and considered sovereign entities by the United States, and the gaming operation must not be directly prohibited in that state. This case is what really opened the floodgates for Indian casinos in the United States, and tribes in Canada followed suit. Today, two Hard Rock Hotels and Casinos in Florida are licensed with the Seminole tribe; these facilities include state-of-the-art arenas and theatres and host some of the world's biggest recording artists and performers.

Following the Las Vegas business model for attracting gamblers, Indian casino operations bolster attendance at their casinos and hotels with world-class entertainment. There are over 400 casinos, but only about 60 of those have expanded into resorts large enough to be viable venues for staging and promoting touring entertainment. Nonetheless, the sovereign Indian tribes expanded what was once unique to Nevada and Atlantic City into a 28-state gambling empire.

The casinos are paying handsomely for the bands to play, more than a concert promoter would guarantee. The band's management will, on that promise, add the date to the tour. Not all of the operations have new entertainment complexes. Along with the less desirable venues are some definitely worth a look. Mohegan Sun in Connecticut, for example, has built a 10,000-seat arena and a theatre to accommodate world-class talent which is attached to a luxurious shopping mall. Mohegan Sun in Pocono Downs was Pennsylvania's first casino, and it recently completed a $208 million dollar renovation project to become the area's leading entertainment complex.

The Hard Rock Cafe chain equips their theatres with the latest lighting and sound systems, similar to the House of Blues chain. In contrast to Hard Rock and Harrah's, which are tied in with Indian gaming license agreements, only 2 of the House of Blues' 24 venues comprised of traditional clubs, arenas, and amphitheatres are hotels and casinos (in Las Vegas and Atlantic City).

You can expect that the growth of Indian gaming hotels and casinos and entertainment complexes will continue. The National Indian Gaming Commission reported growth from $200 million in 1988 to $26 billion in 2007. For the most part, the casino venues are getting it right and improving as entertainment operators, but there are some that should post "enter with caution" signs.

PURPOSE BUILT CONCERT AMPHITHEATRES

Amphitheatres are considered seasonal venues, with seasons beginning in May and continuing through September, possibly October. Just over 100 amphitheatres are listed in the world, primarily in the United States. The oldest amphitheatre in America is probably the Hearst Greek Theatre in Berkeley, California, which officially opened on September 24, 1903. It is an 8500-seat outdoor theatre that still hosts rock concerts. The Greek Theatre in Los Angeles was built in 1929. The majority of the amphitheatres that are considered mainstream concert venues on the summer concert touring circuit were constructed from the mid-1980 to the 1990s, with only a handful being built after 2000.

At this writing the lion's share of all amphitheatres is either owned or operated by Live Nation, with 39 amphitheatres. The advantage for this company is that it also produces many of the artist and package tours that travel exclusively under the Live Nation banner and play at all of their venues. Live Nation is the largest producer of live concerts in the world, annually producing over 16,000 concerts for 1500 artists in 57 countries. The company sells over 45 million concert tickets a year. These figures are exclusive of their other enterprises such as theatrical plays and musical tours.

Audiences enjoy seeing performances outdoors, and promoters and venue owners enjoy certain financial advantages due to not incurring the year-round expenses of utilities and high overheads. For these reasons the outdoor amphitheatre will continue to grow and be a venue of choice.

LED REVOLUTION

The light-emitting diode is a semiconductor diode that emits light when an electrical current is applied. The first practical visual-spectrum LED was developed by a scientist, Nick Holonyak, at General Electric in 1962.[13] His discovery was limited to utilizing the color red.

A former graduate student of Holonyak produced brighter red, orange, and yellow LEDs in 1972, and a blue LED was developed by Shuji Hakamura at a company called Nichia Corporation in Japan. As of this writing, no other color has emerged from this technology. The so-called "white" LED is the result of a coating mix of yellow and blue phosphors.[14]

While the early LED products were used for indicator lights because they did not have a high output of light, they had a very long life of 35,000 to 50,000 hours and were very consistent in output over that life. As their brightness improved, more and more commercial uses could be imagined. It did not take long for the commercial lighting industry to see the potential of LEDs. In 1995 LED luminaires began to be shown at LDI, an annual trade show for concert lighting and sound equipment and designers, in the form of stage lighting. A literal explosion of

concert lighting products is available—from moving lights to walls of LEDs that produce graphic material—and there is the expectation for many more uses to arise in the years to come. It is interesting that for each edition of this book the author has asked industry experts what the next big advance might be, and it has always been suggested that a new light source will be what propels the industry. Currently, LEDs are the newest advance, but at least two other potential sources are being investigated. The point is that lighting designers can't look only at what tools are in hand but must constantly keep up with the advances that might be on the drawing board.

EXPANSION BEYOND THE CONCERT FORMAT

The author observed back in the first edition of this book in 1989 that ways in which these new tools could be used to solve and expand lighting applications in other entertainment fields were only a dream at the time. Yet, it did not take long before many people saw the potential for adding fast, efficient, dramatic lighting to all sorts of productions: corporate presentations, theme parks, television, and film, as well as legitimate theatre. In this edition, we are pleased to showcase a recent Broadway show that has truly embraced the concert philosophy of bold color, high intensity, sophisticated computer lighting control, luminaires, programming tools, and dimming innovations.

THE CONCERT LIGHTING DESIGNER/DIRECTOR

The position of a concert lighting designer or director differs from that of his or her theatrical or television counterpart. One of the main differences lies with their responsibilities. The concert lighting designer is often the only design artist associated with the production; only larger concerts can afford to have a separate scenic artist or video director, so the lighting designer is usually consulted for all such visual concepts. Another difference is that only rarely is

---

[13]Nick Holonyak, Jr., 2004 Lemelson–MIT Prize Winner. News Release from the ECE Department, University of Illinois, Urbana–Champaigne, March 20, 2002.
[14]2006 Finland's Millennium Technology Prize awarded to UCSB's Shuji Nakamura.

there an extended rehearsal schedule. Often a lighting designer has one day to rig the lighting, with no thought of a stop-and-go technical rehearsal. New visualization software (see Chapter 21) has emerged to help designers get a handle on managing the growing amount of moving lights and digital media, even without rehearsal time.

Concert lighting designers must have a highly developed musical sense. Although many are not skilled musicians, they have a natural aptitude for musical interpretation. Because of the ever-changing venues and artist needs, most lighting designers go on the road with the shows they design. A few do leave the show in the hands of an assistant or the lighting board operator, but most designers stay with the tour to personally run the console and call their own cues.

This style of lighting design is an art of immediacy. Lighting designers are not artists who have the luxury of putting paint on a canvas or chipping away at a piece of stone and then standing back to think about their next move for an hour, a day, or a week. They must react instantly, often with no time to document what they did, so they must try to remember and write it down later.

Concert lighting design is an intuitive art. Although very intricate cues can be programmed into highly specialized lighting consoles, there is still a sense that what you are doing is *of the moment*. There is no script and no cue sheets to make notations on. Ways of noting lighting cues have been developed but in a much different form than the standard theatrical method. Chapter 6 will detail several methods. Preparation and organizing clear drawings and charts are key. Every day brings new locations and a new set of problems to solve. Adaptability is a must. One of the most important lessons to learn is:

There is no such thing as a bad decision; the only wrong decision is to make no decision.

If you are prepared for all conceivable problems, then you can deal logically and calmly with the everyday stresses. Although many innovations have been tried over the last 50 years of concert lighting, techniques are still evolving. Just when we think that the size of lighting rigs has been pushed to the limit, a new idea is tried that pushes the physical resources of the media and particularly the physical structure of the buildings, as well as our imaginations, further. Just when it looked like no more moving luminaires could be brought to market, digital moving luminaires appeared. Just as the light sources seemed to max out, LEDs invaded the entertainment market. This is a business that never stands still, and you as a designer can never be complacent. The need to stay on top of what is new is a must. Artists always want the newest or the biggest or anything different, good or bad.

I do not believe in an analytical approach—this business is one quarter art, one quarter science, one quarter intuition, and one quarter adaptability. Teaching by doing, experimenting, and learning what others have tried can be of the greatest value. That is why this book is presented in four main parts. Section I is a discussion about what you need to understand about the work: the business and physical sides as well as the creativity it takes to succeed. The authors have a very strong conviction that designers need to be well versed in business to get ahead in the real world. Section II investigates some of the tools currently being used. Section III takes a look at some interesting designs done by great lighting designers working today. They reflect what was done to solve both the creative and business needs of their specific touring projects. These needs are, in our view, inseparable. Also, for the first time we will look at how artists perceive their role. Finally, Section IV, the Postscript, is the author's assessment of where we are going in the future.

2

TOUR PERSONNEL AND UNIONS

Anyone following the business of rock & roll touring has probably become familiar with the term *roadie*. I would like to counter the stereotype that this term has established and discourage its use in the future. The term *roadie* carries a certain degrading connotation, having its roots in the tradition that spawned another infamous rock & roll term—*groupie*. Many road crew members in the early years may have been little more than male groupies (family and friends of band members) who simply wanted to hang around with the band, but the industry and its requirements have changed dramatically in the past decades. Rock & roll as a whole has become more sophisticated and technically more complex. As a result, the persons charged with the care of the band's equipment possess technical capability and communication skills. These skills are absolutely necessary for today's touring groups and the ultimate success of a tour.

Although people like Michael Tait in England and Chip Monck in the United States were able to start very successful careers in the 1960s without theatrical training, it is much less likely today. Road crew members have evolved into trained technicians with specific expertise in electronics, musical instrument repair, lighting, sound, video projection, camera technology, and the allied theatrical arts. The untrained hangers-on of the early years have been replaced by dedicated, trained, touring professionals. Call them *equipment managers*, *technicians*, *graphic artists*, *directors*, or *managers*. The explosion in technology for both sound and computer lighting means that the people on the road today have

received extensive training, usually by the rental equipment company, before they can go on the road. You can't just pick someone up off the street; there isn't a place for "grunt" labor anymore.

But, sadly, even after 45 years, we are in many ways still bridging new frontiers, and there are few standards or formulas in this business. Crew size, wages, titles, responsibilities, and equipment complexity are factors that vary with the nature of the show, the whims of the artists and their management, and the financial limits of the tour. The titles and duties listed herein are general definitions as applied to rock & roll touring and are subject to adaptation.

ROAD CREW DUTIES

At a minimum, a touring artist has a manager, a road manager, one or more truck drivers, band technicians, and lighting and sound technicians. As an artist's earning power increases, the production becomes more elaborate and the technical staffs increase in size. In addition to the aforementioned basic staff, a touring act may also have a production manager, equipment managers, security personnel, pyrotechnician, rigger, audiovisual specialist, set designer, moving luminaire operator, audio engineer, staging company, video director, master electrician, tour accountant, costumer, carpenter, and any of the other standard theatrical titles. As shows become more complex or egos grow, the crew becomes top heavy with "manager" titles. Dan Wohleen, General Manager of the Maricopa County Event Center in Scottsdale, Arizona, and formerly the production manager at the Greek Theatre in Los Angeles, related one tour roster that had the following:

- Tour Director
- Tour Manager
- Road Manager
- Production Manager
- Stage Manager
- Equipment Manager(s)

Dan said that he sees few tours with a specific stage manager. The tour manager has become the title of power. A road manager has little to do with the production side of tours that have extensive lighting and sound rigs but does maintain a very close relationship with the band members looking out for their personal needs and travel. A look at Figure 2.1 confirms how many positions may have to be filled to get a tour on the road. This particular Metallica tour, way back in 1996–97, listed no fewer than 90 people on the road. Unfortunately, the magazine from which this is taken, *Performance*, is no longer in operation, and its replacement, *Pollstar*, does not print crew and staff information for tours.

BACKSTAGE

METALLICA

Touring: June 27-Aug. 4 (Lollapalooza);
Europe: Sept. 6-Oct. 3; U.K.: Oct. 5-16;
Europe: Oct. 18-Nov. 27; U.S. mid-December-summer '97

Artist Management
Q Prime Inc.
729 7th Ave., 14th fl.
New York, NY 10019
Peter Mensch, Cliff Burnstein
Tony DiCioccio, Sue Tropio
Brian Celler

Tour Manager
Tony Smith

Assistant Tour Manager
Patrick Ledwith

Tour Accountant
Kenny Silva

Production Manager
Dan Braun

Stage Manager
Gary Perkins

Production Assistant
Helen Campbell

Booking Agency
QBQ, Fair Warning-Wasted Talent

Agents
Dennis Arfa, Adam Kornfeld
John Jackson

Drums
Lars Ulrich

Vocals/Guitar
James Hetfield

Guitar
Kirk Hammett

Bass
Jason Newsted

Bass Tech/Equipment Manager
Zach Harmon

Guitar Techs
Andy Battye, Justin Crew

Drum Tech
Flemming Larsen

Sound Company
db Sound (U.S.), SSE Hire (U.K.)

House Soundmixer
Big Mick Hughes

Stage Monitor Mixer
Paul Owen

FOH Sound System Tech
Tom Abraham

Monitor Sound System Tech
Jim Homan

Sound Crew Chief
Bruce Judd

Sound Crew
Roy Parrot, Steve Dando
Niall Slevin, Steve Steiner
Chris Gilpin

Lighting Company
The Obie Company

Automated Lighting
Studio Color

Lighting Designer
John Broderick

Lighting Crew Chief
Ian Cameron

Lighting Crew
Steve Roman, Storm Sollars
Jeff Gregos, Mike Hanson
Terry Smith, Frank Mirabal
John Duncan, Gary Waldie
Jeff Wilson

Status Que Programmer
Ben Richard

Generators
Showpower

Showpower Crew Chief
Carlos Oldigs

Showpower Crew
Ian Smith, Joel Richards

Pyrotechnics
Pyrotek Special Effects Inc.

Pryro Techs
Doug Adams

Pryrotek Crew
John Arrowsmith, Phil Dibello

Set Design
Mark Fisher

Set Construction
Tait Towers

Faulty Towers (U.K.)
Brilliant Stages

Head Riggers
Scott Ward, Bobby Savage

Riggers
Ken Mitchell, Michael Gomez
Chuck Melton

Stage Carpenters
Joe Campbell, Kendall Carter

Freight Forwarding
Rock-It Cargo

Bussing
Senator Coaches
Phoenix Bussing Service (U.K.)

Bus Drivers
John Curtis, Robin Painter
Micky Byers, Gary Wright

Trucking
Upstaging
Transam Trucking Ltd. (U.K.)

Truck Drivers
Martin Ford (lead), Kevin Barnes
Ted Horner, Gordon Brackenridge
Ken Scott, Peter Cook, Alister
Brackenridge, Rod Wilson, Allan
Kearsely, Dale Cole, Davy Jones
Steve Young, Martin Wright, Steve
Dawson, Simon Couldry, Richard
Rivert, Matt Clark, Chris Harte

Air Charter
Platinum Tours

Flight Crew
Jack Roman, captain
Dan Miller, captain
Lisa Mooney, flight attendant
Bill Moody, flight attendant

Band Security
Gio Casparetti, Graham Court

Venue Security
Bob Bender

Tour Pass Security
PERRi Entertainment

Accounting
Siegel, Feldstein, Duffin &
Vuylsteke

Radios
AAA Communications

Onstage Wireless Systems
Showcom

Director of Hospitality
Janine Vogrin

Administrator
Eric Colby

Dressing Room Executive CEO
Allie Amato

Met Club Schmooz
Mike Davis

Tour Book Itinerary
Smart Art

Catering
Snakatak

Caterers
Frank Cribley, Glynn Bramhall
Mairead De Barra
Dawn Harris, Paul Matthews
Mick Thorton, Anne Crawford

Travel Agency
Traveltech/Richard Joseph
Trinifold Travel/Dave Brock (U.K.)

Record Label
Elektra

Merchandising
Giant Merchandising/Bruce Melick
Andrew Scott, Amir Butt

FIGURE 2.1 Metallica tour personnel. (From *Performance* Magazine, 20 September 1996, 26. Reproduced by permission of *Performance* Magazine.)

Artist's Manager

Like their namesakes in the theatre, managers are closest to the performers, often handling their contracts, bookings, and money, as well as acting as the performer's confidant in personal matters. Managers seem to come in an endless variety of styles: Some are deeply concerned with the physical production, some are only happy if the artist is happy, and others don't want to spend any money but still demand a topnotch production. Luckily, this combination of mainframe computer and surrogate mother is usually too busy in the office or managing other artists to tour with the artist.

Road Manager

Technically, the road manager is in charge of the tour. The primary responsibilities of the job are to keep the artists happy, functional, and performing; to see that the show goes on no matter what; and, on all but the big tours, to settle the box office and carry cash for the expenses. Many road managers have hopes of becoming artist managers someday, having worked their way up from crew positions. The road manager will be heavily involved in the pre-tour planning of schedules and tour arrangements and probably has final approval in hiring road crew members.

Tour Director

This position exists to give relief to the road manager on the biggest tours. The tour director may or may not interact with the artist. Being a proven professional in logistics and clearing the way of any obstacles that may stand in the way of a successful performance is what gets this person hired and, if he doesn't keep things running smoothly even if it is not his fault, fired. It is a highly stressful job that takes an ability to watch out for potential bumps in the road 24/7. Many are not very diplomatic and are more prone to the classic World War II general's attitude of "You will do as I say, no questions asked." People can get upset, true, but I have found that often they do not have all the information regarding a crisis and cannot see the whole picture, and there isn't time to take a vote!

Tour Manager

If the tour has one of these it means the road manager has been relieved of one of his biggest headaches—travel logistics. Sometimes this means coordinating private jets for the star while herding a fleet of buses (custom coaches) and trucks from town to town. Also, they make sure the drivers know where they are going and when to arrive at a hotel or hall, they arrange limousines for the star, and they even provide for the comfort of spouses and friends who need to be flown into a show and whisked in and out of the backstage area.

Production Manager

This title generally parallels the normal theatre definition. Anything that goes into the planning or the tour design and staging is this person's responsibility. The production manager reports directly to the manager and works with the artist's financial adviser. Budgets may be set at the beginning of planning, but expenses seldom fit into a neat mold. The production manager coordinates between the designers and suppliers to obtain bids and make the best deals. In some cases, all financial negotiations take place between the manager and the financial advisers, and the production manager is out of that loop. This, in my experience, is a disaster, as the manager and financial people rarely understand how a tour design must be executed on a day-to-day basis. The production manager will also advance the tour, either by himself or with the master electrician, audio engineer, and possibly the stage carpenter. They will discuss with the promoter and venue manager's representative issues of the rider, confirm requirements for the stage hands calls, and deal with catering issues. As the lighting designer, you can only hope that you have explained your needs clearly and that your plans reflect everything you need because you may not be consulted nor even allowed to talk to the suppliers who are bidding.

Stage Manager

Concert tours do not generally have a stage manager as we think of one in theatre. The promoter usually provides a person to act as liaison during the load-in and setup. The head equipment manager, or perhaps the road manager, often does the same for the tour. Only for large or complex shows will there be a need for a tour stage manager who acts as setup supervisor and all-around troubleshooter and would call the cues for the show.

Tour Accountant

A new face on tours began to appear in the 1980s in the guise of a trained accountant, who not only collects box office receipts and pays out *per diems* to the crew but also provides cash or pays bills incurred by the artist on the road. The tour accountant is also the one to come to when expendables have to be acquired while the tour is out. Depending on the deal the suppliers made with the manager, these expenses may include color media, lamps, and other replacements or repairs that arise. But, the real reason tours have accountants is that they can handle the really big money—T-shirts and poster sales. Seriously, the significant monies that are involved in an arena-type show and the very complicated splits of gross almost dictate that a specialist is needed full time on the road. Tours can take in millions of dollars, and a road manager who is busy dealing with a cranky artist may miss a clever ploy of a promoter in the settlement that could easily cost the manager his job. Dan Wohleen had the following to say about the need for accountants:

> The increasing complexity of the building deals [includes] percentages below a break even for the house and added percentages at 80% and sell-out. There are many areas where the band can have monies taken away: catering, overages on the house crew, union meal penalty, extra security, band demands for more tickets, extra limos, etc.

> There was a time when artists thought that all the little things, such as limos, gifts, and special meals, were the result of the local promoter's being a nice guy or a fan of the band. The truth is that all of those costs come out of the artist's pocket in the settlement. There are so many ways to split the take—minimum guarantees, promotional tickets, additional advertising, catering costs, extra ushers or security … the list goes on. The person doing the

box office settlement for the artist must be trained because a definite, specialized set of skills is required. Even with training, Dan went on to say, the road manager or tour accountant often is not actually in control because the house, as in Las Vegas, has the advantage. He knows of one venue that had actually added a row of high-priced seats that did not show on the seating plans. They got away with it for a long time before one road manager actually stood on stage and counted the seats.

Tour Security/Bodyguard

As the title implies, security personnel protect the artist and keep unknown or unwanted persons off the stage and out of the backstage areas. Most bodyguards are very personable and act more like valets than anything else; few carry guns, but most are experts in the martial arts. Frankly, everyone feels a little safer with these people around. It is an unfortunate fact that there are a few people out there who want to harm performers. There is a tremendous investment in the artist, and a lot of people are counting on the artist's ability to perform to maintain their livelihood. Therefore, it is in the crew's best interest to also keep an eye open for strange people hanging out. Keeping the artist safe, happy, and secure is everyone's job.

Public Relations Representative

Often the public relations (PR) or press agent works for an outside firm hired by the artist or record company to handle publicity. PR people are most concerned with image and publicity. The performer needs to stay in the public's eye, so it is essential that a coordinated effort be made to ensure not only that members of the local press have the artist's name on their lips but also that the performer's image is a positive one. While PR people can be a pain in the neck for the crew, the artist's image and thus the crew's bank accounts are very dependent on how well they do their jobs.

Production Designer

When you go to see an artist performing in an area or stadium there is a good chance a production designer was involved in the concept of the show. This person may be the lighting designer or a scenic artist who has made touring design a specialty. For shows in these venues the scenic elements must fit with the lighting rig, the trusses, luminaires, and effects. A train wreck is just around the corner if no agreement has been reached on the total design. The interesting thing is that the lighting designer of a smaller touring artist may also be the production designer by default. The lighting designer is most often the only "creative" person around to interpret what the performer wants the show to look like. While you can say, "I am a specialist and don't do sets," realize that this may be your way into touring, so get out that sketch pad and practice.

Lighting Designer

The lighting designer (LD) strives to meet all of the usual demands of creative lighting but with an added twist: Unlike theatre, which offers a script and a directorial concept for guidance, rock artists rarely have a concept, a program, or a director. At best, the director is the artist himself, whose onstage point of view offers little in the way of objectivity. The LD may also double as the lighting director (see below), running the control console personally and being involved in the physical work of erecting the lighting rig during the setup. Rehearsal time before going on the road is limited, and more often nonexistent, so the LD must be able to improvise at the first few shows.

Over the years, the position of LD has changed. There was a time beginning in the 1980s when some top LDs became "briefcase" designers. They did not perform physical labor; they were not around for the load-in and left before the load-out. Although there are still plenty of LDs like this, the LD staying with the crew and supervising the load-in and strike is again in fashion. Then there are LDs who do not continue on the tour past a few opening dates after they have made sure that what was programmed is working and has the artist's blessing. They return when new material is added or when a special need arises, such as a live taped performance. They make the appropriate adjustments and are off again, handing over the tour to the lighting director as outlined below. There can be two reasons for this. First, the artist may be

comfortable working with a particular lighting director, often a longstanding member of the crew who the artist does not want to see put out of work. A lighting designer with a national reputation is hired just to design the tour, and to have the designer continue on the tour would be cost prohibitive. Second, a certain class of designers (e.g., Marc Brickman, Jeff Ravitz, Patrick Woodruff, Peter Morse) does not want to tour because they make their livings as highly paid conceptualists or designers, not as road crew.

Lighting Director

The touring lighting director goes on the tour to actually run the show. Lighting directors may run the control console, or they may have separate board operators to handle that task while they call the follow spot cues and monitor the overall production. Most of the time, the lighting director will be involved with the actual load-in and setup, handle the light focus, and check the program, as well as coordinate with the house person who will handle the house lights and make sure the crew is on schedule. Lighting directors can find themselves in a very difficult position because they are discouraged from changing the lighting program; however, if an artist insists on changes, the lighting director would have to go against the boss's orders not to change anything without the designer's approval. Loyalty, if the person has been with the band longer than the LD, is probably going to make the lighting director comply with the band's request. Therefore, it is up to lighting directors to make sure they have clear lines of communication and the authority to deal with day-to-day issues.

Lighting Programmer

This is a skilled lighting board operator who may or may not travel on tour. Many of the best only program tours in previsualization programs with designers by their side, and they attend rehearsals to run the board. They are usually so highly paid that they will work on large projects such as a Super Bowl halftime show or the Grammy or American Music Awards television shows more readily than on tours. Some have the desire to become LDs but many are content because they are highly paid to do what they do best—program.

Master Electrician

This is a title that is being seen on tour more and more. It is the more formal theatre title used for the lighting crew chief, although touring shows still use the crew chief title to define the head of the sound department or set crew. Therefore, the designation of master electrician sets these people apart. In 2007, the Entertainment Services and Technology Association (ESTA) officially began certifying master electricians. Essentially, this person has overall change of not only the touring lighting crew but also the house electrical crew. Master electricians oversee the flow of work involved in setting up the dimming, getting the correct power service run to the location, correctly and safely assembling the trusses, and installing the riggers so the hang points are correct when a flying rig is used. They also gather the crew needed to focus the lighting under the direction of the lighting director. During shows, they are lighting watchdogs, constantly on the watch for anything that breaks or goes wrong. Sometimes they will even fix things on the fly while the performance is in progress. Lighting is dangerous from an electrical, fire, and personal injury standpoint. Good master electricians keep their eyes open from the moment the truck doors open until the doors are shut and locked each night.

Lighting Technician

The lighting technician is an electrician who does the physical work of assembling the lighting truss, directs stagehands to run all of the cable, and maintains the moving luminaires and all technologies such as smoke machines, LED luminaires, and special effects that are added to the stage set. They also make sure that the console is properly positioned and that the intercoms to the follow spots are tested.

Moving Luminaire Operator

The moving luminaire operator is a position that has come about since the advent of automated luminaires. Because of the technical complexity of the luminaires and the need for people who have been trained to run the computer consoles, this position is a highly skilled one. Some operators move on to become LDs in their own right. Currently, a rig may contain as many as 10 different moving luminaires. This means the operator must be familiar with and up to date on the attributes of a very large number of luminaires (see Chapter 11).

Scenic and Costume Designers

As more rock acts recognize the value of an interesting set, scenic designers are finding more work in rock & roll touring. Durability and ease of assembly are prime considerations for touring sets; although most sets are fairly simple, super star groups will stage an elaborate production that requires a lot of scenery. Costumers are not currently needed for many artists, but one very good area with the potential for costume work is Las Vegas-style shows. The latest crop of pop divas and heavy-metal acts also go in for exotic costumes, but the general apparel worn on stage by many rock performers is less than a coordinated ensemble.

Equipment Managers

Equipment managers, sometimes referred to as *backline technicians*, handle the basic band gear; set up the amplifiers, drums, keyboards, and other instruments; and remain ready during performances to deal with a broken drum head or guitar string, generally aiding the band. Guitar and keyboard technicians have evolved into highly skilled computer specialists. Many keyboards have sophisticated computers and MIDI controls so the position requires extensive knowledge of specialized components. Although lighting and sound personnel may work for companies contracted to the artist, equipment managers are individuals who generally work directly for the band and are not under outside contract to any company. Whether the artist keeps them on a retainer or not, many will return tour after tour to work with the same artist.

Crew Chief

Most often the crew chief will be the head lighting technician or another crew member who is designated as crew chief, especially with a large crew that needs to have a master organizer and doesn't have a production manager. The position requires someone who can coordinate all departments for a smooth flow of daily scheduling and possibly collect *per diems* on behalf of the rest of the crew. (*Note:* The author is personally against any crew member being responsible for other people's money, especially cash.) It's a good bet that savings in local crew overtime will pay the crew chief's salary. Some might refer to this position as the stage manager or the production manager. In other instances, each department, such as lighting and sound, will have a head person with this title (e.g., crew chief of lighting or head sound technician).

Rigger

The human fly called a *rigger* ascends the heights inside arenas, field houses, ice rinks, and anywhere else necessary to secure hanging points for flown lighting and sound systems. The early riggers came from ice shows and circuses and found that the fast pace and reduced setup time of concert tours offered a unique challenge to their daring abilities. Because almost all lighting rigs are now being flown, this position is essential. House riggers are most often just as skilled as their touring counterparts, and they have the advantage of knowing the venue intimately; however, riggers who tour can get rigs up faster because they already know the weight and balance needs and the sequence required to get the show up efficiently.

Sound Engineer

Also known as sound system designer and front of house (FOH) mixer, this crew member controls the mix that the audience hears from the middle of the audience in an equipment area known as the FOH position. Because the acoustics of many venues, such as arenas and convention centers, were not designed with sound reproduction in mind, sound engineers must use not only sophisticated electronic equipment to balance the sound but also their own expertise and experience to give the live audience a great experience. Using a variety of processors and effects, they constantly listen to the overall blend and make decisions about the individual volume of instruments, voices, and manipulating effects. Live and studio recording equipment once differed greatly, but today the same tools used in the studio are also used to enhance live sound. Often, shows are recorded and made available to the artists so they can make changes or improvements to the

song sets. FOH consoles and effects equipment are often specified by the FOH mixer and used to adjust relative levels as necessary to meet the demands of the music. The sound engineer and monitor engineer are generally responsible for the daily setup, testing, and strike of their equipment.

Monitor Engineer

This role is essential for live music and is a hot-seat position. Performers depend on the monitor engineer to produce custom mixes for them. Part practitioner and part psychiatrist, this crew member has the challenge of pleasing the musician by producing an individualized sound mix. Not only do monitor engineers have to be a good at their job, but they also have to interact with artists during very tense moments—while they are performing, when egos and emotions are out of control! Monitor speakers are still used, but for the most part have been replaced with an in-ear monitor system (IEM) system or a combination of both. Similar to hearing aids, they are basically a pair of headphones that are custom molded for the musician's ears. The IEM system reduces outside noise and protects the artist's ears from damage by using controllable gates that prevent the sound mix from reaching exceedingly painful levels. This position requires familiarity with the latest digital consoles and audio processing, being a master of communication, and being able to tolerate high-stress situations.

Sound Technician

Like the lighting technicians, sound technicians do the physical work of assembling and maintaining the sound system.

Pyrotechnician

The pyrotechnician is an experienced technician who handles flash pots, smoke pots, explosives, and similar effects. Local ordinances are usually very strict about the use of such effects, and pyrotechnicians usually must be licensed by the state in which the show is performing. Although a few acts with especially heavy effects carry their own pyrotechnicians, most acts have the promoter hire someone who is locally licensed. We have all seen footage of

Great White pyrotechnics gone bad at a club in West Warwick, Rhode Island, on February 21, 2003. A highly trained specialist is *always* needed whenever these effects are used during a show.

Audiovisual Specialist

Multiple-screen slide, film, and video projection, as well as light-emitting diode (LED) screens, are on the increase in rock & roll touring, and this has created a need for a separate specialist who sets up, maintains, and operates the audiovisual equipment on the tour. Almost all outdoor concerts now travel with a video crew to provide large-screen video magnification of the performance. With the advent of LED modules or blanket systems, very large images can be produced that match the high illumination output of today's moving luminaires. It is becoming more common to see some type of video screen display used even in smaller venues.

Wardrobe, Dresser, and Dressing Rooms

Even if the artist is not performing in a traditional, theatrically designed wardrobe, there is a need to keep the artist's stage clothes clean and ready for them when the show goes on. A wardrobe person on tour is no longer uncommon. This person may or may not double as a dresser for quick changes by female artists. The third title in this group may seem misplaced; however, if there is a wardrobe person, it is likely that setting up the dressing room according to the artist's demands in the contract rider can be handled by the wardrobe person. If not, the duties fall on a lower road technician, or a local person is hired and supervised by the production manager or road manager. The actual setup will most often be done by the local promoter to the specifications of the contract rider, but remember the admonition offered under the tour accountant section. Much or all of the expense here could come out of the settlement, so don't be quick to ask for things beyond what is called for in the rider.

Laserist

Laserists operate and maintain laser equipment. They are qualified by training and will have a federal variance (license). Laser equipment is seen less on tour today than at the height of its popularity in the 1980s, but it is gaining popularity now because of its ease of use and new higher power. New technology lasers are solid state, meaning that they require standard wall power, no cooling water, and far less room than previous versions; a standard 5-watt laser is about the size of a shoe box. Audience safety is of paramount importance, particularly with regard to where beams are allowed to fall and their not exceeding certain power levels. The laser's variances are subject to inspection by federal and state authorities. Local laws may require inspection by a health and safety officer before the performance, and the inspection takes time to arrange.

Video Director

This would seem like a straightforward position—a person who places the cameras and calls the shots that will appear on the screens—but often it is much more complicated. A large amount of recorded media must be placed on the screens, as well as live shots. Some tours even add live satellite images and answer questions called in by the audience right on the screens. Throughout all of this, coordination with the lighting is critical. Calling the shots on the tour is one thing, but is the video director in charge of the overall direction of the lighting and video? Not in most cases. Obviously, this is a case-by-case issue, but lighting designers should be prepared to discuss and stand up for their lighting designs as equal partners in the production design. This means that lighting designers had better have studied some video and system integrations software (see Chapters 16 and 26) to know what is possible and how it can be used in the show.

Video Crew

The video crew can have several components, including the normal positions of camera operator, video controller, switcher, and graphics operator. Other positions might involve producing documentary footage before the show that can be cut into the performance or included later as part of a television special or music video. While it is exciting to be traveling with a famous artist, the video crew must keep things in perspective, as should all crew members in lighting, sound, costume, and scenic departments, as well as equipment managers. You are there to serve the show, not be the show.

Road Chef

More common in Europe, caterers with ingenious names like the Rolling Stones, Just Desserts, and Meat and Two Veg can travel with a tour. American artists don't normally carry chefs, with the exception of those requiring special diets or who have more discerning tastes. Road chefs are a combination of chief cook and bottle washer, confidant, massager of large and fragile egos, and provider of a sense of "hearth and home" for everyone from the lead singer of the band to the last member of the lighting crew. Road chefs have to be accomplished and imaginative cooks, in addition to having accounting skills, being good at logistical problem solving, being able to work under a lot of pressure, and being good diplomats and listeners with the stamina of a mountain goat and the patience of a saint. Oh, and a sense of humor is essential!

(*Author's note:* Because most arenas in the United States have signed exclusive contracts for food services at their sporting events, an outside caterer or chef cannot work on the premises. ARA, Inc., has the largest network of such contracts. Some artists, however, have found a way around the rule—most notably the Grateful Dead, who had their own caterers travel with them. For theatre touring, this is not as big an issue. Most theatres do not require food services on a constant basis, so there is rarely an in-house exclusive contract.)

Truck Driver

Many drivers and trucking companies specialize in hauling for concert tours (see Chapter 7). The drivers live by one rule: Get the equipment to the next hall no matter what! They are usually young and non-union and carry the necessary Class A driver's license for tractor–trailer rigs. They may or may not help with the load-out each night, but most do because they want to make sure the load is secure in their trucks. There are major trucking companies with union drivers who do a great job, but the percentages are against them. Good truck drivers are highly valued, because drivers who are late to the hall cause many crew hours to be wasted. In the worst case, a delayed truck could cause the show to be cancelled. Drivers and rigs can be part of the transportation package supplied by a supplier or a separate trucking company that specializes in tour transportation.

Unions

Generally road technicians range in age from 21 to 30, but we are seeing a growing number of old-line legitimate theatre technicians being drawn into touring by the high wages. These professionals are often 40 or older; riggers in particular seem to be older men as a result of the training and years of experience required for competency in the field. It is also true that as many superstars turn 40 and 50, even 60, so are the long-time touring crews who started in the early days and have stayed with it. It is fair to say that they have earned the coveted title of "road dog."

I know of no rock acts that are staffed by union stagehands. Sound and lighting companies that provide road technicians do not have International Alliance of Theatrical Stage Employees (IATSE or simply IA) contracts. Wages, as always, are at the core of the controversy. The anti-union side feels that it cannot afford to pay the union scale and that the union, once installed in their ranks, could control future wage increases. What these people fail to consider is that a worker who is working under the IA banner would also receive health and welfare benefits. Yes, that means a percentage added to the salary, but it also relieves the company of providing a health plan. The workers are normally only covered when they are touring, assuming they are not full-time employees. As a proud IA member, the author must say that to deny crews such benefits is not right. I don't know if any of the lighting and sound companies provide health insurance for their road crews, but, if so, great.

As could be expected, promoters, producers, and managers are not very interested in paying additional wages anytime. House union crews work side by side with the non-union road crews. Would the IA like to have all stage hands be members? Yes, and I have a letter from the current IATSE president, Matthew Loeb, saying that the union will continue to pursue organizational efforts on all fronts. No matter what side you take, a proven technician is paid what he demands. Some artists and managers appreciate the fact that experienced personnel can save them money in overtime and damaged equipment, which more than makes up for the premium pay.

One faction of the IA that does get contacts signed is the United Scenic Artists (USA), the union that generally represents the interests of theatrical designers in the United States. This union is actually now part of the IA. It is somewhat unique in that its members are designers, not day workers, and the contracts are very different. Because the average rock & roll lighting design fee is far above the USA/IA minimum scale, many members do not fight management that is unwilling to pay into the union contract, which has a clause for vacation, health, and welfare contributions above the fee. Most designers form independent contractor businesses that are able to provide health and medical insurance, as well as a retirement plan. Because designers seldom have a 12-month contract with a concert artist, it is better for them to provide their own insurance. Many designers also go on tours and work the actual shows, but standard contract wage arrangements do not exist in the USA/IA contracts for touring and would have to be worked out on an individual basis. A West Coast USA/IA representative said they are always willing to craft an agreement with their members as long as it follows minimum guidelines.

Industry Certifications

Certification for lighting designers in the architectural field has been available since 1991; it is administered by the National Council on Qualifications for the Lighting Professional (NCQLP).[1] More recently, similar certification recognizes skill sets for technicians in theatre and concerts positions. It has been a misconception for many years that gaining membership in one of the IA locals gives the person some sort of certification, but this is completely false. There never has been any type of licensing in theatrical fields for technicians or designers. The United States Institute for Theatre Technology (USITT) had

the idea to develop a certification program but was never able to direct their energies or enough of their budget to the effort until the Entertainment Services Technology Association (ESTA)[2] decided to take a shot at it. It took many years to determine the legal implications to the organization as well as how testing should be accomplished. Right from the start, the IATSE offered its full support and contributed both people and money.

ESTA was formed of USITT members who were primarily suppliers of products and services to the entertainment industry. Today, members include dealers, manufacturers, manufacturer's representatives, distributors, service and production companies, scenic houses, designers, and consultants. The ESTA standards program is the only American National Standards Institute (ANSI)-accredited technical standards program for the entertainment industry.

In March of 2003, the ESTA board established a certification program for entertainment technology technicians. That same year, ESTA was joined by IA, USITT, the International Association of Assembly Managers (IAAM), the Themed Entertain-ment Association (TEA), and the Canadian Institute for Theatre Technology (CITT), and the next year the Alliance of Motion Picture and Television Producers (AMPTP), Live Nation, and PRG management offered their time, money, and support. A lot of volunteer committees worked years developing programs as well as bringing in the money required to develop the testing.

The real distinction was that this program would provide *certification*, not a certificate. A certificate only implies attendance, not an assessment process that recognizes an individual's knowledge, skills, and competency in a particular specialty, according to an ETCP brochure, which goes on to state: "Becoming certified indicates mastery/competency as measured against a defensible set of standards, usually by application or exam. It is awarded by a third-party organization that has set standards through a defensible industry-wide process resulting in an outline of required knowledge and skills. Certification typically

---

[1]National Council on Qualifications for the Lighting Professional (NCLQP), P.O. Box 142729, Austin, TX 78714-2729; (512) 973-0042; www.info@ncqlp.org.

[2]Entertainment Services Technology Association (ESTA), 875 Sixth Avenue, Suite 1005, New York, NY 10001; (212) 244-1421, http://esta.org.

results in credentials to be listed after one's name, has ongoing requirements in order to maintain, and the holder must demonstrate that they continue to meet requirements."

The first program to go online, so to speak, was the rigging certification. It is broken down into two categories: arena and theatre. A person can earn certification in either or both categories, depending on prior experience and passing the ETCP exam. Initially, no one was sure how such certification would be received in the industry, but it was quickly embraced by facilities, employers, and the union. The certification helps to reduce the liability costs of facilities and employers because it proves to insurance companies that the employees working in this potentially dangerous field can be shown to have at least 30 points on the eligibility scale, between hours of rigging experience and education as well as passing a rigorous exam.

The next certification program to go online was for master electricians, and it has met with the same success. Certifications are tested and administered by the Entertainment Technician Certification Program (ETCP),[3] which was founded by ESTA to administer these programs. More programs are in the planning stages.

The next program planned is Essential Skills for the Entertainment Technician (ēSET).[4] This will be an introductory level, not a proven skill level, exam. The idea is to test people on a defining core body of knowledge for numerous disciplines practiced by the live entertainment technician, from basic stagecraft, including safety, terminology, and professional protocols, to lighting, rigging, audio, etc. According to the ESTA Foundation which was formed for legal and tax considerations by ESTA with the initial focus of the program is disciplines that directly affect the health and safety of performers, crews, and audiences. Fundamental knowledge has been

developed. For several years, the author has been the co-chairman of this committee charged with the development of a body of knowledge (BOK) to provide the foundation that all people working backstage must have to work safely and efficiently. At this writing, a beta version of the BOK is available online for comment. Go to ēSET under the www.estafoundation.org/eset.htm website to see and comment on the definitions. One has already been adopted for this book—luminaire, which replaces fixture to define a housing containing a light source. Why the need for basic, entry-level testing? The reasoning here was that people may enter the business with a variety of training backgrounds—from union apprentice to high school theatre tech to college undergraduate to master's degrees at a prestigious theatre program at a major university. Yet, there is no way of assuring an employer that any of these methods of preparation is better than another. This form of testing allows for a level playing field.

The future plans of ESTA are to create an intermediate or journeymen level between essential skills and master rigger or electrician. Other skill categories are in development. To find out more specifics as to prequalifications for testing or details on what the tests contain, contact the organization.

The United States is not the only country looking at these issues. The Professional Lighting and Sound Association (PLASA)[5] in England is developing a rigging certification (http://plasa.org). They refer to it as a "skill card," which is an ID card that indicates that the candidate has been awarded a certificate for one of four levels: trainee rigger, rigger, rigging supervisor, and rigging managers. The big difference here is that, once agreed to, the government owns these standards and incorporates them into the qualifications by PLASA. The assessment standards are more rigorous than ESTA has set, but then again this program has the weight of being a government-authorized credential.

---

[3]Entertainment Technician Certification Program (ETCP), 875 Sixth Avenue, Suite 1005, New York, NY 10001; (212) 244-1421; www.etcp.esta.org.
[4]Essential Skills for the Entertainment Technician (ESET), 875 Sixth Avenue, Suite 1005, New York, NY 10001; (212) 244-1421; www.info@estafoundation.org/eset.htm.

---

[5]Professional Lighting and Sound Association (PLASA), 38 St. Leonards Road, Eastbourne, East Sussex, BN21 3 UT; +44 (0)1323 410335; www.plasa.org.

I t is relatively easy to get a group to say, "Okay, do our lights!" It is not so easy to keep from getting ripped off. One problem is that most designers go into a meeting eager to show the group how much they know, so they spill their creative guts. Do not be so naïve as to think the manager is not mentally taking down every concept you throw out, even if he doesn't react verbally. All too often your ideas show up on stage, but you do not!

This is an old story. Because it really does happen, the United Scenic Artists (USA) and similarly the International Alliance of Theatrical Stage Employees (IATSE or IA for short) unions have specific rules that no member puts pen to paper or presents an idea to a prospective client until a contract is signed. Excellent rule, but you need two sides to play the game. Rock & roll has only one side—the manager or producer of the artist. The other side that supports the designer or crew does not exist. Sure, the USA and IA would love to have designers work under their banners, but, frankly, they did not realize the economic potential for their members early on. They were not alone; most of the adult world felt that rock & roll was just a fad. An economic plan geared to the rock & roll designer's needs and the setting of industry standards are a long way from being a reality.

What are the economics of rock & roll touring? Who makes the money and how much? Should you work for a company or an artist directly, or be independent (freelance) with your own consulting company—which is best? These are questions that you should consider before you walk into that first meeting.

There is no governing body setting fee standards in rock & roll; instead, there is a range of fees, which have been static throughout the years because of an increase in the supply of people wanting to enter the business. The coming years don't hold much promise for dramatic changes. The economics are such that it is all relative to how much you are in demand as a designer. The few mega-tours and their designers will continue to make the big money relative to the other 99% of designers working in the field. However, that is no different than in any other consultant or design field. The idea is to create the demand so you can get paid whatever you want.

FREELANCE

You can follow three roads to lead you to work. The first, the independent or freelance way, is simple. All you have to do is find a client and convince the artist, road manager, girlfriend, manager, accountant, business affairs manager, and several close friends of the artist that you can do a great job. Realistically, there is a step before this—how do you get past the secretary in the first place? When I was first starting out, I would try to get one of the top managers on the telephone. If you accomplish this, maybe you should go directly to the presidency of a record company; why stop at being a mere lighting designer!

After you have set up a meeting, the question becomes "How much should I charge, and where do

I get the equipment?" It is rare that a group will hire you to design a show without your also estimating a budget for the equipment it will take to do your design. Therefore, you must convince an equipment rental company that you have a client on the line and you need their best price. Cross your fingers that the company does not go directly to the artist's manager and cut you out. I am not implying that this is a regular occurrence, but it has happened. Some of the things the rental company will need to know beyond how many dimmers, lamps, and trusses you will want are length of tour; personnel requirements; equipment list; who covers hotel, trucking, and travel expenses; and deposit and payment method.

COMPANY EMPLOYEE

Artists' managers do not want to be bothered with payments to multiple companies or to a lot of individuals. A *package* approach is most commonly used for touring today, especially by promoters such as Live Nation. A company that can supply sound, luminaires, trucks, travel arrangements, and other services is in a very strong position. Often, these companies sell their services like packaged home theatre systems: good surround speakers, bad CD/DVD player. Artists only need to pay for any sets, costumes, dancers, and musicians they want to have at the dates.

Herein lies one of the reasons for a designer to hook up with a company. Actually, there are several excellent reasons, including:

1. *Accessibility*—The company has been at this game a while and already has clients. It can get you past the first steps outlined for the independent, and then you can get down to work.
2. *Learning rock & roll*—Designing for rock & roll is a new ball game to most college-trained technicians. Although your education is a good base, you still are not ready to design Madonna or U2 shows. The best way to learn the ropes is to work for a company already in

the business. Working for someone else might not be your goal, but it can help you get your feet wet while you get a paid introduction to the field.

You will lose a little freedom when you join a company, because there will be rules and procedures and a boss who may appear to interfere with your creativity—that is, if you're allowed any creativity. At first, you'll haul cable around the shop and fix connectors and, dare I say it, sweep up. Oh, don't be such a baby; Yale and Harvard lawyers start out at big New York law firms in the mail room.

But, most of these companies do small local shows, and this is where you'll get your first chance to design. Take it. It may be a fashion show or a college dance, but it is experience in the professional world, so don't knock it. The other big advantage is that your management and sales responsibilities are eliminated (two things we will discuss in more detail in this chapter), and this is a big relief for a new designer.

Another way to be an employee has recently become available. A number of designers have formed design firms with two or more principals leasing their services out to artists. Visual Terrain, Inc. (a firm I founded with Jeff Ravitz) was one of the first outside or architectural or facility design lighting firms to go after the touring market. A different set of entry skills is required for younger designers; instead of starting out in the shop cleaning and storing equipment you might be drafting or assisting one of the partners. Then, as you gain experience, you could be given small one-off shows, and, if you show some style and business ability, who knows—one day your name could be on the masthead.

DIRECT CONTACT

The third method of employment is to work directly for the group or artist. Although the trend of groups owning their own equipment has faded, Supertramp's equipment spawned the lighting and sound production company Delicate Productions in Camarillo, California, which remains in a highly marketable position to this day. Working directly for the artist does have advantages, but in the long run you may feel trapped just working with one artist. You could also become involved in the infighting that plagues so many artists' personal organizations. Still, this way you do get a weekly check, and if you are into hanging out with a star (we call them *paid friends*), then you have that opportunity. It is still my least favorite option for starting in the field. I have seen too many designers, equipment managers, and road managers be asked to do "other" distasteful tasks during their off time, because the managers think that they are paying them, so why should they complain? They have a job, right?

PAY

Because there are no union guidelines, your income range is really wide. From a survey I sent to concert touring rental companies and freelance designers in the late 1980s, I was able to extrapolate the following figures. If you work for one of the major concert lighting companies, you could expect a road salary of about $350 to $500 per week while on the road, depending on your position on the crew. In a 2008 survey, those figures had moved to $500 to $800 per week; however, many people out there have upwards of 20+ years of road experience, and they can demand higher salaries for the expertise and proven reliability they bring to the production.

The increase in salaries for company staff crews has not kept pace with the IA union salaries, which generally run about $37 an hour in the major markets, which is $1480 a week for 40 hours. Over the years, nonunion entertainment workers have seen a 25% increase in pay but generally no health insurance. At the same time, IA has provided a 38% increase with health insurance and welfare. Back in the shop, between tours, the pay is usually less. Most people are amazed to learn that IA road contracts ask for an amount that falls very short of what the nonunion touring designers are getting. Occasionally, a good head electrician, whether union

or nonunion, will earn about $1500 a week, but this takes some climbing up the ladder.

Touring concert lighting directors can and do make $3500 to $7000 a week, but as I have said there is very often a big difference between what people say they get and what the truth is. For a one-off show, a freelance designer can make $800 to $1500 per day. In addition to the road pay, name designers will receive a flat fee of $5000 to over $50,000 upfront for preproduction meetings and the design time invested in designing a tour. They also receive a daily fee for the road days of rehearsal. Good lighting board operators can earn $2500 to $4500 weekly, and with a lot less stress. For one-off shows, they can make $500 to over $800 per day.

As a designer, you must also be a good salesperson. Your product is your design ability. Your ability to push your fee as hard as possible and not give in to a low-ball mentality is sorely tested. Just remember: *There is no loyalty to anyone, and everyone eventually loses.*

PER DIEM

You should also receive a *per diem*. This is nontaxable income that the Internal Revenue Service (IRS) allows for meals and incidental expenses (M&IE) for each day you spend away from home on business.[1] The maximum currently is $58 per day in major cities; smaller cities have lower rates, such as Kansas City at $45. The rate varies and even accounts for seasonal changes, so this is a little tricky. Your employer will not want to mess with making individual calculations so they may settle on a safe rate lower than the maximum; however, you can keep a diary (the IRS even offers you a form on their website) to maximize your deductions. Most accountants advise that if you get $35 or more per day then you should keep receipts. If you wish to view *per diem* regulations, see IRS Publication 1542, available only online.[1] Consult a tax expert as to exactly

---

[1] For *per diem* rates (for travel within the continental United States), see the IRS online publication at www.irs.gov/pub/p1542.pdf.

what you must do to maintain the tax-free status of this money. Yearly tax changes can eliminate many deductions, so keep informed each year in order to take advantage of whatever tax breaks there are.

Because most concert tours do not allow time for the crew to exit the building and search out a restaurant, a lunch meal is generally provided. The contents can be specified in the artist's rider. Breakfast may also be provided, although if you want more than donuts and coffee that will also have to be spelled out in the rider. Dinner provided onsite is also common. This brings up an interesting issue. If your employer is paying you a *per diem* and is still providing one, two, or three meals on show nights (not travel days), how does that affect the *per diem* from the viewpoint of the IRS? Well, we could sit down with an accountant and figure out how many days a week are show days and how many are travel or off days and come up with a figure of so many meals provided which should rightly be deducted from your *per diem*. But, that's too complicated, so the average *per diem* currently is about $35 and everyone should be happy. You get some free meals and still have some cash to do laundry, buy snacks, and maybe a drink in the hotel bar after the load-out is complete.

Besides the *per diem*, your hotel and travel expenses should be covered by your employer or the artist. Some companies and artists try to work deals whereby you get a flat *per diem* amount that also covers your hotel. This is a common practice on theatrical bus and truck road shows that are going to spend weeks in the same city but I am against this practice for one-night tours. Your day is already too busy to try to make such arrangements, and you are forced to stay at cheap hotels while the band is probably staying at five-star hotels. I do not believe that the technical crew should be treated like second-class citizens, and neither do most artists. It is only accountants and business managers who don't

understand the rigors of touring; they are just looking at numbers.

EQUIPMENT COSTS

Tim Murch, currently with PRG (Los Angeles) put it this way: "When it comes to putting a tour together, the lighting people are the last to know. The lighting designer can't do his job until the scenic elements are together." Even if a budget is known, specific elements are rarely computed accurately. Tim continued:

Then when the cost of the scenic construction comes in and trucking and crew is added in, the lighting gets what's left over. The accountants don't have a clue what we do and nine times out of ten, they push the lighting service company to cut their rates rather than kill the LD's creative input.

Spy Matthews, of Delicate Productions, said:

All of a sudden a client we have worked with for years hires a new attorney and a new nine-page rider shows up at the office just days before the tour is to leave. We are expected to sign it and get it back to them in 24 hours. Hardly enough time to read it, let alone get any legal advice of our own. We are totally screwed.

Like salaries and fees, there are no official guidelines to equipment costs. A company can get $20,000 a night for a tour (the mega-tours do not follow any formula), while another will only get $10,000 for a similar-sized system and crew. Although I cannot be sure that what companies tell me is always the truth, and trying to account for bragging rights factor, the figures given below represent norms set this past decade and are not likely to change well into the 2000s:

Small	50 to 80 luminaires, 1/2 to 1 truck	$750 to $1000 per show
Medium	100 to 150 luminaires, 1 to 1-1/2 trucks	$1250 to $1500 per show
Large	150 to 300 luminaires, 2 to 3 trucks	$1500 to $2500 per show
Stadium	300 to 400 luminaires, 3 to 4 trucks	$2500 to $3500 per show

Note: All of these costs take into consideration a 20-unit moving luminaire package plus conventional luminaires, increasing to 100 or more moving luminaires for the large and stadium packages. All figures are based on a guaranteed five shows per week.

CONTRACTS

Because the USA and IA standard contract forms do not normally apply, you are left to your own devices. I had a $20,000 lesson in contract writing in my early days. I got a group to use my services, so I wrote up what I thought covered all the points of our agreement, and a representative of the group signed it. Troubles developed within the group, and they broke up. It took over 3 years to collect my money, and I had to pay my lawyers' fees.

The answer is not necessarily a long legal contract. Part of the problem comes from the extremely fast pace at which this work is done, as mentioned by Spy Matthews earlier. On average, a group finally gives you the go-ahead a few weeks prior to the start of the tour. If you use a lawyer to draw up a contract, you can wait at least a week or two for the finished document to be delivered to you. After the contract is presented to the group, it will go to their lawyers and the process starts over again, which could mean the tour will be over before the contract is signed!

I have found a happy medium with *letters of agreement*. These look rather informal, with no whereas or wherefore—just plain language that tells all the details and duties of each of the parties. But, even this informal paper must contain the basic components of a legal contract to be valid. The best thing to do is to confer with an attorney who will give you a list of things that must be covered to protect you properly. I devised a checklist for myself of points that should be covered in all my letters of agreement. Do not consider this a legal reference, because each state has conditions that should be verified by an attorney; however, the following will help you understand the basics.

Who Are the Parties Involved?

It is not as easy as it might appear to ascertain the responsible parties. Often the person with whom you have negotiated the agreement has no legal standing or power to execute a contract on behalf of the party who will be responsible for your payment.

What Are You Going to Do for Them?

You must write a job description. Be specific and include even things you assume they should know you will do as the designer, such as provide plans and color charts, work with the equipment supplier to assemble the equipment, call the cues, and travel with the show.

What Are They Going to Do for You?

You want the client to pay you, of course. How much, how often, and by cash or check should be made very clear before you start work. Are they paying travel expenses? All of these things must be spelled out clearly if you do not want to have it come up later in court. Payment schedules that are simple and straightforward are best. Accountants and tour managers like amounts that get paid regularly rather than many add-on charges that cannot be determined before the tour starts. Remember to ask for program or air credit for yourself if shots are used of your show design for a television special or music video. If you also ask to be paid additional fees for such use of your tour design, it will not go over well and could be a deal-breaking point. Most managers will not sign such a clause; they feel it puts them in a bad negotiating position. You will most likely get credit, but they will not want it to be locked into a written contract. The only time to be sure to be paid is if you are a nontouring designer who has to be called out on the road to modify plans for a show that is going to be taped; that is clearly a negotiation point.

What Are You Providing?

You are providing a finished light plot, yourself as touring LD, a staff, and/or equipment. Explain exactly what, in detail, you will give them for their money.

When Will the Tour Begin and End?

You need in writing the starting date of rehearsal and the dates of the first and last shows, so if the tour is cancelled you have justification for a claim for loss of income.

An Optional Paragraph

Your final paragraph can include the standard legal line about suing the other party if it does not live up to the agreement and that any legal fees are at their sole cost and expense. Actually, the expense is set by the court and does not come close to today's high legal fees. It does show, however, that you do know something about the law and it should keep them on their toes.

AUTHORIZED REPRESENTATIVE

Finally, have the *authorized representative* sign and date a copy. Make sure the person signs "On behalf of XYZ Productions" and returns a dated copy to you. Because the Uniform Commercial Code has not been adopted in all states, there are many variations and degrees of force with which the courts will hold a letter of agreement as legally binding in this simplest of forms.

Consult an attorney in your own state to set up a personal legal plan for your specific situation and needs. The expense may seem high, but not as high as the expense of a lengthy court battle.

If there is a single area in which most designers fail, it is business. Read some books on contract law, business law, and accounting. It can save you great expense and trouble later on. Do not, however, try to be your own attorney or accountant. These are areas where a little knowledge is very dangerous. There is just too much to deal with, and it takes years of training to become an expert. Because most people are eager to work, they tend to jump in with both feet before they take a look under the water. Even if you are going to work directly for the artist or the production company, get an agreement in writing that spells out these simple points. You will not be sorry, and the time you spend will be worth it.

CONTRACTING THE CREW

As theatre, film, or television lighting designers, normally we do not become involved in the stagehands' contractual part of the production. In theatre, a union head electrician would be hired for the show, and he would assist the union in arranging with the producer for the crew. Concert tours take a slightly different road: Virtually none of the shows has a traditional unionized road crew, so there are no department heads to discuss crew needs in the preproduction meeting. You will find that the artist's manager is looking to you to know how long the setup will take and how many local stagehands it will require. Actually, the crewing, physical stage needs, and other items to be supplied in each city in which the concert will play are usually determined jointly by the lighting director, audio engineer, road manager, and production manager (if there is one) unless there is an extensive set, in which case the scenic construction company's representative should be consulted also.

So, how do you get your requirements for stagehands known in the different towns? There are two ways, both of which should be used: the *yellow card* and the *contract rider*.

For a concert appearance, the booking agent sends a general contract, which states the flat performance fee or percentage splits of gate, deposit, billing, and other financial conditions, as well as date and time of show and length of performance, to the talent buyer or promoter, who signs and returns it. The rider, which covers the specific performance needs of the artist, is usually put together by the road manager after consulting the lighting and sound people who are doing the show.

THE CONTRACT RIDER

The *contract rider* is usually considered a supplementary agreement. It is based on the original contract and is incorporated into the contract by reference to it. The problem is that, if it is not part of the original agreement, then it is, in effect, a new contract and therefore there must be a separate and distinct passage of consideration from the offeree. What is it that the artist offers the promoter to accept the rider?

The law generally considers something called *trade usage* or custom. This means that if you can prove in court that it is a widely used and generally common practice in your business to send riders containing certain information and demands for equipment and services, after the original contract has been signed, a court could accept the rider as a valid part of the agreement.

If you are in a position to send a rider to promoters, send it registered mail, return receipt, so you know that it was received. At the very least, send it via an overnight package service such as FedEx or UPS. People tend to take this form of delivery as more important than a regular airmail letter and will usually see that it gets to the recipient as soon as possible. The issue of Internet transmission is still a gray area. Many e-mail services have a way for the sender to receive confirmation that the recipient did in fact receive the e-mail. I do not know of any court case that has tested this method by holding the receiver liable for not living up to the contents of the e-mail.

RIDER ITEMS

Three general areas must be covered in the rider:

1. *Artist's requirements*
   A. Piano, tuned to A440; specify size of piano
   B. Piano tuner to tune piano prior to sound check and be available after sound check for touch-up
   C. Amplifiers, organ B-3 with one or two Leslies, electronic keyboards
   D. Dressing room needs, how many people in each
   E. Limousines required, airport and hotel pickups
2. *Food*
   A. Soft drinks for crew during setup
   B. Breakfast (if early load-in), lunch, and dinner for crew
   C. Food trays in dressing rooms for artist, with cheese trays and fruit and whatever liquor and beer is requested (many artists specify brand names)
3. *Stage requirements*
   A. Time of load-in
   B. Number of stagehands required
   C. Number and type of follow spots, possibly position (front, rear, side)
   D. Power requirements for lighting, sound, and band
   E. Stage size
   F. Rigging requirements

In addition to this basic group of items, the rider must include the specifics that pertain to the actual production. These include any items that could incur costs for the promoter or the people or services expected to be provided, such as balloons or a seamstress. Special attention should be paid to alerting the promoter if truss-mounted follow spots will be used. Some stage-hands will not operate them, and a half hour before the show is no time to learn that a crewperson hired for follow spot duty will not climb the truss.

Remember, all of these requests of the promoter will come out of the artist's settlement, so make sure management is aware that the band may be incurring hidden costs.

The rider shown in Figure 3.1 was developed by Michael Richter.

Other sections in the rider deal with handling the press, contract fees, payment method, billing, comp tickets, and other needs of the artist.

THE IMPORTANCE OF THE RIDER

Write everything into the rider and be as clear as you can, because if you believe that everyone knows that you need power on stage to do the show then guess again! People with money promote concerts, not necessarily people who know what it takes to mount a production. It can happen that the promoter is extremely naïve and so taken with show biz that he thinks everything is magical, including how a show is set up.

The main reason why I place so much importance on the rider is that it shows the promoter that competent production personnel are on the show and gives a sense of security that the coming production is together. Confidence is half the battle.

Second, it helps you get your act together. To write a clear, full, and accurate rider, you must do your homework, and that helps to anticipate problems before you hit the road. There is no substitute for good planning. If there is ever a time that Murphy's Law will come into play, it is on a concert tour of one-night stands. You cannot avoid problems completely, but you can be better prepared if you have spent the time in the preproduction stage.

Reo Speedwagon Tour 2008   Production At A Glance

This document is for 2008 headline tour dates in the U.S. only.

Production Manager: Michael Richter (619) 277-2614 cell

What we have:
- One Fully loaded semi
- Two tour buses
- Complete band gear
- Complete monitor system
- FOH audio console & effects
- Audio snake
- A set (inclusive of Marley, risers and steps)
- Merchandise

What we need:
- A clear clean stage of any and all obstructions when we arrive to unload our truck
- Stagehands-14 at load in and 16 at load out to unload our truck and 1 electrician to tie in shore power for our buses (if it's available)
- 4 spot operators for show call
- At least 4 deckhands for show call if there is a support act on the bill
- One english speaking runner with a LARGE CLEAN car from load in to load out with a cell phone or pager
- Audio-stacks and racks to properly cover the venue, front fills, a drive rack and system techs to deal with it
- Power for our backline and board groups
- Lighting rig per our plot with system techs
- 3 clean dressing rooms with private bathrooms and showers (with HOT water in them)
- 1 Production office with high speed internet access, a phone line, two tables, four chairs and power
- Catering-Hot meals (B-14, L-14, D-21) dressing room catering, stage drinks, and after show food per our rider (all to be discussed with Michael Richter)
- Lots of ice for the buses
- Showers (if not in the venue then 2 hotel rooms will need to be provided to us to use for showers)
- 4 dozen bath towels
- 2 dozen hand towels for the stage (dark colored if possible but not mandatory)
- 12 bars of bath soap

FIGURE 3.1   REO Speedwagon technical rider (2008).

FOLLOW-UP

The major problem is not what to put into this rider and to what degree of detail, but will it get go to the right people in time? This is why a follow-up or hall advance should be made in conjunction with the rider. Dan Wohleen, formerly a production manager of the Greek Theatre in Los Angeles, California, said this is still a big issue. He found that he often had to track down the road crew to obtain updated information. He said it is pretty silly not to think that by contacting the venue in advance that you are helping to ensure that things are ready for your arrival.

After making sure that each promoter gets the newest, most accurate rider, it is very important to follow up to see that the promoter gets it to the person whose job it will be to arrange for those things the rider requires. If your artist is not making his or

Notes for you:

- We do not require a crowd control barricade (if there is one in house it's ok but it's not necessary)
- Please be aware that we require a high quality stage, PA and lighting system, not "just enough to get by".
- We cannot accommodate any opening acts on our audio consoles. Any opening act FOH audio and monitors must be provided for separately.
- We are fine with DJ's or yourself making onstage announcements but they must happen at least 15 minutes prior to the scheduled start time of the show. We do not want the band introduced but you may say "coming up soon" is REO Speedwagon. If you intend to do this please give me ample advance warning so I can let my sound guys know it needs to be dealt with.
- If the show is outside, be sure to have at least two full rolls of Visqueen (plastic sheeting) and four 10 x 10 pop-up tents available for our use.
- If we can get access to the venue 1 hour prior to load in time for showers we would really appreciate it.
- We have monitor world and 1 large guitar world stage left.
- We have two large guitar worlds stage right.
- Please have your caterer contact me directly as we make changes to the bus stock on a regular basis to cut down on waste. (We prefer they shop for the buses on the day of the show after we generate a list of things we ACTUALLY need).
- Our FOH engineer is Neil Schaefer and can be reached at 248-219-8316 or neilschaefer@mac.com
- Our Lighting Director is Paul Dexter and can be reached at 818-599-0593 or paul@masterworkslighting.com
- Our Tour Manager is named Wally Versen and he handles all ticketing, travel, & meet and greets. 602-616-1134 or hitwally@aol.com.
- We have a merch guy on the road with us named Bill Stone 602-721-3142 or b-stone@mindspring.com

FIGURE 3.1 (Continued)

her first headline appearance, then there is already a rider floating around out there. Because it is usually not attached to the contract, it is quite probably the old one. Good follow-up can take care of this.

Your problems are not over just because you've written the rider. Chances are that you will get the rider back with changes, deletions, and notations. At that time, check with the booking agency and the artist's management to make sure they are aware of any such changes or deletions. Often, a booking agent receives a modified rider, signs it, and sends it back to the promoter and then never tells the road crew. Producing your best effort if the stage is 5 feet too narrow for the ground-supported truss or if there is only one follow spot instead of the four you had planned in the design takes a lot of physical as well as mental adjustment.

THE PROMOTER'S VIEW

Interviews with several promoters revealed their feelings about the contract riders they received. They brought up their constant annoyance with out-of-date riders, but conceded that it was possible that the booking agent or a management secretary sent out a rider that had been lying around without first checking to see if it was the latest version.

From the promoter's standpoint, essential information that needs to be covered in the rider includes stage size, power requirements, number of follow spots, security needs, and band equipment not being provided by the artist. Asked what is most often left off the rider, these promoters specified keyboards, which the artist expects the promoter to rent for the performance, and the request that a piano tuner be there.

Of the riders they said they see, 60% are clear and complete, but the remainder are confusingly written or simply are not complete. Many promoters voiced complaints about catering requirements. They felt that artists insist on too much food and beverages. Most artists just do not consider the promoter's costs to meet the rider.

It was also pointed out that, although most riders make clear how many stagehands will be needed, they do not break down the time—for example, 4 hours to set up, 2 hours for sound checks, 3 hours to strike and load out the show. Because some union fees have different rates for these periods, it would help if the artist gave the average time required for each part of the day's activities.

Although not actually part of the rider, insurance is a big expense for the promoter, and the fees are based on the past history of all the shows the promoter has done. That means that the promoter must split the cost between the shows being promoted that year. Some artists have begun to take out their own insurance because of their past good records, thus obtaining lower rates if they have not had problems with equipment or, more importantly, audiences. This is no small sum. On a per-show basis, the fee can run $2000 to $15,000 for liability insurance on a concert. If the promoter does not have to pay this, he has more money to put into the physical production.

In summary, the promoter wants clarity, reasonable requirements, and up-to-date riders. Too often the rider spends more time on additional frills than on services required to put on a good show. Requests for pool tables backstage and limo pickup for girl-friends, etc., can mean the difference between the production having that fourth follow spot or not.

SMALL PRODUCTION—LOW BUDGET

What about the artist who does not carry luminaires or sound on the road? This is usually a new rock & roll artist, but also includes many Vegas-type acts as well as many jazz and comedy performers. These artists, though, are becoming more and more aware of the need for production values, and they are adopting the rock & roll mode of production.

The rider must be written so it can be understood by the electrician or promoter's representative who will be given the paperwork to prepare the show. Do not be fooled; the standard theatrical template may not be as universally understood as you have been led to believe. Somehow you must produce a light plot that it is reasonable to believe the promoter and facility can re-create. Just leave enough leeway so you can be flexible when it comes to the physical limitations of the facility.

If you do not carry the lighting equipment, it is best to give colors in general terms (e.g., light red and moonlit blue), rather than specify a #821 Roscolene, because several very good color media are available, and sometimes not all brands are available in every town. The same holds true for instrumentation. Will you get all PAR-64s when you need some ellipsoidal spotlight? Try to make clear the area each lamp will cover and its function, so if a substitution has to be made the local supplier or electrician can give you something that will come as close to your needs as possible. Make sure you specify a working intercom between follow spots, house lights, light board, and yourself. Although good intercoms are the rule you should still specify as follows:

> The intercom shall be of the proximity boom mike type and double-headset type and shall have the capability to talk back at all positions.

The actual form of the rider need not be drawn up by a lawyer, but good English and clear presentation of your needs are musts. Plus, road managers will probably put together the final documents from the material they have been given by the sound crew, lighting, road technicians, rigger, band members, and others.

After the rider is completed, you are only one third of the way done. The remainder of the work is follow-up. Of that, one-half is tracking down the person who *should* have the rider for each venue, and the other half is explaining and working out the compromises. You could say that the contract rider is a waste of time, if in fact you need to work out all these compromises, but by covering all the bases you should improve your chances for success.

Ideally, a single promoter will be doing the whole tour (it is happening with more and more frequency) and then you can get everything set once and not worry—much.

THE YELLOW CARD

A *yellow card* show is one that is negotiated with the IA local, usually in the town in which the show rehearses or previews. They work out the needs for all the road and local crew in advance. The form that is sent to each IA local happens to be yellow, thus the name commonly used in conversation. It specifies how many persons in each department— props, carpentry, electrics (sound is still considered part of the electrical department in many venues)— are required for load-in, setup, show run, and strike. The IA local's business agent then sets the calls for members based on these requirements. The card also shows how many people in each department are traveling with the show. You can be sure that the business agent of the local will check to make sure each road person listed on the yellow card has a valid IA card as well as a pink contract (a road contract issued by their local).

I have never heard of a true rock & roll tour having a yellow card. A yellow card cannot be issued unless *all* road personnel are IA members and hold valid cards and pink contracts. Just because someone holds an IA card does not mean he or she can get a road contract, as certain requirements must be met. So make sure, if you are in a position to hire, that you ask the applicant if he or she can get a contract to go on the road, not just "Do you have a union card?"

The chances that the rock & roll sound company on the tour has been unionized are even slimmer. None of the major sound companies is currently unionized. This effectively blocks the issuance of the yellow card, even if the lighting company's employees and the carpenters and riggers are union members. So why even tell you about the yellow card if it is not possible to obtain one?

There are always exceptions to the rules. I have put this section in, even though the IA has not penetrated the rock & roll touring market, because the rental companies can now supply all kinds of touring productions: dance, opera, car shows, theatre bus, and truck productions. In these markets, it is much more likely that a yellow card could be issued.

PREPRODUCTION

Before you can sit down and start a light plot, even for a play, you need to know the physical criteria of the location. For the concert lighting designer, there are some added twists. It is impractical to look over floor plans for the 40 or more facilities in which the artist will perform on a tour. But, because facilities will vary in width, height, and power availability, it is difficult to design without this information. Where do you get your facts? Usually it is the road manager or, possibly on a large tour, the production manager.

The chances are slim that a city-by-city, hall-by-hall, in-person survey will be possible a month or two before the show hits the road; the schedule usually arrives a week or less before the first show, if then. So, we must deal in broader classifications—that is, theatres, arenas, college gyms, outdoor festivals, racetracks, etc.

A checklist of the basic information required includes:

1. Type of halls to be played (theatre, arena, outdoor)
2. Budget (per show or weekly and what it must cover)
3. Artist's requirements
4. Stage limitations
5. Crewing
6. Opening acts
7. Prep time available (before tour and on the road)
8. Rehearsal time available (before tour with luminaires)
9. Contract rider, as it existed on the last tour (see Chapter 3)

TYPE OF HALLS

If the manager can at least narrow it down to the type of halls to be booked, you have the most important piece of information; however, this will be clear to you only after you have played a variety of buildings. You must use your own judgment in limiting your staging, basing it on general categories such as theatre or arena, knowing full well that variability from structure to structure is great even within these groups. The types of performance spaces artists play range from arenas to clubs, rodeos, college basketball arenas, and glass field houses, and anything in between.

Some reference books are available. Also, *Pollstar* is a weekly publication available online that provides reviews of concerts, tour schedules (Figure 4.1), and many annual directories such as artist management (Figure 4.2) and venue information. Unfortunately, the number of other publications in this area has dwindled over the past years, leaving *Pollstar* alone in this category of magazines today (Figure 4.3).

One organization that can be helpful, the International Association of Assembly Managers (IAAM), which used to be called the International Association of Auditorium Managers, provides a number of annual guides to venues. It provides probably the most complete listing of stadiums, auditoriums, arenas, and theatres, but these listings are designed primarily to provide hall and facility contacts for ticket sales, catering, and building services. While lacking detailed technical data, they do give hall type, floor or stage size, seating capacity, and other general details, as well as names and telephone contacts at the venues. In truth, almost all venues now post

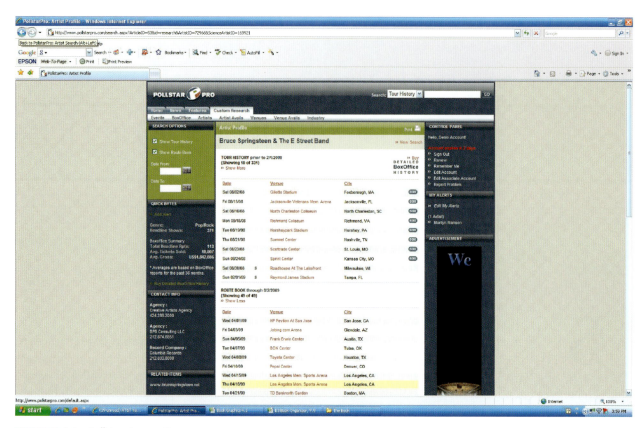

FIGURE 4.1 *Pollstar*—tour routing.

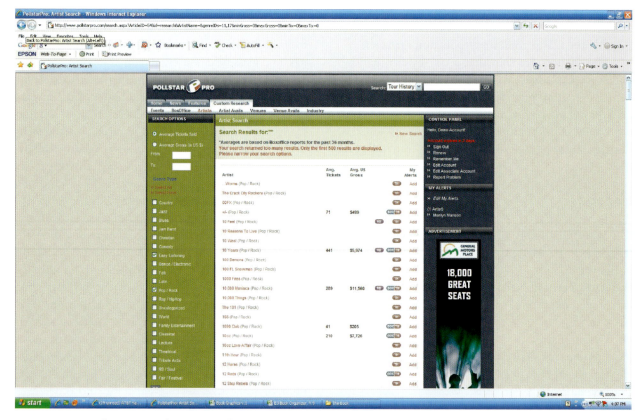

FIGURE 4.2 *Pollstar*—artist's directory search.

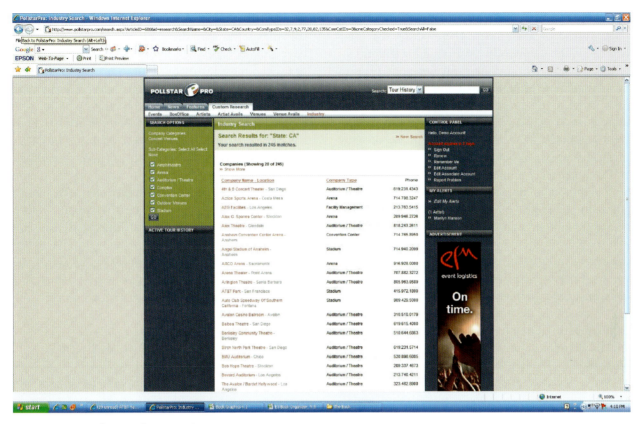

FIGURE 4.3 *Pollstar*—industry search, venues.

plans, facility technical data, and contact information online. If you know the name of the facility, with a little effort you can get a head start on gathering information, but the annual directory makes easy comparisons possible.

The IAAM also operates a sister organization in Europe the provides similar venue information. Other activities of the IAAM center around their mission statement: "To educate, advocate for, and inspire public assembly venue professionals, worldwide." To that end, the group offers a number of courses leading to certifications in such areas as risk management, emergency planning, and venue safety and security, in addition to certifying facilities executives. The IAAM does this both in the United States and through the European Academy for Venue Management (EAVM) run out of Innsbruck, Austria.

Legitimate theatres offer a degree of uniformity, because Broadway road companies for plays and musicals are most often booked here, but this uniformity is limited to some 30 or 40 theatres across the country. A concert tour plays everything from an old movie house to a symphony hall and it is still called a theatre tour. A theatre tour for a rock band could go from the Capital Theatre (an old movie house) in Passaic, New Jersey, to the Minneapolis Symphony Hall (no overhead pipes), to the Arie Crown Theatre in Chicago (very deep stage with excellent grid system), to the Constitution Hall in Washington, D.C. (a domed stage with no grid), to Pine Knob Music Center outside Detroit (another covered stage with open-air lawn seating). This is why trusses and other structures are trouped by concert artists even when doing a theatre tour.

When we talk about arena tours, we can expect even less in the way of theatrical facilities—that is, pipes and stage. Arenas are for the most part simply large airplane hangars with seating.

The newest trend is for a hybrid between theatre and nontraditional concert hall. Nokia has built two such state-of-the-art facilities in New York and Los Angeles. They are truly unique and incorporate trusses that slide to different positions up and down stage instead of fixed pipes. There is talk that they will be building more of these around the country.

If you are working at a facility and want to get updated information regarding the technical rider or plans of an artist, it is now quite easy to track a show that is coming to your facility via the artist's website. Their Internet sites can be used by stage managers, production managers, designers, and technicians. Another way to stay up to date is by logging on to the sites of some of the facilities where the artist has just performed. The technicians or stage manager will be able to add very specific information that may not appear on the rider. This can be the most useful information you get. Ideally, the artist's lighting designer or stage manager maintains an active website to keep everyone aware of any changes in the tour technical issues or band requirements. Certainly if they are on tour and you are the housemen, contact them via e-mail and ask questions.

At a few exceptionally well-equipped theatres around the country, the house staff cannot understand why we do not use their luminaires and counterweight systems. What they do not understand is that the elements of timing, consistency, and repeatability are the keys to a good concert. A theatre production usually rehearses in each new town, so there is time to make adjustments in hanging positions, focus, and dimmer levels. A touring concert has less than 12 hours from load-in to curtain. The concert designer, as well as the artist, must be assured of what will be seen during the performance.

Budget

A few elite designers refuse to let budget restrictions encumber their creativity. They prefer to obtain as much information as available and prepare the design the way they believe it should look. They then present it to the artist and try to convince management that the idea is worth any expense.

Maybe I am too much of a realist, but I like knowing at least the approximate budget. Designers who develop the budget after the design is submitted

are, on the whole, the name designers. They have sold themselves with the understanding that they will have a blank check. Let us consider the other 99.5% of us who must justify our cost to management.

It is wise to ask for a budget when you first talk about the project. More likely than not, you will receive blank looks because a budget has not even been considered, or they will throw it back to you in an attempt to see if you will give them figures lower than they paid last time. One way to obtain a rough figure is to give a range per show and observe the manager's reaction. This is not to say that, if you return with a terrific idea that will cost more than the budget, you will not be able to talk them into using it. The budget should simply be a median point to be used as a guide; you can come in under budget or even over budget if you feel it is worth the effort (see Chapter 3).

Artist's Requirements

The initial meeting should be with the artist. The only way to get a feeling for how an artist wants to look is by meeting the artist, not the artist's management (see Chapters 23 to 25). You will be creating something that will affect the artist's performance significantly, and you must be in tune with the artist's desires. Concerns such as stage movement and placement of band equipment must be discussed.

One of my first clients was Billy Preston, who was often called the fifth Beatle. I was called in at the last minute to do his tour. At the time I only met his road manager, not Billy. The road manager assured me he had been with the artist so long that he knew his every move and requirement. The first show looked like I had designed the lighting for folksinger Joan Baez rather than the high-energy pop music he performed.

A rapport with the artist must be established. Confidence in you as another creative person who wants to make the artist appear in the best light (no pun) is essential. It is the theatrical equivalent to the theatre designer's relationship with the director. The artist or band's will is paramount in the development of the ideas process. It is the one thing that you can count on that will actually help you to separate one design from the next. The challenge is to foster a relationship that will build trust in you as a person and as a designer with solid ideas. This can be accomplished with the way that you present yourself and your ideas and how you communicate with the artist, which must be unassuming and respectful.

STAGE LIMITATIONS

Many physical limitations must be considered in the mounting of the show. Ask yourself questions such as the following: Is a backdrop being used that must be lit? In arenas, will there be seating in the rear? In outdoor shows, will the wind factor preclude backdrops and some scenic devices? What are the sightlines and width of the stage? Dan Wohleen, now managing director of the Maricopa Country Center, remembers when "an artist had changed from rear to front projection after the tour started and had not informed us. I lost 800 seats to the change and it came out of the artist's pocket." In another case, the booking agent had sold the act with 270-degree seating availability to the promoters. When Dan got the rider, it was quite clear that there would only be 120-degree seating available. Dan contacted the tour production manager and was told that "someone" had not informed the booker of the change. Luckily, this was discovered before seats went on sale.

What do you do when the band shows up and says it has decided to let someone videotape the performance? Ignore for a moment the additional costs this will incur, but what about that "little" 5-foot platform they will need to place a camera dead center in the expensive seats? That happened to Dan: "It wiped out 120 seats. People were mad at the house, not the band."

Because you cannot control the performance spaces in so many locations, the design must anticipate physical staging problems before they develop. Once on the road, you must meet the day-to-day problems that arise. Time is a constant devil, so preplanning and giving some forethought to possible solutions to every conceivable staging problem are essential. As an old college professor told me, "You plan for the worst, expect the unexpected, and you'll get something in between."

CREWING

Will you hire the crew personally or be contracting a lighting company for personnel and equipment? How many technicians will be traveling on the lighting crew? How much time is available for setup? How many local stagehands will there be to help out? If required, are riggers to be provided by the promoter or are you taking a head rigger or a full rigging crew with you on tour?

Good, efficient crewing is vital. You can design the most spectacular show in the world, but it will be for naught if it is not ready by 8:00 p.m. Remember that time is the unrelenting demon of touring. A touring lighting crew of two technicians is average, one for dimming and one for truss supervision; three or four can be found on large shows, with a moving luminaire technician specialist as part of the crew to keep them operational. They usually handle the lighting setup with the assistance of four to six local stagehands. However, a large number of special effects will take time to set up and can slow things down, as do heavy scenic elements. By 1997, moving luminaires had evolved from being a novelty to being the main element in many touring lighting rigs. As the PAR-64 luminaire became the signature luminaire in early concert design, the moving luminaire took on the status of king of the hill; however, this wonderfully versatile luminaire carries with it a great deal of baggage. A member of the lighting crew must be a trained maintenance technician. These units are far from bulletproof, even when following the manufacturer's instructions for how they must be operated under road conditions. Weather (e.g., heat and rain) can damage moving luminaires. They are the true temperamental stars of the lighting galaxy. Another factor is that artists or their managers may have the mistaken impression that they make life easier for the designer, when in fact it takes longer to program all of the individual attributes for each luminaire. The lighting consoles also become exponentially more complicated to operate.

Who handles the road supervision and repair of the set, scrims, drapery, and such? A road carpenter is still rare on a rock & roll tour. As the designer, you are likely be responsible for all of these elements. That is not to say that you personally will have to take hammer and nail to a broken flat, but you will probably need to make sure everything is kept in working order.

Opening Acts

The opening act is often slighted. Opening acts are at the bottom of the food chain most of the time, and little consideration is usually given to their needs for stage space or lighting. The headline band's equipment technicians usually refuse to move one piece of their stage gear. This is often out of spite, because it was done to them before they became headliners. But, it is a fact of the business that you must deal with the issue.

Opening acts with the same management company as the headliner will probably receive more consideration, but not always. Stars' egos can be enormous, and artists quickly forget how it was when they were starting out themselves. Do not take for granted that the opening act will be given *carte blanche* to add special lighting even if they pay for it. In fact, do not agree to allow any equipment to be added to your rig until you have had it cleared with management in writing. If the opening act is permitted anything, you may still want to place restrictions on the extent to which they can utilize your design.

There is a tendency to package a single opening act with the tour. In theatre, we look at the show and how its parts are integrated into the whole. In concert touring, 99% of the time we design for *one* element of the show only. Keep in mind whom you work for, and do not try to be a good guy to the other artists; you could lose your job for that effort. You might be surprised to learn which artists are vindictive and downright nasty to opening acts even if they were themselves opening acts not too long ago. It is payback time in their minds, somehow, and you can get caught in the middle if you are not careful.

Prep Time

The first consideration in prep is the time allotted to design and physically put the equipment together, check it out, and be ready to rehearse and hit the road. If you are going to use one of the companies in the business of supplying tour lighting, the time required can be cut to a couple of weeks, or in a pinch much less, because they have much of the equipment already packaged and can prepare it quickly. A substantial delay occurs when the trussing must be fabricated. If time is tight, do yourself a favor by checking out truss plans that can be accomplished with in-stock units. The company may have special pieces that you can adopt or designs you may not have considered. Also, remember that how busy the shop is will be a determining factor in preparation time and very possibly cost.

The second area to consider for preparation is the time allocated each day to load-in and be ready for the sound check. A show that requires a long time to load-in, hang lights, and focus must be planned well in advance so the promoter can provide adequate assistance. Local stagehands and access to the hall must be arranged for as many hours as it will take to accomplish all these things. Planning takes on even more importance with mega-tours such as Madonna, Faith Hill and Tim McGraw, and Bon Jovi, for which the load-in could take two to three days.

Ability to judge the setup time is an acquired skill. You can consult with the lighting company in the early stages to discuss your concepts and let them give you their ideas as to crew requirements and setup time. It takes time to move equipment and personnel from one city to the next. A show that is booked with 600-mile jumps in between will never make a 9:00 a.m. stage call. Figure on an hour of driving time per 50 miles of travel, but areas of the country where superhighways are not the rule will lower the speed, as will travel through large cities to get to venues such as Madison Square Garden in New York City. The time of year (e.g., winter in the Midwest) is also a factor. Unfortunately, booking agents are notorious for their complete lack of consideration on this point.

All the innovations that have come along—especially previsualization software such as WYSI-WYG ("what you see is what you get"), a three-dimensional (3-D), real-time programming tool (see Chapter 19) that allows a designer to create cues and record them on the touring lighting console in addition to printing out color snapshots of looks to use as a storyboard—can be very helpful and a real time saver when it comes to the number of days necessary to set up the rig. There are also proprietary visualization programs that come with several computer lighting consoles. While this is discussed in detail later, it is also a big factor in scheduling rehearsals and the space and length of time necessary to set up the rig.

REHEARSAL TIME

Most top acts produce a full production and approach their tours as if they were theatrical shows. The more rehearsal time available, the tighter the first show; however, a long rehearsal with luminaires is rare. The availability of large stages and the very high cost to the artist of a lengthy full rehearsal place limitations on this phase of the preparation. Here is where your pre-planning and ability to make quick decisions are put to the test. Rock bands are not actors who know how to freeze while a lighting look is being set and then pick up the action again. They will play through the song and possibly repeat it, if they wish, but I cannot stress enough that a true rehearsal such as theatre companies enjoy is nonexistent for a touring rock show.

Jeff Ravitz has a system where he has an assistant keep time. He allocates a certain number of minutes to each song. When the time is up, it's up, and he moves on to another song. If he doesn't keep strict time, the tour may open with great lights for the first three songs and then a dark stage. Computer-based lighting consoles, largely due to the reliance on moving luminaires, take time to program a cue, not like the old manual boards where you could bring up channels easily without recording the cue (in other words, "wing it"). Jeff tries to at least build the shell of each song before the tour hits the first city. "I often go into the first city with only a basic look and the absolutely necessary cues within the song, then over the next day or two I fill in the holes. There just isn't enough time allotted by the management anymore. They think it's magic!" (See Chapter 6.)

TIMING

Timing is the key to efficient production of a concert tour. Most problems during setup grow to become major disasters, not because of faulty equipment or damage that cannot be repaired, but because you only have 15 minutes to an hour to fix it. As if that is not bad enough, you usually find yourself in a strange town, often close to or after 5:00 p.m. Time is your biggest enemy. Effective handling of your setup time will give you more time to discover problems and, when needed, to get parts and make repairs in time for the show.

VARIETY OF VENUES AND ARTISTIC STYLES

A career in concert lighting does not limit you to one type of music. A great -ety of artistic styles is being performed in concert venues throughout the country each day, with over 600 musical artists and bands on the road at any one time. The peak seasons are April to May, July to August, and October to November. If you take a closer look at these 600-plus tours, you'll find that only 6 to 8 of them are superstars, and only a couple are mega-tours at any one time. About 40 of these acts comprise the extra large tours, and about 50 to 100 use touring lights and sound. Another 150 to 200 are just getting a foothold and may take out a designer and rent equipment locally or regionally. Another 30 or 40 are casino circuit acts playing Las Vegas, Reno, Lake Tahoe, Atlantic City, and a large number of Indian gaming establishments have attracted name artists in recent years with large paychecks. This group of venues can allow for some very creative design decisions, so do not brush them off. Many of the facilities are extremely well equipped, and they are always looking for something new. The rest of the touring acts are playing as opening acts or simply cannot afford or do not care about lights and sound.

The range of places that are used for performances is vast and can create not only frightening challenges for the designer but also wonderful opportunities. Once the preliminary areas are covered—type of hall, budget, artist's needs, staging, crew, opening act, prep time, and rehearsal—you can sit down, listen to the music, and start to rough out a light plot. The frustration created by designing before you know most of this information can be utterly debilitating. Learn as many facts as possible and be positive in your concepts but be prepared for changes. That is a known: There *will* be changes!

They come in many sizes and wattages, but generally concerts use the largest and highest wattage PAR-64 (1000 watts). F(N) or very narrow spot (VNSP) lamps are used for special and accent lighting. The PAR-64 and the less used but valuable family of PAR luminaires (see Chapter 18) are the workhorses of conventional luminaires used in concert lighting. Being relatively trouble free and devoid of parts that can jam or break, these units have the most effective, dimmable light source available among the standard theatrical and film luminaires. For 45 years, the PAR lamp has continued to be the most reliable source for "punch" lighting.

The efficiency and type of beam spreads available for the PAR-64 luminaire should be viewed in a couple of ways. There seems to be a preoccupation with efficiency because the availability of a sufficiently large power service can often be a problem. If you are going into an old theatre or college gym, access to auxiliary power can be a headache; consequently, you want as much light as you can get for as little power draw as possible. The PAR fits this bill best of all. The sealed-beam lamps offer a range of beam spreads that give you coverage from very narrow to wide flood fields. These lamps can give you concentrated beams of light that will project even dense color a great distance. For example, compare the beam spread at a 30-foot throw between the very narrow lamp—3.7 × 6.4 feet—and the medium flood lamp—6.5 × 15.5 feet—as measured at 50% intensity. While on the subject, those same 1-kW lamps produce 560 fc and 150 fc, respectively, at that distance, compared to a 1-kW Fresnel, which can only produce 195 fc at spot and 40 fc at flood focus at the same distance.

Air Light

Air light, often referred to as *air graphics*, became an increasingly prevalent term in the late 1970s and early 1980s. There was an undeclared race among touring rock bands to reach new records for how many trucks they had out on tours and how many of those trucks carried truss, lights, and staging. At the time, Van Halen proudly claimed that they carried 1300 lights, including every PAR light bulb and each one on 8-light audience blinders. Air graphics are possible

when there is a lot of haze in the air. This contained smoke gives narrow-beam PARs the air surface necessary to trace a light beam from 16 to 24 feet above the stage floor, but without physical staging to light up or reflect light. Smoking was still allowed, and smoke machines or hazers were not carried until the mid-1980s. Big arena tours sought to achieve a smoky club atmosphere, and some lighting designers even requested that a facility's powerful air-conditioning units be shut off just before a show so fat beams of light could be seen. As a way to avoid having to mount more and more rows of lights to illuminate the performers or staging, air light became a practical way to animate lighting scene changes using color from different beam angles and create movement in the air. Designers arranged banks of light purely for effect, and air graphics became a large part of the overall design of many tours and television specials. By the time moving luminaires came along in the early 1980s, air graphics had become a finely tuned craft, and moving beams through the air added greatly to the excitement that air light creates (Figure 5.1).

Moving Lights (Luminaires)

A big change in the way we speak of theatrical lights has occurred recently. For years, people referred to them as anything from *lamps* to *fixtures* to *lights*. The term commonly used today for lights that remotely change position is *luminaire*. This is a word that Vari*Lite originally coined for their computer-controlled lights, and it has now entered into the lexicon of the entertainment business. Recently, the Essential Skills for the Entertainment Technician Committee of the Entertainment Services Technology Association (ESTA; see certification discussion in Chapter 2) adopted the word to replace *fixture* to define all fixtures, not just moving lights. As of this writing, the industry has not fully embraced this change for a long standing term, but we believe with the Committee's endorsement the term will start to be used by younger technicians and designers.

If you have a moving luminaire package, you might be well served by using them more for repositioning and color or pattern changes than for flash or

FIGURE 5.1 Example of "air light" effect. (Photograph by Lewis Lee.)

to add sizzle at the end of a song or the band's bows. I once saw a Bruce Springsteen concert designed by Jeff Ravitz. I had been told that the lighting package was all moving luminaires. But, as I watched the show, I never saw a luminaire move. I actually thought the program had somehow malfunctioned. Then, as Bruce was leaving the stage after the last number, the lighting exploded in a riot of color and movement. There was no way the audience would not applaud and bring him back, even if he wasn't "The Boss." Jeff had used the moving luminaires to change color, pattern, and position when they were not on so each song had a different look with a minimum of luminaires, and he saved the big bang for the end. That showed great showmanship on his part.

Computer-controlled moving luminaires are everywhere. At last count, there were over 35 manufacturer's producing club and touring models worldwide.

OTHER LUMINAIRE CHOICES

Other available choices are beam projectors, ellipsoidals—Lekos, ellipsoidal reflector spotlights (ERSs), or, as the Australian, English, and European

companies call them, profile spots (see Chapter 18). Fresnels can also come into play, as well as the various types of striplights and luminaires designed around multiple PAR lamps. Recently, light-emitting diode (LED) light sources have given us a whole new category of lighting, as have the digital luminaire instruments that can show images as well as colors. The beam projector, also an open-faced luminaire, does just about the same job as the PAR-64, but with a very low efficiency factor (see Chapter 18).

The ERS or ellipsoidal has not been given the status on tour that it enjoys in the theatre, but more and more we are seeing it being employed for concert lighting. Some people consider the author to be this luminaire's biggest booster. I once did a tour with over 180 ellipsoidals and only 4 PAR-64 luminaires. This does require a high degree of confidence in your design; however, this is true only when you have a show in which you are not confident that the talent will hit the rehearsed marks. Artists who cannot hit marks do more than hurt your design; they are wasting their money. If the effect cannot be seen because of their inability to be consistent in their movements, then the luminaires are not needed and are just so much excess baggage. Electronic Theatre Controls (ETC), Inc., was the first company to revolutionize Mercury (Hg) the standard ellipsoidal in 1992. Strand Lighting and others followed later with their own versions. ETC's concept coupled a new lens design with a new lamp design. The lens was sharper, and it had fewer aberrations around the edge and cleaner shutters. Moving luminaires can also project patterns (*gobos*), or you can choose from a variety of lens spreads or even add an iris to narrow the beam. So, consider the cost difference. If you want the pattern to be used multiple times, why tie up an expensive moving luminaire?

The Fresnel luminaire doesn't have a great following in the concert market because of its weight and limited efficiency. Other luminaires include the film Mini-Brute, an early application of the PAR lamp designed for broad fill lighting for film and location shooting. Mercury medium-arc iodide (HMI) Fresnel luminaires do find special uses from time to time. Generally, you will find that concert equipment manufacturers have produced their own designs for grouping PAR-36 luminaires into single luminaires usually used as audience lights (see Chapter 18). Another variation of audience lights, created by Doug Brant and Justin Collie of Artfag LLC, has a strobe built in; it is called the Litepod and is manufactured by Wybron, Inc. (see Chapter 18).

In the final design, virtually any luminaire can be used if the designer has the vision to see what effects it can produce and can use it accordingly. Always keep in mind that the instrument or effect must work night after night. Nothing is worse than when an effect the artist comes to expect does not work.

PLACEMENT OF LUMINAIRES

Concert lighting reverses basic light direction used in theatre. Whereas theatre puts great importance on frontlight and least on backlight, concerts completely reverse this concept. Two factors are responsible for this. First, there is little or no possibility of consistently getting a balcony or front-of-house (FOH) position to use for a concert tour in 30 or 40 *found spaces* (spaces used but not specifically designed as performance areas). Second, lighting in concerts is viewed as providing effects and accents rather than the theatrical function of facial visibility. Thus, the

idea is to produce all the front light with one or two follow spots and to focus the fixed lighting on providing special effects and complementary colors for stage washes, breakup patterns, and backlight to surround the singer or players with color.

This use of backlight is known as *accent lighting*, to distinguish it from the theatrical style. Light is concentrated at strategic points on the stage to punctuate the music with deep colors. This method of lighting truly fits the maxim:

> It's not where you put the light; it's where you *don't* put the light that counts.

Keeping this in mind during your designing efforts will serve you well in not overdoing the design. Too many shows load in 200 luminaires when 100 will do. I had a problem with a group that wanted me to add more luminaires to the tour because they were cutting back on the set. The logic was that if they had less scenery, more luminaires could fill in the void. Wrong. You cannot light what is not there, except when using air light. I already had an adequate number of luminaires in the plot to light the band and didn't desire anything more except effects and projections, but they had just eliminated the things I needed as surfaces for projection.

The quantity and placement of luminaires must be directly related to the placement of band and vocals. It is unreasonable for a manager to request 80 lamps, or 200 lamps, for his client; my standard reply is, "Okay, I will use that for the budget figure, but I must know more about the artist's needs." I see no point in renting 80 (or 200) lamps and then trying to figure out what to do with them, but I have been put in this position on several occasions.

The most effective designs I have seen used fewer luminaires. The timing of the cues, the color selection, and the organization of effects were far more powerful than adding additional luminaires would have been. Good design in any field is not based on quantity.

COLOR

In my own designs I try to think in terms of colors, not quantity of luminaires, before I begin to draw a light plot. What I am trying to picture in my mind are *looks*, planned patterns of light direction and color that will be used one or more times during the show. The term *looks* started out as a television term, but now most designers use it to describe the patterns created by the lighting.

Primary and secondary colors are generally best. For a general plot, try for four colors as sidelight and five colors plus white for backlight on the lead artist. Bands usually get four colors for backwashes and three colors for sidelight, plus two back specials—one warm and one cool color, such as Lee filters #105 (orange) and #137 (special lavender)—to accent their solos. I prefer not to use a follow spot every time a short guitar break happens. A nice hot backlight stabbing through the darkness, while dousing the follow spot on the singer, can be very effective. When I am able to get front specials into the design, I add a warm special and a cool special to each solo player, plus at least two general front or top band washes.

How different colors affect the audience's visual interpretation of the music is *key* to how a designer interprets the music. You are strongly advised to read one of the many excellent theatre lighting textbooks in print for a full discussion of light and the human eye. My favorite book on the specific subject of how light is received and interpreted is *The Beauty of Light*,[1] by Ben Bova, a physicist, although it is currently out of print.

CIRCUITING AND DIMMING

The luminaire complement is tied to the dimming available, except with HMI luminaires. It is fashionable for designers not to concern themselves with such trivial matters. After all, you were probably taught to leave that to your master electrician or gaffer; however, reality dictates that the availability of equipment most often forces you to pick from a choice of dimmer packages, usually in groups of six channels of a particular wattage: 2.4 kW or 6 kW. Not many 3.6-kW, 8-kW, or 12-kW dimmers are found on the

road. The 1-kW dimmers are usually packaged in larger groups, a common quantity being 72 per rack (see Chapter 12). Always leave some dimmers with unused capacity so that, if and when you have a failure, you can gang channels. Better yet, leave two of your largest capacity dimmers as spares in the first place.

LAYERING

Besides basic position and angle, look to another idea for most of your visuals: *layering*. This is the process that creates depth and separation by using different shades or saturations of color. Except for effects, I rarely bathe the stage in a single color except between songs. That gives no visual depth. With a conscious consideration of color, coupled with intensity, you not only can direct the audience's attention to a specific part of the stage but you can also put the total stage into perspective. Too often, I see shows that rely totally on the follow spot, with its dramatic shaft of bright light, to accomplish this task. That is the easy but boring way out. The source of light is important, but so are the hue and tone of the colors used on stage. Many readers will believe they have to put a lot of different colors on the stage all at once, but for a more sophisticated effect you can layer in a single color by hue variation. Also, do not forget that the absence of light also contributes to the overall effect.

I first realized how important this idea can be during my early television lighting assignments. In video, the camera cannot show depth of field; lighting directors must accentuate depth by means of backlight and differences in intensity between the foreground and the background. That same approach can be applied to concerts when the audience can be as far away as 150 feet from the stage. In theatre, layering can be applied to subtly draw attention to a particular part of the stage. Here, again, not just the choice of luminaires and angle but also the mix of colors you use can make a better, clearer design.

LAYOUT AND SYMBOLS

Lest anyone think that with the drawing of the light plot the design is finished, let me talk about the next step in my approach to design. I have often marveled

---

[1]Ben Bova, *The Beauty of Light*, John Wiley & Sons, New York, 1988.

at beautiful computer-aided drawing (CAD) light plots, only to be disappointed by the execution of the show. On the other hand, I have often seen light plots scribbled on graph paper or the proverbial envelope that are exciting and show great style; however, even when circumstances have made proper preparation impossible, the additional effort and time wasted explaining such crudely drawn plots cannot be defended. The designer owes it to the technicians to give clearly understood and accurate instructions, and they are best conveyed by the properly drawn light plot. Figure 5.2 provides an example of a concert light plot for a festival; it doesn't follow the USA layout rules but is quite effective for touring designs.

As you study theatrical lighting, it will become apparent that there are several schools of thought regarding technique. Some styles are created by teachers at a particular institution of learning, or you might learn the so-called Broadway method perpetuated by the USA/IATSE union examination. The Broadway style prevails, but some West Coast designers have made changes that reflect a more open thinking. (Keep in mind that if you do wish to practice your craft on Broadway you will need to learn it as required for the exam.) Add in the distinctive style of the British school of design that has been injected into the United States and you have an evolving system that no one can agree upon. The United States Institute for Theatre Technology (USITT) took years to agree on standard symbols for lighting instruments.

Many concert and even successful theatre designers have had no formal schooling (see Chapter 23, where four top designers talk about their educations). Everyone will either succeed or fail based on what they put forth; no one method works for everyone. This is a highly creative field that always awards style and ingenuity. Partly because a portion of these designers are not formally trained and therefore don't know the established rules, they tend to simplify the plot so anyone can understand the color, circuit, and control channels at a glance. The major differences between the two styles of drawing (the traditional *versus* the modern) can be categorized as:

1. *Templates*—The traditional method uses both top and side symbols, while concert designers use only top symbols or even simple circles.

2. *Electrical hookup chart*—Traditionally, this chart is similar to electrical engineering drawings, where a line is drawn joining all luminaires of a circuit. Concert designers tend to use a number inside a symbol to represent the patching.

3. *Presentation*—The East Coast, or more traditional, method uses separate sheets of paper for the many different details needed to complete the design. The concert and Las Vegas approach is to include all the information on one sheet.

Over the years, I have evolved a method I believe is even clearer. Some variation of this method can now be seen on many concert lighting plots. The light plot shown in Figure 5.3 of a 2008 REO Speedwagon tour has all the information on one sheet that is needed to complete the color and hanging of the show.

I have always had trouble printing numbers clearly on drawings, especially when trying to get the color number, circuit number, and control channel number all in the same space. Although a computer drafting program eliminates this issue, a move toward adding more symbols to the drawing is founded on an idea used by Len Rader, who for many years was the head electrician at the MGM Grand Hotel in Las Vegas. Every light plot that came to him was different, and he said a lot of time was wasted trying to get the crew to understand what the plot represented. So, he started redrawing the plots and keeping them on file so that when the artist returned the load-in was much simpler. On one sheet, he had the circuits, dimmer assignments, and color in a form that everyone on his crew could understand. I adapted his form, which I found easy to understand, and I began using this style all over the world with great success. The plot shown in Figure 5.4 is from a Las Vegas show with Smokey Robinson.

The rectangle (symbol for circuit number) behind the luminaire symbol is blank. When the show is not touring with dimmers, it must be left up to the house to assign their dimmers. Existing house circuits are assigned by staff electricians to best facilitate the hang in a particular house. Luminaires are pointed in the desired direction of focus rather than straight

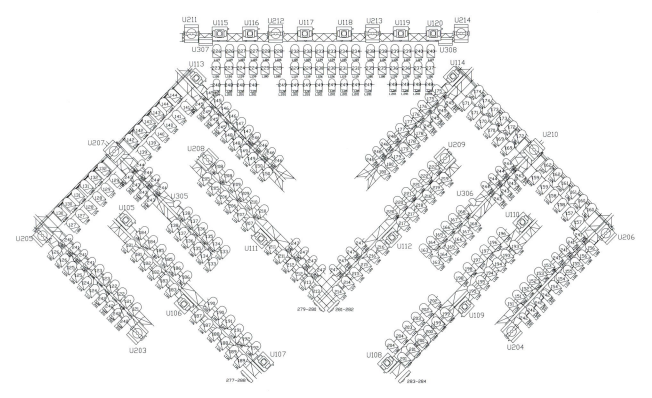

Instrument Count

Venue:
Designer:

Type	Lens	Count
ACL		96
Atomic 3000		8
Coemar CF1200 Spot		14
Mac 2000E Profile	Standard	20
PAR 64		480
PAR 64 Stubbi		24
Thomas 8 Lite Par 36		20
		662

Colour Count

Venue: (All Layers)
Designer:

Venue: Show:
Designer: Assistant:

Colour	Type	Count
L105	10" Colour Frame	80
L106	10" Colour Frame	76
L119	10" Colour Frame	84
L126	10" Colour Frame	82
L139	10" Colour Frame	58
L152	10" Colour Frame	40
L180	10" Colour Frame	60
L201	10" Colour Frame	96

SwedenRock 2008

Lighting Design: Peter Lumini
Venue: Festival Stage

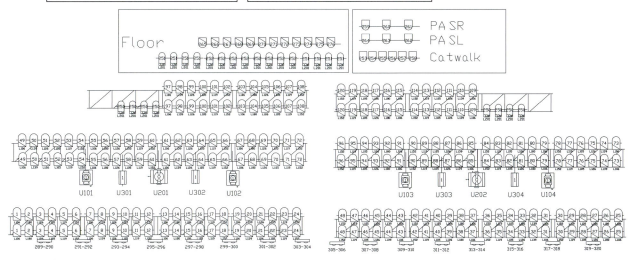

FIGURE 5.2 Sweden Rock Light Plot (2008). (Designed by Peter Johanssen.)

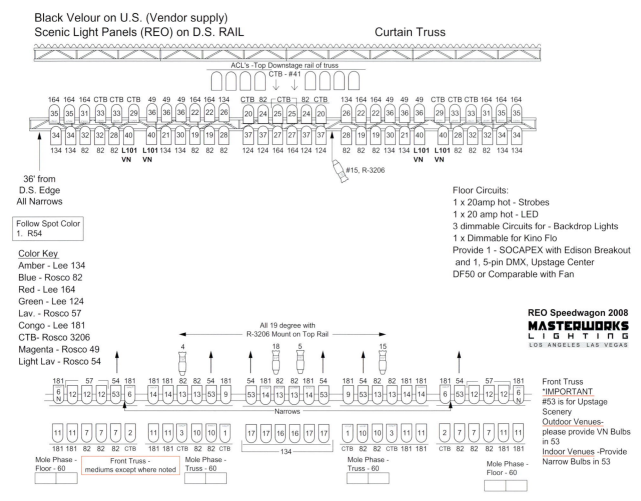

FIGURE 5.3 REO Speedwagon two-truss lighting Plot (2008). (Designed by Paul Dexter.)

side-by-side, as the USA exam and the more traditional method require. The 5-foot marks allow for a quick reference to luminaire placement along the pipe.

As mentioned earlier, computer drafting programs have enhanced the clarity of these drawings. Many CAD programs are available. Some are designed specifically for theatre, others for concerts and event production or even architecture. VectorWorks by Nemetschek, North America, Inc., is without a doubt the most popular software among Broadway, theatre, and concert designers. Although you will develop a favorite, all such programs have a common link. The work can be changed time and time again without starting over. This has been a boom to designers who need to make several versions of a tour plot. A fuller discussion of CAD programs is presented in Chapter 21.

HANGING

The final step in your design is not deciding whether to mount the luminaires on a pipe or on a prerigged truss. What looked good on paper may be junk in the air. As the designer, you not only must work closely with the supplier to ensure accurate reproduction of the design but must also be open to suggestions concerning changes that will improve the repeatability, packaging, and focus during the tour. If it is a *one-off* gig, the access to luminaires must be well thought out because of scenery or band placement. It will be too late once the load-in starts to try to solve a major problem. It might be wise to use a truss so the luminaires can be focused by someone walking or crawling on them (with the proper fall protection

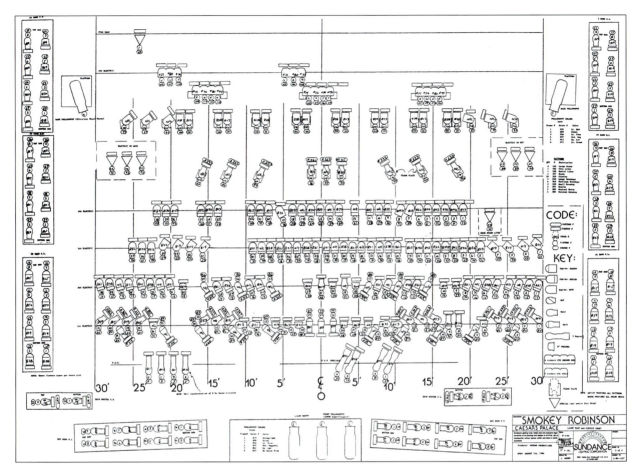

FIGURE 5.4 Light Plot for Smokey Robinson. (Designed by James Moody.)

system in place; see Chapter 9). Restrictions as to circuit availability, drapery obstruction, and trim height must all be considered well in advance. I never leave the theatre during load-in on a one-nighter. Designers not attentive during rigging do a great disservice to themselves and the crew and show a lack of respect for the crew's efforts on behalf of the show.

Often the house stagehands will have suggestions that will help simplify the rigging and focus. It is like hunting or fishing in that the locals know the woods and best fishing holes. The designer must keep an open mind to suggestions and not become defensive. It not only helps to have the crew feel part of the show because you listen to their ideas but also shows your good sense to know that someone else can look at the problem and possibly come up with a better solution.

SAMPLE LIGHT PLOT

The light plot shown in Figure 5.5 illustrates a typical rock & roll band setup. It is made up of all conventional lighting luminaires and no moving luminaires. Replace the keyboards with a steel guitar, and you have a country band. This could even be a jazz group. The back truss and sidelight tree design is the starting point for all concert design. Heavy backlight is the trademark of concert lighting.

The band backlight washes could be split so the drummer is separated, as he has no backlight specials. Some people split the sidelight left and right so they can have, for example, a red sidelight from stage left and an amber color from stage right. The real controlling factor is the number of dimmers available.

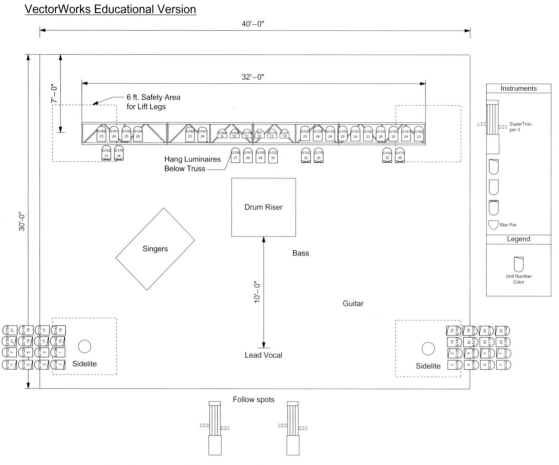

FIGURE 5.5 Basic concert light Plot. (Designed by James Moody.)

A more ideal control of these luminaires would be

- 4 band backlight washes (4 luminaires each)
- 4 drummer specials (1 luminaire each)
- 6 band specials (1 luminaire each)
- 6 lead-singer backlight specials (1 luminaire each)
- 8 band sidelight washes (2 luminaires each)
- 8 lead-singer sidelight washes (2 luminaires each)
- 36 total of channels of control

That quantity hits the nail on the head. Most consoles come in 24, 36, 48, 60, 72, or 96 channels of control. I insist on spare dimmers and control channels on the road. That way you can adjust at the last minute if the failure of a dimmer or a control channel occurs. I also make sure I have firmly fixed in my mind, well in advance of the tour, what control I could give up in order to solve a problem.

Preparation is critical. Using this plot as an example, first I would give up sidelight circuits, and then I would gang drummer backlight to band washes. In either case, the dimmer capacity must be high enough to accept this additional load, and advance planning can cover that point. Always have a few dimmers that can accept additional loads. If some moving luminaires were added to this plot, you should already have thought out what they might be able to compensate for if you were to lose a critical conventional luminaire.

The colors I would select change, depending on whether the lead singer is male or female and even whether they are black, brown, or white. Without going into the basic theory of color as it applies to theatrical lighting, it should be easily understood that any color (actually the absence of a portion of the visible light spectrum) projected onto a color will affect how the human eye perceives that color. Therefore, the designer must be sensitive to skin tone to most effectively illuminate the artist. The following color chart would be an acceptable starting point for most acts. Note that I have not designated specific color media. At this point, we are more concerned with the broader picture; exact color numbers are not yet important:

- *Band and drum backlight*
  - Red
  - Blue
  - Lemon
  - Green
- *Band backlight specials*
  - Light pink or lemon
  - Lavender or sunrise pink
- *Band sidelight*
  - Amber
  - Blue
  - Blue/green
  - Lavender
- *Lead-singer backlight*
  - Red
  - Lemon
  - Light blue
  - Lavender
  - Magenta
  - No color
- *Lead-singer sidelight*
  - Red
  - Amber
  - Blue
  - Lavender
  - No color

COLOR CHANGERS, MOVING LUMINAIRE, AND EFFECTS

A few color changers (the scrolling type can give 6 to 32 colors) can make even the simplest plot more flexible for an advanced designer. (Color changers are discussed further in Chapter 18.) The possible combinations that this multiplicity of color brings to the design would be much more advantageous on the

road than the increased quantity of luminaires and dimming that would be required to achieve the same effects. Remember that shipping space is critical.

Moving luminaire and truss-mounted follow spots could also add to the flexibility of the system without increasing the physical truss configuration. The pizzazz that is possible from even a simple layout can be used over and over again. Add the toys only after you have mastered the straightforward plot or you might get in over your head. Complexity can add confusion, and the time it might take you to make a decision can spell disaster.

VARIABLES

Lighting fascinates me because it never looks the same twice, even with the reliability of a computer console. The atmosphere of the room changes from hall to hall, and people's perceptions are slightly altered. Smoke in a night club or at concerts makes the light stand out more. Light from exit signs or candles on the tables can change the darkness level (black level, in TV parlance) from show to show.

Other variables are voltage to the lamps and changes in position and elevation. When you move the show from one hall to another, the voltage could change. Not many halls have exactly 120 volts at the input service—it often varies between 108 and 120 volts—and then when dimmers are added they will drop another 2 to 4 volts on the output side. String 50 to 100 feet of portable cable to the lamp, and it is rare to have 120 volts at the lamp filament. Therefore, the color temperature changes each day, and the color won't be exactly as planned. With many colors, this will mean visibly altered colors.

Moving the trusses or truss upstage or downstage a couple of feet makes a difference in the angle at which the light will strike the performers; this alters the shading and contrast of your lighting. Another variable is in the follow spot positions from hall to hall. Are they straight on or at a 45-degree angle to the artist? Every variable will make a difference from show to show and alternate plans must be considered in advance.

Twelve designers can take the example in Figure 5.5 and create 100 different looks. That excites me,

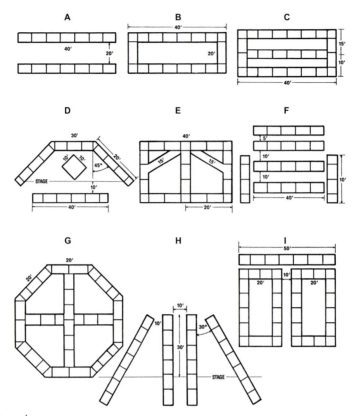

FIGURE 5.6 Examples of truss layouts.

and I think concert lighting releases the same creative imagination as theatre. We must take 50 artists, all of whom perform with similar staging, and make each look different.

One final word of advice: Experiment, experiment, experiment. I do, even with a show that is already touring. If I am running the board, I may try different things. If you are not sure how your client will react to unannounced changes, you might want to play with the changes at sound check to get a reaction before adding them into the show. Sometimes it is better, sometimes it is a toss-up, and often it is worse than my first idea. But, I still try and I will keep on trying to bring clarity and definition to my design.

MORE COMPLEX DESIGNS

After you become proficient at a simple, straightforward plot, you can move on to some more grandiose designs. This field of design loves the outrageous.

The creative freedom begins with the structures you design. Trusses come in all lengths, shapes, and load capacities. At this point, simply think of geometric patterns and designs. The examples in Figure 5.6 are only a few basic ideas for possible grids. More than making a pretty design, the objective is to choose light angles that create mood and drama.

Although there is the practical limitation of load capacity, I have seen trusses that defy structural logic. In Chapter 13, we provide the technical information required to help you decide on a final truss layout. I would caution that before becoming married to a design you consult with a rigger or at least the lighting supplier for advice on how your dream can be accomplished. However, at this pure design phase of our discussion, you can see how the structures become a real element of the look. Keep in mind that, in almost all of the touring systems, the structures will be in view of the audience. The layout of the structures can be both functional and an aesthetic element of your show.

6

CUEING THE MUSIC

While doing the layout and color chart, you must be thinking ahead to how the cues will work with your plan. Does the layout give you the degree of flexibility you want? Does the color chart work as a palette in harmony with the music? A separate design skill now enters the picture. The designer must take the raw data from the light plot and make the design come to life in the same way a musician does with notes in the score.

From the first time you hear the artist's music, you should be mentally planning the choreography of your lighting, breaking down the music into cue points. Songs are like movies and novels; they have formulas. If you listen to a song all the way through first before writing anything, you will hear the formula or flow of the sound. The usual formula can be identified as intro, verse 1, verse 2, bridge, and chorus, which repeat, followed by verse, bridge, chorus, and a buildup to a guitar solo or instrumental melody, bridge, chorus, and end. Each and every song has some variation on this formula—not exactly as described here because some portions will be reversed or added in a different order—but most commercial pop songs have what is referred to as a *hook*. This is generally the title of the song and is the part of the song that fans remember and sing. Often, the hook has an anthem style and is considered to be the chorus. The bigger the anthem, the bigger, more intense, and more animated the lighting and effects should be.

CUES, SONG PUNCTUATION

The concert lighting designer's job is to complement the song and enhance the experience, not overshadow it. I have seen concert lighting where every song looked exactly the same with lights flashing to every drum beat and guitar riff. It was so nauseating that my favorite part was the blackout at the end.

The artists spend considerable time writing their songs, rehearsing them, producing and recording them. This labor of love has structure and deep meaning to the artist and loyal fans. It is not possible to take short cuts in writing and producing music and have it be successful to the degree that people will pay to see their favorite artists in concert. Concert lighting is a visual translation of that music, and the lighting requires an equally professional approach to the process of creating cues, just as much as much it did to create the music. Sometimes a song is easy to translate into lighting, such as a ballad or an anthem, and a sometimes it isn't, such as a complicated rock song that is thematic with extreme highs and lows, even tempo changes. Regardless, each song requires attention to detail so it will not be anticlimactic and you will have something left in your bag of tricks for the ending. The only way to accomplish that is to be patient and take the necessary time to structure the song and schedule effects for the entire song set.

I approach new music in the same way I was taught to read a script. You read it three times: first for the fun of reading it, second to find the written stage directions that affect lighting cues, and a third time to find potential cues hidden in the dialog. I approach a song the same way. The first time, I listen to it for the sheer enjoyment, and the second time I try to pick out the cue points, accents, verse and choruses, and tempo changes. The last time I listen for the "color" of the song: happy, yellow and orange; moody or melancholy, blue and lavender; fiery and hot, red; soothing, green and blue–green. Listen to music with your eyes closed, and try to visualize the colors that might be associated with it.

PICKING THE CONSOLE TO MATCH THE CUES

Analyzing a song helps you determine the type of lighting console you will need. Manual consoles are almost a thing of the past, but many hybrid consoles allow manual operation with recorded cues that can be accessed when needed. I was one of the first to take a computer console on tour, the early Light Palette by Strand Lighting, Inc. My reasoning at the time was that the artist, John Denver, had a very large song list to draw from and he was performing without an opening act, thus making for almost 2 hours of music. The computer allowed me to design each look using smooth, subtle cues that were absolutely repeatable and to write them directly into the memory. I devised a way to retrieve each song no matter what the order.

FIND YOUR LOOKS

My years of experience allow me to take a few shortcuts now, but I am sure I still take these steps subconsciously:

1. Listen to the song; try to pick up a lyric, a musical phrase, or the dynamics that make the general statement of the song.
2. Translate the song into a primary color.
3. Find the high point of the song (it may not be the end).
4. Find the repeating portions of the music, such as the choruses and verses.

This process will probably lead to four or five looks: the opening, the chorus, the verse, the solo

spot, and the ending. Changes between verse and chorus may be repeated several times. Another cue may occur at the *turnaround*, a musical device found in most pop music that allows the songwriter to repeat a melody by interjecting another musical phrase between similar themes. Are the cues going to be bumps or slow fades? It does not matter what they are. What is important is that they each act as musical punctuation not just flashing lights.

After doing this for all of the songs, look at the song order, or *set*. See if you have the same color patterns for songs that are played back-to-back, and do not hesitate to change the colors to avoid repeating a look in two adjacent songs. Certainly, a look can be repeated later in the set. If the same color simply must be used, try to change the direction the color comes from; for example, use an amber backlight for the first song but an amber sidelight or an amber follow spot for the next song, with white as a backlight.

CUE PLACEMENT

How do you remember where the cues go? Most artists do not follow a script, such as a play or an opera does; however, when lyric sheets are used, the traditional theatre stage manager cue book method could be used (Figure 6.1). Some touring lighting designers, including Jeff Ravitz, and almost all of the Las Vegas show designers refer to lyric sheets (Figure 6.2) for cues, much as a play does. This is a theatrical form not normally used in concert touring but sometimes handy in longer running shows, such as Las Vegas. Although it is not general practice in Las Vegas for the lighting director to leave after a day or two of a 2-week run, the board operator could follow the lyrics, call the follow spots, and execute the board cues from those notations. I have done this on two occasions when the shows were on long, open-ended runs, and it worked.

THE CUE BOOK

Usually the lighting designer runs the console, except on the biggest tours. If not, the person doing so is traveling with the show and is as familiar with the

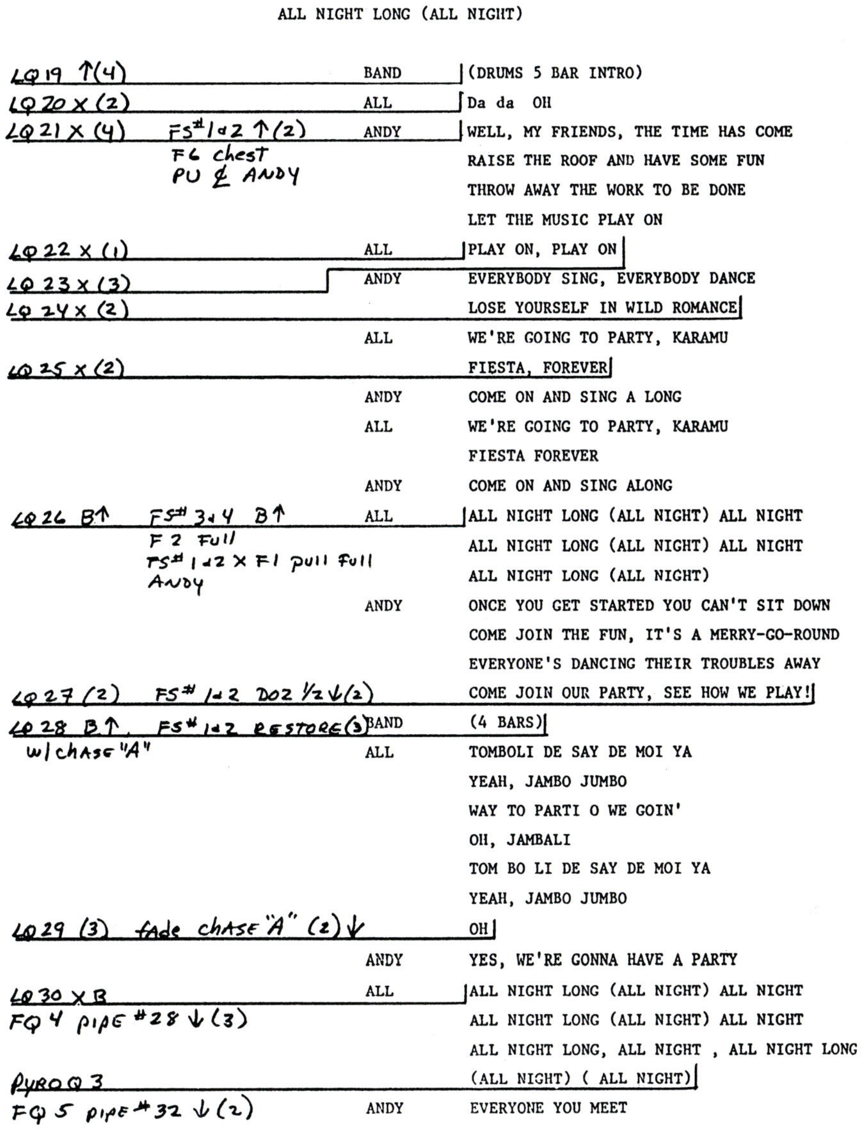

ALL NIGHT LONG (ALL NIGHT)

```
LQ 19 ↑(4)                              BAND    |(DRUMS 5 BAR INTRO)
LQ 20 X (2)                             ALL     |Da da  OH
LQ 21 X (4)    FS#1&2 ↑(2)              ANDY    |WELL, MY FRIENDS, THE TIME HAS COME
               F6 chest                          RAISE THE ROOF AND HAVE SOME FUN
               PU & ANDY                         THROW AWAY THE WORK TO BE DONE
                                                 LET THE MUSIC PLAY ON
LQ 22 X (1)                             ALL     |PLAY ON, PLAY ON|
LQ 23 X (3)                             ANDY     EVERYBODY SING, EVERYBODY DANCE
LQ 24 X (2)                                      LOSE YOURSELF IN WILD ROMANCE|
                                        ALL      WE'RE GOING TO PARTY, KARAMU
LQ 25 X (2)                                      FIESTA, FOREVER|
                                        ANDY     COME ON AND SING A LONG
                                        ALL      WE'RE GOING TO PARTY, KARAMU
                                                 FIESTA FOREVER
                                        ANDY     COME ON AND SING ALONG
LQ 26 B↑    FS# 3&4  B↑                 ALL     |ALL NIGHT LONG (ALL NIGHT) ALL NIGHT
            F 2 Full                             ALL NIGHT LONG (ALL NIGHT) ALL NIGHT
            FS#1&2 X F1 pull Full                ALL NIGHT LONG (ALL NIGHT)
            ANDY                        ANDY     ONCE YOU GET STARTED YOU CAN'T SIT DOWN
                                                 COME JOIN THE FUN, IT'S A MERRY-GO-ROUND
                                                 EVERYONE'S DANCING THEIR TROUBLES AWAY
LQ 27 (2)   FS# 1&2 DO2 ½↓(2)                    COME JOIN OUR PARTY, SEE HOW WE PLAY!|
LQ 28 B↑    FS# 1&2 restore (3)         BAND    |(4 BARS)|
    w/chase "A"                         ALL      TOMBOLI DE SAY DE MOI YA
                                                 YEAH, JAMBO JUMBO
                                                 WAY TO PARTI O WE GOIN'
                                                 OH, JAMBALI
                                                 TOM BO LI DE SAY DE MOI YA
                                                 YEAH, JAMBO JUMBO
LQ 29 (3) fade chase "A" (2)↓           OH|
                                        ANDY     YES, WE'RE GONNA HAVE A PARTY
LQ 30 X B                               ALL     |ALL NIGHT LONG (ALL NIGHT) ALL NIGHT
FQ 4 pipe #28 ↓(3)                               ALL NIGHT LONG (ALL NIGHT) ALL NIGHT
                                                 ALL NIGHT LONG, ALL NIGHT , ALL NIGHT LONG
Pyro Q 3                                        (ALL NIGHT) ( ALL NIGHT)|
FQ 5 pipe #32 ↓(2)                      ANDY     EVERYONE YOU MEET
```

FIGURE 6.1 Lyric cue sheet, including a method for noting fly and other cues. (Designed by James Moody.)

cues as the designer is. So, if the designer is with the show and knows the music inside out, why bother with a formal cuing method? There are two good reasons: the possibility of illness or accident and the fact that the designer is most likely working for several clients, and without total recall it is difficult to remember all of the cues for each show. Also, after a few years of touring, an artist can build up an extensive repertoire of songs, not all of which are in the current show. As the tour goes on, old songs may be substituted or new songs tried out. A lot of mental agility is required to stay on top of all that music. John Denver had over 50 songs that the band rehearsed and could play on any show, and he often substituted without warning. Jeff Ravitz told me that on the most recent Bruce Springsteen tour he had a cue book that was 4 inches thick!

Another way to generate a visual plan for a song is to count out and note each eight bars, just as a choreographer counts out the measures (1 and 2 and 3 and 4 …), and then write the number down and mark any musical changes. Jeff Ravitz, who includes

Ringo Starr and Bruce Springsteen among his many clients, says that he cannot possibly keep up with all the cues from all the songs, so he uses this method.

Take a sheet of paper and listen to the song. Count out each eight bars and then strike a line, put another line after the next eight bars, and so on. If something changes before you get to your count of eight, strike a line and insert the number of bars above it. This way, even if you must call the cues after months or years, you should be able to count out the measures and get the timing right. It is also a great method for knowing when the end of a solo that does not present a nice clean tag is coming to an end—that is, if the artist always plays the song the same way. In any event, it is a good way to jog your memory and helps you visualize the song.

In Paul Dexter's form of the cue sheet (see Figure 6.3), he generally writes a few words of lyrics after

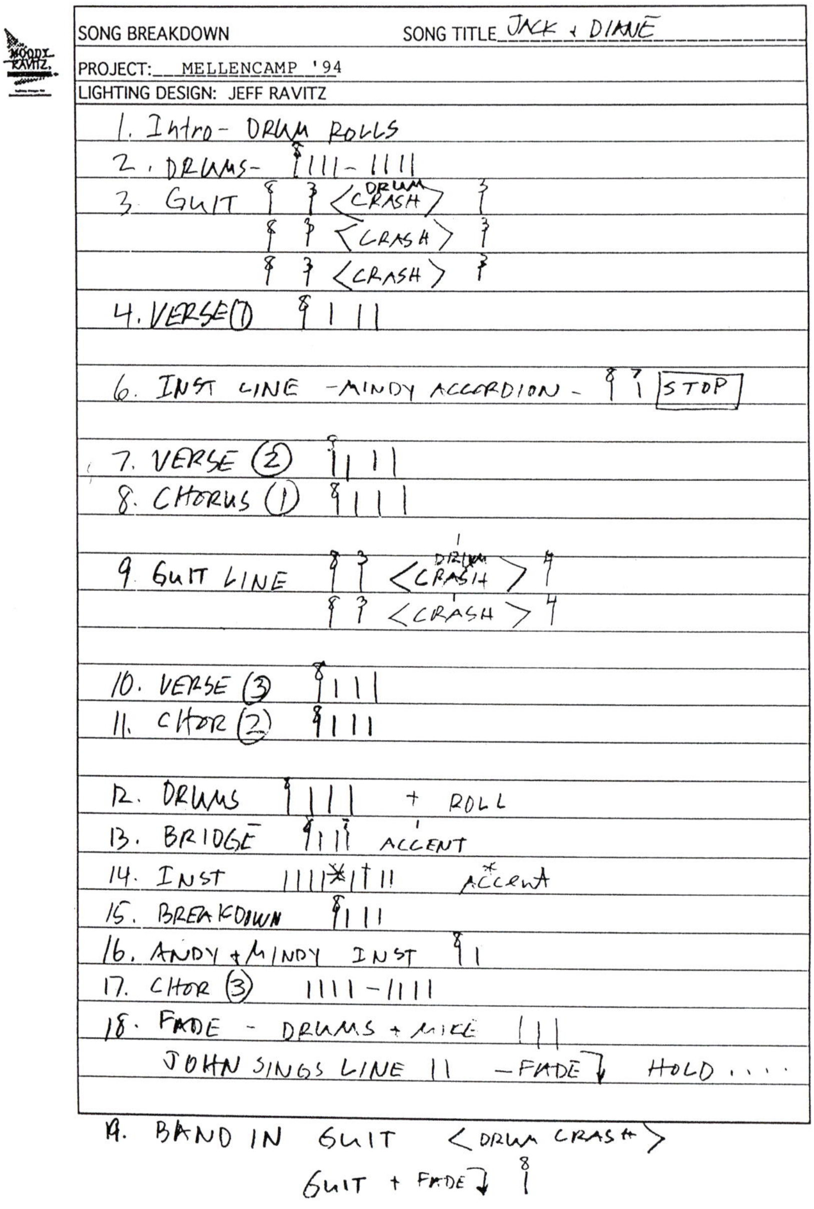
SONG BREAKDOWN SONG TITLE JACK + DIANE
PROJECT: MELLENCAMP '94
LIGHTING DESIGN: JEFF RAVITZ
1. Intro- DRUM ROLLS
2. DRUMS- 1111-1111
3. GUIT <DRUM CRASH>
<CRASH>
<CRASH>
4. VERSE (1) 1111
6. INST LINE -MINDY ACCORDION- STOP
7. VERSE (2) 1111
8. CHORUS (1) 1111
9. GUIT LINE <DRUM CRASH>
<CRASH>
10. VERSE (3) 1111
11. CHOR (2) 1111
12. DRUMS 1111 + ROLL
13. BRIDGE 1111 ACCENT
14. INST 1111*1111 ACCENT
15. BREAKDOWN 1111
16. ANDY + MINDY INST
17. CHOR (3) 1111-1111
18. FADE - DRUMS + MIKE 111
JOHN SINGS LINE 11 -FADE HOLD....
19. BAND IN GUIT <DRUM CRASH>
GUIT + FADE

FIGURE 6.2 Eight-bar notation cue sheet. (Designed by Jeff Ravitz.)

each of the labels to remind him (and others) that a change is going to occur—for example, from the chorus to a lead-in for a guitar solo with a lyric such as "time for me to fly." He sees key words on a page that serve as a warning before a change. He said that, "The cue sheet in this basic form is a place to write new discoveries and notes and elaborate exactly in the spot where the effect is going to happen, such as a backdrop change or Kabuki drop, a truss movement or projection cue." No matter what is supposed to take place visually, it can be organized with the cue sheet.

SIGN OF THE SOUTHERN CROSS

	Verse	"…we feel"
	Bridge	"…Crystal Ball"
2	Inst. Melody	
2	Breakdown - -Sustain (fades to dark)	
	Verse	…"seen around"
2	Chorus	- - - - - TAG "…fade away"
2	Inst. Melody	
	Breakdown Sustain	
Verse	(Inst. Hits)	"…sail the ship of time"
	Chorus	- - - - - TAG …"fade away"
	"I can't accept it anymore"	
5	Crescendos - - - - - 4 x 3 hits to Guitar Solo	
	- - Drums - -	
2	Inst. Melody	
	Breakdown - - fades	
	Verse	"…never seen around"
1	Chorus	- - - - - TAG to (2) Inst. Melody
	Verse over Inst. Melody	
4	Inst. Melody	4 X 3 Hits - - Drums -- out

FIGURE 6.3   Cut sheet example. (Designed by Paul Dexter.)

THE CUE CARD

I use 4 × 5 index cards as forms. Full-size 8-1.2 × 11 pages are too big to place conveniently on the console and scan quickly. What kind of information you place on the card is the trick. I have a format I have used with a few variations for many years. Figure 6.4 shows the form for a manual console, and Figure 6.5 shows a form for a computer console. Although they may not seem to differ much, the detail in the Preset column (column 4) is simpler for a computer, whereas a total breakdown of dimmer numbers and intensity is required for a manual board.

Cue Number

I use A–B–C instead of numbers for cues, so if I am calling board cues along with follow spots no one can confuse cue 1 with follow spot 1 or frame 1. Also, I do not believe in consecutively numbering cues straight through the show, because the song order can and often does change during the course of the tour.

SONG TITLE		MANUAL BOARD		SHOW ORDER (11) (Mark each night in pencil)	
CUE#	TIME	GO	BOARD	F.S.	
A	↑ (3)	Music	4/6 10/F 12/8 14/7 20/F	1/F6 PU Full	
B	X (3)	Verse	1/F 2/F 6/8 12/4 ↑ 22/F 23/8		
C	B ↑	Piano solo	27/F	B.O.	
D	↑ (1)	Verse	Same as B	Restore	
E	X (2)	Chorus	9/F 10/F 14/2 18/6 26/9		
F	B.O.	End			
		RESTORE			
G	↑ (3)			1/F N/C Full	

FIGURE 6.4   Manual console 4 × 5-inch cue card. (Designed by James Moody.)

CUE#	TIME	GO	NOTES	F.S.
	SONG TITLE	COMPUTER BOARD		SHOW ORDER (4) (Mark each night in pencil)
10	X↑(6)/(10)↓	Song starts	Watch drummer	
10.1	Add ↑ (4)	Build band		
11	X (3)	Vocal		1/F6 Full 2/F 3 hips
12	X (2)	Chorus		1/F4 B X
13	X (4)	Verse		X 1/F6 (4)
14	X (2)	Chorus		
15	B X	Guitar solo	Stage right	1/B.O. 2/B.O.
15.1	B X	End solo		1/F4 PU guitar
15.2	B X	Vocal		1/PU singer 2/F3 H/S
16	X (2)	Add backlite singers		
17	X (4)	Tag	Lead w/ F.S.	All Fade (3)
18	↓ (3)	End song		
19	B ↑	Restore		

FIGURE 6.5 Computer console 4 × 5-inch cue card. (Designed by James Moody.)

Cue

The cue is the downbeat, instrument, lyric, or whatever you use to indicate when the action should take place. Because the shows are not scripted, I use very simple ones, such as "band starts," "first lyric," "sax solo," "chorus," "end song," or "restore."

Action

The action notations (under the Time column) can change slightly depending on whether you are using a computer or a manual board. A computer board is simply noted—for example, P.S. 11 × 12 (2), which means preset 11 is to cross-fade to preset 12 on a 2 count (2 seconds). A variation can be written as P.S. 10 @ (6) and previous cue @ (10), which means to pile on preset 10 and fade the previous preset on a 10-count fade. Note that the parentheses always indicate time; the number inside the parentheses indicates the seconds. A mark of ↑ indicates a bump up, and ↓ or BO indicates a blackout. Brackets, [ ], can be added for other notations. It is also possible on a computer board to use a channel in the manual mode. This is indicated as M d36/7 (8), which means to manually add dimmer 36 to a level of 70% on an 8 count. On a manual preset board, the notation

could be A × B (3), which means that scene A cross-fades to scene B on a 3 count. The notation A @ B @ (3) indicates to add scene A and fade scene B on a 3 count. Another action would be A @ (3), which means to add scene A on a 3 count.

Preset

The Notes column provides references to the actions or things that remind you of why the cue is done. On the manual board, the Board column indicates each dimmer level specifically—for example, d9/F d10/F d14/2 d18/6 d26/9 × (2), which means dimmer 9 level full, dimmer 10 level full, dimmer 14 at 20%, dimmer 18 at 60% and dimmer 26 at 90%, cross-fading on a 2 count. Similarly, d14/3 @ (10) would mean to fade out dimmer 14 on a 10 count. This could also be used to lower the dimmer level— for example, d14/3 @ (5)—from the previous level to a lower level. Because a fade out and a change in level look the same in the notes, some people underline when it is a level change only.

Follow Spot

The notations under the Follow Spot (F.S.) column can be complex if more than one reference must be considered. As a general example, however, follow spot notations might read (john) F#6 @ (3), which means to fade up on John in frame 6, full body on a 3 count. The use of ↑ @ indicates a bump up, and ↓ @ or BO @ indicates a blackout the same way it does with dimmers. You can develop letter or numerical notes for the size of the circle of light needed; for example, I use "1/2" for a waist shot and "HS" for head and shoulders. You can make up anything that is easy for you to remember when you are looking at your cards.

Miscellaneous

A Miscellaneous column could be added for additional notations such as effects cues, set or curtain cues, warnings, etc. I do not use colored pencils on my cards, because it is time consuming and the colors can be misread under red or blue work lights. You may have been taught that warnings were indicated in one color pencil and go's in another; however, we do not normally have the script format to do that, so it doesn't work for me except in Las Vegas venues. Be aware, though, that the work light may have a color filter over it, and it may be difficult to see what color you have used on your cue cards.

VERBAL COMMUNICATIONS

When calling the follow spot cues, be friendly but firm and do not joke with the operators. If some operators try to take the instructions lightly and make inappropriate comments, then I say that there is not time for that now. Paul Dexter said he waits until 15 to 20 minutes before a show to put pressure on the operators so they stay focused.

The way you communicate to a house light operator, a board operator, or a follow spot operator will directly affect the smoothness of the show and accuracy of the cues. It is truly an art form unto itself. Chip Monck has always been highly regarded for his effective follow spot cuing, especially on the 1972 Rolling Stones tour. He is certainly one of the pioneers in concert lighting and developed many of the cuing techniques still being used.

When I was an air traffic controller, I learned something that is generally overlooked. Speech pattern, meter, and accent are very important. If the operator cannot understand you, he will not be able to do the cue properly. Your diction and local accent can frustrate the operators and reduce your comments to unintelligible noise. This is not to put down any regional accent; it is just a proven fact that how you speak will affect the understanding of the listener. The Air Force says that the Midwestern accent can be best understood by every English-speaking person. Meter is important because the speed and inflection you use in your speech will affect the operator emotionally. Obviously, if you are screaming or talking very fast over the intercom, you do the greatest harm possible, because the operator will tune you out. He will think you are not in control, and that is disastrous. As air traffic controllers, we were taught to speak at a rate of 60 words a minute, and I still find this a good speed for show cuing. I realize that at times you have a lot to say in a very short time between cues, but I would suggest that you take the time to find a way of saying it in fewer words, rather than speeding through to get it all. After all, if you say it so quickly that no one understands what you've said then you might as well not say it at all. All of these points can be worked on in a speech class if you have the opportunity, but you can also learn by speaking cues into a tape recorder and then assuming the role of the operator. See if you would listen to yourself. Or, have a friend who is not in the business listen and try to repeat what you say.

CUEING FOLLOW SPOTS

The object is to simplify, particularly if you are operating the console. You will not always have union or experienced follow spot operators, so if you always maintain a keep-it-simple approach, no matter what, then you can depend on some continuity from one show to the next, even when different people are involved every night.

Most often follow spots already have a number assigned to them by the facility that the operators are used to hearing, but the numbers may also represent a simple position. As a very basic example, 1 and 4 might be the lead singer, 2 the guitar player, and 3 the bass player. In this case the following spots will be assigned to operators with very little position deviation during the entire show.

Some people like to meet with the operators by the side of the stage before the show to point out the stage positions and lay out how the follow spots will be numbered or called. Other designers prefer not to meet the operators face to face before the show, only asking that they be in their positions 15 to 20 minutes prior to the show. With the house lights on, you can ask each of them to direct their lights on to the area where they will be assigned. This is when the designer can adjust the iris sizes, balance the intensity, and review some of the rules.

Because the designer has so much to look at and keep track of, most tell the operators to fade out on their own if the performer walks off the stage; they should not follow the performer and wait for him to return. If the performer is getting a drink, talking to technicians about the sound levels, or changing guitars, the operator should automatically fade out.

You can use the *key word* method of follow spot cuing. This is a method of calling cues by using words such as GO or OUT or FADE to prompt someone to react as planned, instead of relying on visual or mechanical signaling devices to call the cues. The pre-show speech, whether delivered with the operators standing around you or over the headset when they get to their positions, might go something like this:

> Good evening, ladies and gentlemen. I'll call the follow spots as spot 1, spot 2, etc. [*Indicate which light you are referring to; for example, northeast corner, left or right of stage, man in red shirt, whatever makes it obvious which position is which.*] I give all cues as READY ... and ... GO. Do nothing until you hear the word GO. It will indicate a color change, blackout, fade up, or pickup. Anything else you hear is information or prepping for an upcoming cue. All cue warnings are given as, for example, STANDBY spot 1 in frame 6 waist on Fred for a 3-count fade-up ... READY ... and ... GO. The counts are in seconds, and all fades should be evenly executed throughout the numbers.

Verbally count off so they can hear at what speed you count. Lay out the placement of all players, including items that would help the operators remember each musician. Some people use letters for areas of the stage. If the artist moves around the stage a lot, do such a plan in four or five areas across the stage and two or three areas deep. To close, I say:

> Thank you, ladies and gentlemen. I'm looking forward to a great show. Take your color and head for your lights; check in on the intercom when you are in position.

Some designers use the operators' first names to call cues, but you should do whatever is most comfortable for you. Using their names makes it more personal and eliminates one number that could be confused with all the other numbers being heard by the operators. During the show, I tend to talk a lot to the operators, mainly to make them feel as though they are a part of the show and not just robots who turn lights on and off. And, if you find that the operators like to chatter a lot, keep up a running dialog so you can call a cue anytime it is needed. But, be careful not to confuse them as to what is your chatter and what is actual cue information. I find this approach to be better than telling people to "shut up." Give people credit for their skills, and they generally respond with their best effort. If they have a general idea of what is about to happen, they relax and the tension is relieved. Certainly, this is not always the case, and I have been in positions where the less said the better; your own judgment must be used in each show.

SUMMARY

Whether you use cue cards or other forms, the important point is clarity and being frugal with cue words and notations. Cue cards are not necessarily designed so that another designer and board operator could walk in and take over right away. The cue cards are a summary of my personal preferences; they are only there to assist me in running a smooth show. When I use board operators, I do not write out their cards. I use mine to talk them through and let them do their own cue notations as they choose.

In the final analysis, the smoothness of your show will be directly related to your ability to cue both board and follow spots in time to the music.

Your choice of colors and angles will affect the audience, but not nearly as memorably as a late or missed cue. The rock & roll audience has become very sophisticated and realizes that the lights are a very important part of the show.

There has been a change in how some shows use follow spots. It is not unusual to see very rapid color changes and utilization of all six available frames with saturated colors. Because more video is being used, and the audience views performers on larger-than-life image magnification (I-MAG) screens, there is a tendency to keep a constant light on the performers at all times during a song at an intensity level that balances with the stage lighting. No matter where the camera is focused, you want the performer to look good. Even when no cameras are present, you may find this a comfortable method to use because it is a reliable way to maintain continuity. If you involve six or even eight other people in your show on the follow spots, chances are greater that something is going to go wrong. The less you rely on follow spots to do anything other than give the performers key light (front light), the better!

7
ROAD LIFE

Time becomes the most important factor when a truck is late. A show normally loads-in at noon and gets a sound check done by 6:00 p.m. (on average, depending on production complexity). When the truck does not arrive until 4:00 p.m. or 5:00 p.m., can the show be ready? It had better be! Here is where your organizational skills and efficient use of local resources are really put to the test. Problems on the road can be handled on a day-to-day basis if you stay calm and face each one with your mind open to several alternative solutions. Make your decisions as quickly as possible. The worst thing you can do is to delay making a decision. There is no right or wrong, but in retrospect some decisions are better than others. In the end, though, the show almost certainly will start at 8:00 p.m., with sound and at least some lighting. How you accomplish the tasks you are assigned determines whether or not you are a successful road designer; if not, learn from experience and move on. The road is not for everyone, and in most cases the road will choose you. Read Karl Kuenning's memoir, *Roadie: A True Story (at least the parts I remember)*. This a great view of one person's experiences of being on the road in the 1980s.

POWER SERVICE

After timing problems, the second most important problem can be finding a good power service to use. This is especially true at colleges. Most schools do not have field houses equipped with a bull switch or other power connection devices for use by road companies,

although this is not generally true of schools that have recently built a basketball arena. If the facility does not have a dedicated company switch for special events, power may have to be taken off the house light panel. That is very bad electrical practice and, in most instances, a code violation. But, when there is no other power within 400 feet of the stage, you make do. This is an area that requires a discussion all its own. The alternative is to have the promoter hire a "frequency-regulated" generator. This just means the generator has been fitted with electronics that keep it operating at a constant speed so the voltage isn't drawn down when an additional load is added. The best solution is to be as knowledgeable as possible; read up on the subject, or better yet talk to a licensed electrician who does temporary service connections for a living. Put in the artist's contract rider that a house electrician must be on hand at load-in. Even though you probably do not need power for an hour or more, it gives you time to discover if there is a problem and then time to solve it without delaying your schedule. If rigging is to be done, the crew needs power for their rigging motors as soon as they start so as not to delay the load-in.

FOLLOW SPOTS

Unless the venue is a dedicated performance venue, locating follow spots tends to be left up to people who are unfamiliar with the importance of their location to shows—for example, the person hired to erect a scaffold tower. Whether it is plain laziness or they run out of tower before they reach a platform that is no higher than the stage, more often than not this issue is about as important as the positioning of the portable toilets. As part of your advance procedure and as a follow-up with the venue or promoter rep, request that the platform be as high as possible (16 feet or more) and better yet on a scissor lift. Performers do not like to stare into follow spots that look like a freight train headed their way; also, a follow spot bouncing all over the set and offering no isolation can ruin your great stage looks. If the facility does not understand your power requirements, it usually does not have good follow spots or operators, either.

A good road technician can diagnose what is wrong with a follow spot without being at the light and is able to tell the operator how to correct the problem. Talking operators through the cues is another problem. Because the follow spots are the one lighting element not generally carried with the show (except those mounted on the trusses), the efficiency of the unit is tied to the operator's ability. Unlike Broadway shows, where several days of rehearsals with follow spot operators are accomplished before opening for previews, a concert starts with only a brief talk with the operators (see Chapter 6). When you give the first cue, you will see how the operator handles the light as well as the efficiency of the unit. There is no time to replace or repair the unit. I can safely say that follow spots have caused me more problems than any other element of a show. They are the great unknown factor.

STAGES AND CEILING HEIGHT

Stages themselves can be a major problem. Although the size may not be known before arrival, it is usually the way the stage is built that becomes the real issue. When the crew arrives to find an unsafe stage, either because it is not braced properly or because it is uneven, the delay in getting it corrected takes away valuable time from the load-in. Size variations should be considered in advance so the set and lights can be adapted—within reason. Large public arenas and auditoriums have sturdy portable stage structures. It is when the stage is built by people unfamiliar with the devices that are to be placed on it that problems

can occur. Concert artists generally bring complete portable lighting systems and, if ground supported, these structures add a tremendous weight to the stage. I have actually been told that the house crew thought everything would be okay because the stage had been big enough and strong enough for the high school graduation ceremony.

Ceiling height can also be a problem (see the discussion in Chapter 1 on Indian casinos). Problems usually arise in clubs and small facilities that were designed as multipurpose rooms. The ceiling may be 20 feet high, but if the stage is 6 feet high then you can only put lights up 12 to 14 feet above the stage. You must also consider what the ceiling is made of. Can it take the heat that the lighting luminaires give off? A low ceiling height is an enormous problem for backlighting, because the lamps will be at about a 20-degree angle to vertical (normally it should be 35 to 45 degrees). The light will spill out into the audience a considerable distance, making it difficult for people in the front sections to view the performer.

SPECIAL EFFECTS

Projected effects or special effects such as smoke, fire, and explosions can be an accident waiting to happen. Projected effects usually require a bigger or deeper stage. What happens when the stage just cannot be 40 feet deep, the screen cannot be moved upstage any farther, or the projector cannot be rigged where it is supposed to be placed? The show should carry both rear and front screens for projection. (Actually, a show carrying these effects should have had sense enough to check out these issues in advance.) Many of these problems have been solved with the advent of light-emitting diode (LED) screens (see Chapter 16), which do not require much depth and are very bright; also, digital luminaires have zoom features from 35° to 70° or more which will make composition for variances in distance. Pyrotechnical effects are another serious issue. Most cities require special licenses or permits to have flash pots or open fire on stage. Some cities flatly forbid the use of fire or pyrotechnics on stage. Where permitted, these effects must be handled with extreme caution. Generally, a licensed technician must be in charge.

ROAD LIFESTYLE

A lot has been written about rock stars and their drug use, sexual activity, and spending habits. Because the technician spends 16 to 20 hours a day with the artist, some of these excesses can rub off. The problems begin when the technicians and designers forget their purpose on the tour. It is not to be buddies with the stars nor to see how many pickups they can make at the concert. The good technician or sound engineer should consider the mounting of the show as the real reason for being there. Drugs and booze are both problems. A drunken technician is just as likely to make a mistake and hurt someone as a drugged one. Personally, I have had more trouble with drunken crewmen than drugged road personnel.

Most all of your hours on the road, particularly for lighting crew, are spent at the gig. The lighting crews are the first in and generally the last out. Depending on the size of the system, there is very little time for breaks, especially if it is a medium to large touring system or you are a crew of one that has to adapt a show to every venue. The work is very challenging. The hours can be a strain, exacerbated by a troublesome technical day or another crew member who isn't pulling his weight. For the most part, people who don't want to be there soon leave or they are fired, leaving only those who really want to be on the road and are staying for all the right reasons: They are passionate about what they are doing, and they have come to accept the lifestyle, embrace the good days and endure the bad, and learn to do it better the next time. The road life cannot be compared to any other situation because the circumstances are so different from any other profession—you are in a fairly autonomous position traveling with others who are like minded and with unusually famous individuals. That combination makes for a whole lot of fun, memories, and friendships that you can count on forever. It is an identity within an exclusive club. Another book I recommend is Dinky Dawson's *Life on the Road*. He started his rock & roll career as Fleetwood Mac's roadie in 1968 and became one of the most sought-after audio engineers by many top artists into the 1980s.

Outdoor performances—such as at state and county fairs and music festivals—bring a whole other set of issues and problems, and crews can sometimes feel as though they've gotten the short end of the stick. Such issues as food and restrooms come to mind. There are so many variables associated with outdoor stage shows that, more often than not, it is a lot more difficult to put on the show than if it was in a controlled indoor environment. Following are just a few reasons why rock shows shouldn't be outside.

Weather is the most obvious reason, of course— hours of relentless winds, exceedingly hot temperatures of 90° or above with matching humidity, a gusty cold front passing through bringing rain with it (and you didn't bring a jacket). Significant amounts of time are spent covering up equipment with rolls upon rolls of tarps at the first signs of rain while you're freezing and looking for a warm, dry spot.

Sitting on long bus rides gives ample time for crew banter. One day, with the movie *My Cousin Vinny* on in the background and a crew that had seen all possible types of gigs in the stretch of a week, the conversation was at a sarcastic high. It was determined that a tour could have shows in three categories: *primary* markets, which would be major cities such as Dallas or New York with good facilities; *secondary* markets, which are for the most part good venues but in places like South Bend or Oklahoma City; and then a third category—*dairy markets*! Upon stepping off the bus to greet the new day, you're greeted in turn by the smell of longhorns, horses, cows, and pigs, and the stage is on a race track.

As an example, let's consider the "Stampede Show." Because the authors do not bear a grudge against these fine people, the actual name of the annual event will not be disclosed here. The stage wasn't on the race track, but on the dirt of a rodeo stadium. Pulling the stage to the grounds was planned for 3 p.m., once the horse show was finished, by using enormous tractors with 10-foot-diameter tires. Bear in mind that horse stadiums are frequently plowed during the day to keep the dirt soft for the show horses' hooves; however, the down side is that soft dirt quickly turns to squishy mud when it pours rain continuously for an hour. The rain was only the warm-up act for the hail storm that followed, when we were attacked by hailstones the size of small meteors. This was all taking place while the audience began filing in to find their seats. There was no room to move on stage, but the crew inched through, fighting with rolls of tarp to cover up equipment while stage hands frantically swept water puddles off the stage that threatened to soak amplifiers and dimmers.

Every single reason for not having an outdoor concert all happened in one day—in a dairy market, where the crew cooked in the heat and sweated like a chain gang in the humid afternoon heat but then practically froze that night while slogging through knee-deep mud. "Go forth and make music?" It's crazy. The audience never knew what the crew did so they could enjoy the event. It stopped raining for the show, and 5000 people got their money's worth. The crew? Well, they got back on the coach and did it all over again the next night. "What, and quit show biz?"

A typical workday can be demonstrated by the typical schedule provided in Figure 7.1. Consider that this schedule is repeated five to six times a week for 6 to 12 weeks to really get a feeling for the strain and hardship such touring puts on people. You can also read *Concert Tour Production Management*, by John Vasey (1997, Focal Press), for another view of the working day and schedule.

A person who goes out with a show like this must be able to handle both his physical and mental health. The body can only take the pace for so long. You should prepare as best you can, eat as regularly as possible, and organize your sleep and play time to benefit yourself the most. The body needs play, yes, but on tour most often it will need some alone time in bed. Living out of a suitcase can be much harder than the actual physical pace of the work. I know many road technicians who do not have a permanent address. Most humans need a place to call home or someone waiting for them to return. The road technician often has neither.

Even as I say these things, I know that most readers of this book will believe that the excitement of being part of this piece of our culture and theatrical history outweighs all of the hardships. Maybe that is what keeps us going. The newness of each day is at the top of my list. The author does not believe he could be happy doing a 9-to-5 gig. Even now, as a

Road Timetable

	Daily Timetable for a fly/drive tour	Daily Timetable for a bus tour
6:00 A.M.	Wake-up call	
7:30 A.M.	Depart for airport	
8:00 A.M.	Arrive airport; check bags and turn in rental car	Rigging call
8:30 A.M.	Flight departs	Lighting Crew Breakfast
9:00 A.M.		Lighting/Video Load in
10:00 A.M.		Audio Load in
11:00 A.M.	Arrive next city	Stage Set Load in
11:20 A.M.	Rent cars and get bags	
12:00 noon	Arrive at hall (check into hotel, maybe), begin setup	Band Gear load in
12:30 P.M	Grab a quick lunch on the way to the venue	Crew Lunch
1:30 P.M.	Focus lighting	Focus and set Floor lighting
2:30 P.M.	Audio Line Check	Audio Line Check
3:00 P.M.	Check lighting programming	Program lighting
4:00 P.M.	Band Sound check	Band Sound Check
5:00 P.M.	Reset band equipment for opening act	Opening act stage set up
5:30 P.M.	Opening act sound check	Opening act sound Check
6:00 P.M.	Re-set stage	Catered dinner starts and preset stage/effects
6:30 P.M.	Crew and Bands Dinner	
7:00 P.M.	Open house	Doors open for audience
8:00 P.M.	Show starts	Show Begins
11:00 P.M.	Show ends	Show Ends
11:00 A.M.	Load-out begins	Strike and Load out
1:30 A.M.	Load-out complete, go to hotel	Shower at venue, climb into bus bunk

FIGURE 7.1 Possible daily timetable.

college professor, the days vary and a lot of production work occurs at night. The job certainly is not for anyone who wants security and the typical 9-to-5 job. Each day begins and ends in work-related activity. Very few minutes are spent not tied to the group or the people who are traveling with the show. You had better all get along or tempers will eventually flare.

TRANSPORTATION

The methods of transportation have changed. In the early days of touring, it was in to fly everywhere, but the economy has caught up with even the wealthiest stars. Tours now take their cues from what many country/western artists have known for years. Custom-built buses and motor coaches outfitted for sleeping and riding in comfort are more practical. The time spent running through airports and the questionable on-time arrivals of flights have been replaced with the ease and security offered by a bus. After all, a rock star has never been killed while riding on a bus; injured, yes (Gloria Estefan, for example, was injured when her tour bus crashed). Of course, the same cannot be said about flying, as aircraft and helicopter failures have been responsible for the deaths of Buddy Holly, the Big Bopper, members of Lynyrd Skynard, Ricky Nelson and his band,

Stevie Ray Vaughn, promoter Bill Graham, and John Denver (who was flying his own aircraft).

Of all tours since 1978, 90% have used private coaches (we tend to call them "buses" but they are hardly your typical Greyhound bus) or motor homes as the primary means of crew transportation. If the truck can make it with the equipment, so can the coach. Touring has returned to the days of the old "bus and truck" touring theatre companies. The difference is in the comfort and class of the bus interiors. They are completely custom designed and fitted with sleeping sections as well as front and often rear lounges that have everything to keep the band and crew occupied during the trip. This is why coaches are so popular. The crew can maximize their sleep time, and the bus doubles as a crew lounge and even provides a place for a quick nap during the sound check. Normally, a bus can sleep 8 to 12 people comfortably. Figure 7.2 shows a line-up of Senator Coaches on a Def Leppard tour.

Figure 7.3 shows a custom coach built by Florida Coach, a company that has been providing custom coaches for artists for over 20 years. At any one time, they might have 30 such units on tour. The interior of a Provost H3-45 coach is designed for the comfort and safety of a touring band and their crew. Only the shell, engine, and drive train come from the manufacturer; the custom coach builder does the rest, either according to his own design or often to specifications provided by the artist who will use the

FIGURE 7.2 Senator Coaches on Def Leppard tour (2008). (Photograph by Paul Dexter.)

FIGURE 7.3 Florida Coach exterior (Provost H25-45). (Photograph by Florida Coach, Inc.)

FIGURE 7.4 Florida Coach rear lounge. (Photograph by Florida Coach, Inc.)

FIGURE 7.5 Florida Coach front lounge. (Photograph by Florida Coach, Inc.)

bus. Three commercial bus manufacturers are generally used: Eagle, Provost, and VanHool. In Figure 7.4, the view is into the rear lounge of a Florida Coach, while Figure 7.5 shows the front area. Most are outfitted with microwaves, refrigerators, televisions, DVD players, Wii consoles or other electronic games, and stereos. Aft of the area shown in the photograph would be the sleeping area. A list of custom tour coaches can be seen in Figure 7.6.

Another big part of transportation is the trucks that haul the lighting rigs as well as sound, band gear, projection/video gear, sets, and costumes (Figure 7.7). The trucks in the figure are all lined up at a loading dock ready to disgorge their cargo or load it all up to head for the next gig. Companies that specialize in this work have drivers who know they need to be on time and to make sure the load is secure and handled with care. Companies such as Roadshow, Upstaging, and one of the oldest, Clark Transfer, have in one way or another saved a lot of lighting designers from a last-minute focus because the truck was not late. Figure 7.9 shows a group of several companies lined up for a European show. One of the companies that has been in this business the longest is Edwin Shirley Trucking (EST), which also has facilities is Australia.

Egotrips	www.egotrips.net
Florida Luxury Coaches, Inc.	www.floridaluxurycoach.com
His Majesty Coach	www.hismajestycoach.com
Senators Coach	www.senatorscoach.com

FIGURE 7.6 Custom tour coaches.

FIGURE 7.7 Trucks at the loading dock. (Photograph by Paul Dexter.)

Clark Transfer, Inc.	www.clarktransfer.com
Egotrips	www.egotrips.net
Roadshow	www.roadshowservices.com
S.O.S. Transportation, LLC	www.sostransportation.com
Stage Call	www.stagecall.com
Upstaging, Inc.	www.upstaging.com

FIGURE 7.8 U.S. trucking companies.

AN ESTABLISHED FIELD

Touring is now a well-established theatrical field. Touring has proven to have lasting entertainment and cultural value. The economy will always have its ups and downs, which will affect what the ticket buyer can afford to spend on entertainment, but it has already been proven that concerts hold their own, even in slow economic times. People now associate entertainment as a must-have to escape their routine and not something to cut back on when times are tough. It is a comfort as well as an escape to get lost in the live music of one's favorite band.

Artists who are now superstars will continue to draw huge crowds and, it is hoped, will continue to increase their production values. Artists in the production/show category, such as Britney Spears, The Spice Girls, Hannah Montana, Bon Jovi, and

FIGURE 7.9 European trucks on tour. (Photograph by Paul Dexter.)

Van Halen, are spending even more production dollars to draw crowds. The 1987 David Bowie tour cost a reported $1 million a week to put on the road. Given how much costs have risen over the last 20 years, even if figured at only 5% a year, that same tour would cost about double that today!

New headliners are the ones that will be worst hit by increased production costs, but, whatever the state of the economy, the production values, budgets, and complexity of designs continue to expand in the concert field. With this expansion has come adoption of these techniques in other media. The technician or designer who sees the value in the advances made in concert lighting is better equipped to deal with problem solving in other entertainment venues.

8

WORKING OUTSIDE THE UNITED STATES

 hen U.S. artists decide to go on tour overseas, one of three approaches to satisfying their technical needs is employed. Most commonly, the lighting designer is told by the artist's management that all arrangements have been made through the local promoter and that a local supplier will provide the equipment. You are expected to use what you are given and to "make it work." In a second approach, you may be asked to submit a lighting plot, but from there on you are told that you must accept what they give you. Or, manna from heaven, you get anything you request. This last scenario is not that hard to achieve these days.

Don't forget about the bean counters, though. Depending on the artist's contract, the artist may be paying for the production values out of their cut of the ticket sales. If this is the case, you may have the problem of making the financial people understand that prices may be more overseas for almost all services.

Touring outside of the United States adds tremendously to the budget due to such extra expenses as translators, higher hotel and travel costs, and higher *per diem* rates. Some of this added expense will be paid by the local promoter, but there is no question that the bottom line is raised for foreign tours.

This book cannot detail touring in every country, but I believe that the European Union, South America, and Japan represent the largest foreign markets and that most other countries regions deal with concerts similarly. Major differences among countries are most likely to be for political reasons. Some

countries repress certain types of music and artist images or control air play of their music. Artists have been told that they can perform in a particular country but then are told that they must delete certain lyrics, songs, or physical gestures. Most of these restrictions are imposed on the artist, not the production or the crew, but be aware that in any country you should respect their customs and attitudes, such as no drinking in Muslim countries.

CARNETS

A merchandise passport (*carnet*) is accepted in over 75 countries and territories. This form is filled out and given to the U.S. Customs Service before a tour leaves the United States carrying any equipment, even personal items such as expensive cameras, that will be brought back into the United States. You will need all serial numbers, descriptions of the items, proof of the country of manufacture, and estimated cost. The U.S. Customs Service stamps the forms as verification that the equipment originated in the United States so the artist is not charged import taxes upon their return. Furthermore, when the artist enters a foreign country the form eliminates the *value-added tax* (VAT), duties, and posting of security bonds normally required at the time of importation.

The carnet is used throughout an artist's travels abroad to maintain a paper trail of the movement of equipment and to certify that the artist does not plan on selling the equipment. The form is checked when the artist leaves each country to ensure that all the items that entered the country are leaving, thus eliminating any sales taxes or special tariffs or restrictions on its removal.

Either the tour manager has been through all this often enough to make sure the paperwork is correct or the artist's management has hired a specialty firm that handles this type of freight-forwarding paperwork. It is good if that company also has agents in the country the artist will be visiting to assist with any issues at that end. These companies can also expedite the artist's departure from the country. This issue is near to my heart because on one of my first trips touring in Europe I was told "someone" had taken care of everything, but when we arrived that "someone" was nowhere to be found. Calls to the phone number we had went unanswered. He did eventually show up, but we were delayed in our setup and our hearts were beating fast for the rest of the tour worrying that he would disappear again. Don't let this happen to you.

While you as the lighting designer may not be directly responsible for pulling together all of this paperwork, if you are taking lighting, even your console, it would be best for your peace of mind alone to make sure the equipment is taken care of properly.

Valuable personal items you are carrying in your luggage, such as a camera, should also be checked by customs in the airport (check for their location and hours of operation). A form will be given to you that will allow you to return to the United States without paying duty on the item.

I did work with an artist who put some fake Rolex watches in his amplifier, and, yes, customs did find them. At first the artist denied putting them in there and insisted that one of the crew must have done it. In the end, he did pay the duty but all of our equipment was detained in customs for a few days. The artist also had to pay the extra equipment rental required.

TOURING IN THE EUROPEAN UNION

If you are touring in Europe, you will find virtually everything that you need and are familiar with in the United States. In addition to all of the American-built equipment that has been exported to Europe are very good English and European luminaires, dimming, and consoles. Europe can be a candy store for designers.

In addition to English companies, several of the U.S. tour lighting and sound companies have branches in England; for example, See Factor, Inc., has an association with the English company, Bandit Lights, and PRG have their own facilities, as does Clair Audio (with some lighting). Check with them or your favorite U.S. company early in your preparation for your overseas assignment.

These companies often combine the best of European and U.S. technologies to create their systems; therefore, it is relatively easy to tour in England and the EU as far as equipment is concerned. Just about any piece of equipment is available, although quantities may be limited, and early commitments are usually necessary if you want to be assured of getting everything you want.

Effect of the European Common Market and the Euro

One of the goals of the EU was to allow borders to be crossed more easily, but 9/11 and the war on terror have affected this goal somewhat. Be ready to show your passport at borders, but if you are traveling by tour bus or car you most likely will not be stopped. Air travel still requires you to show your passport, but there are separate lines in the airport for holders of EU passports. British citizens will not get their passport stamped but they are scanned to track their travel movements. Other nationals will get a stamp upon their entry into England. My British contacts say that customs in England are still a bit of a pain, more so than in other EU countries. There is also an EU identity card, but citizens of England do not have them yet.

Those in the United States who think our custom and airport security is random and somewhat capricious are not alone. Englishman Andi Watson said, "As in the States, airport security is a somewhat random affair with the complete confusion of not knowing if you have to take your laptop out or leave it in, shoes on/shoes off. At London Heathrow, the rules can change literally by the hour, and in Paris CDG you have to take every single electronic or electrical item out of your possessions and put them in a special tray including *anything* with a wire in it, adapter plugs, cables, USB drives, etc., etc., which makes doing the sorts of trips you take in this industry a slight nightmare."

The euro was introduced in 1999 as the common currency for EU member countries. All but the United Kingdom, Sweden, and Denmark now use the euro. Even a number of countries that are not members have also adopted the currency. While it started out at a 1:1 parity with the U.S. dollar, the euro has gained substantial monetary status and now most often trades at a higher rate against the dollar. In 2009, the conversion rate was about 1 euro for 1.3 U.S. dollars.

To enter England or countries in Europe with equipment such as band gear, lights, sound, costumes, or sets, you must have your paperwork in order and complete. And I do mean complete. A customs agent having a bad day could decide to go through every piece of equipment (which also holds true for arriving back in the United States). If the agent finds just one small item not listed it could mean a hold in customs for hours or days … even weeks. Don't take the chance. Work with a good freight forwarding company familiar with exactly what a carnet requires (see Figure 8.1).

Cultural Differences

Customs with regard to the way Europeans take their breaks and meals, the hours they work, and even crew availability can vary widely throughout the European Union. That is not to imply that European crews are not as good as, if not better than, U.S. crews. It just means that you must keep in mind that they were raised with a different work ethic and it can take some getting used to on your first overseas tour.

Language also comes into play. You can carry the handy translation book *Theatre Words* (OISTAT, www.theatrewords.com) in one of its four versions: World, Northern Europe, Central Europe, or Italian, each covering the specialized technical words we use, but even such a guide will not be a total solution to language problems. Many promoters do hire English-speaking crews on the Continent; in some cases, they are American servicemen or their dependents stationed in the area. I find the best solution is to make sure the head electrician you get from your British equipment supplier is able to converse in French and German. That is not too much to ask, as many Brits do speak some of both languages.

Working through an interpreter is difficult, because most are not familiar with the theatrical terms we need to have communicated. I also find that the interpreter ends up spending more time with the management personnel than with the crew. I have gotten pretty good at the few words and numbers you need to communicate, and I usually find that enough people speak some English so you can get your point across. Patrick Stansfield worked for the Rolling Stones back when they were virtually unknown in the United States. They toured small clubs throughout Europe. An American, Patrick was continually frustrated with people when he was sure they knew he was an American yet still insisted on speaking to him in their native language. He had a feeling they spoke English but were just trying to put him down. So, he learned a phrase something like, "No, no, let's not speak in Italian [or German, etc.] today. Why don't we practice our English?" in a perfect native accent. He contends he was never challenged. They always switched to English and never knew he didn't speak their language. It would be difficult today to find someone who speaks absolutely no English, especially

Rock-it Cargo USA, LLC	www.rockitcargo.com
Showfreight International	www.showfreightinternational.com
Stage Call Corp.	www.stagecall.com
Rock On The Roll	www.rockontheroll.net
Road Show Services, Inc.	www.roadshowservices.com

FIGURE 8.1 Freight forwarders.

among young people, and in Japan they would fight you if you even attempted to speak Japanese. The Japanese love everything American, and language is the key to their understanding our pop culture.

I did a show in Berlin where it turned out that only one crew member spoke some English. We were playing the International Congress Centrum, which was brand new. The plan was to use only the house lighting crew and equipment for a Mercedes-Benz corporate show. Because we had a 2-day load-in scheduled, I agreed to try to work with the crew. We had a translator, but I knew she would be busy with other people most of the time (and I was right). I was told that the crew all spoke some English. The fact was that the facility's technical director was an American, but he could not be around all the time to lend a hand, as he had a very large facility to run with several other theatres and convention spaces. So, I was left to my own devices.

With a lot of gesturing and the few words I knew, we got the show mounted. The real problem was calling the follow spots. What I contrived was to write out all the cues and spend plenty of time with my one English-speaking man going over each cue until I thought he understood. Then I would point to the cue that was next up and he would read the cue in German to the follow spot operators and I would give the GO or BLACKOUT (*Hauptregler*) or FADE OUT (*Ausblender*) in my best German. The crew tried very hard and did an admirable job. Of course, we had 2 days of rehearsal in addition to the load-in time, so I could take this chance. On one-nighters, I would not recommend trying this at all.

Also in Germany, I found myself trying to convince local officials that it was safe to rig a lighting truss from a catwalk in their beautiful but very old municipal theatre. Every official we talked to passed the buck up higher. Finally, we had to call in the town's *Burgermeister* (mayor). The German promoter was acting as translator and go-between, and everyone was talking so fast that I could not understand what was going on except that it did not look like I was going to get my truss hung. I finally got them to stop long enough for me to motion that we should all go up on the catwalk so I could show them what we needed to do. Once on the catwalk, I asked the promoter to tell them that all of us standing at the very point where we wished to attach the truss weighed more than the lighting truss! We did the show as planned.

Responsibilities

In most European countries, especially in Germany, the technical director is personally responsible for the safety of the theatre. He can be legally held liable, fined, or jailed if there is an accident. Of very great concern to the authorities throughout Europe and England is the rigging. Many of the buildings are very old and highly suspect as to their actual load-bearing ability. Do not be shocked when a rigging plan representing all the rigging and bridle points and their respective weight loading is demanded in advance of the show. In point of fact, the Greater London Council (the GLC, as it is commonly referred to by Londoners) will request such a plan and they will need to approve it several days in advance or there will be no show.

This is a practice I wish was more prevalent in the United States. I believe I have worked with the finest riggers in the world, and safety is always a top concern, but even they would be the first to admit that many buildings have not had a safety test inspection in many years and that structural stress can change from day to day. Any touring designer or rigger who does not worry about every building he or she enters is a fool, and that high level of concern is what keeps us safe.

George Gilsbach, a well-known concert rigger, says that he is actually encouraged when facilities want to see his rigging plan. That they are knowledgeable about the potential risks is reassuring. Gilsbach has found that more and more buildings are asking for plans. The ESTA rigging certification program (see Chapter 2) is also helping to connect riggers and building structural engineers in more and more buildings.

Power Supply

Power problems are not serious if you are using one of the British concert equipment suppliers. They are prepared for the various power setups in each of the countries and have adapters to do the job; however, if you insist on bringing your own lighting console from the United States, make sure your console can accept 220-volt inputs or request a transformer to plug it or you will be out of luck. Many rock & roll lighting consoles do have adapter systems built in so they will work on 110 and 220 volts, as do the more recent versions of moving luminaires. Any band gear with motors or fans will require 50-cycle adapters, which are expensive, so it is better to rent that equipment (electric organs, computer-based keyboards, etc.) in England.

TOURING IN SOUTH AMERICA

The market for artists touring in Mexico and Central and South America is tremendous—not only Latin artists but also English-speaking artists can draw huge crowds. These people are as keen to see the flash and color of large touring shows as we are. Univision and Telemundo television networks beam speculator music specials and award shows throughout South America. These shows are heavily dependent on color and moving luminaires, so it is no wonder the people want to see those same effects live. Glitzy, almost Las Vegas types of productions have always been a trademark for Latin shows, so it is no wonder this area pushed for American-style rock & roll lighting earlier than anywhere else. Demand has brought many U.S. and English lighting and sound companies to Latin America to sell their products, start joint-venture companies, or establish their own entities, as Vari*Lite did. Equipment availability has increased over the years to the point where it would be fairly easy to have a design translated to a company there that closely matches the comparable U.S. rig.

Tricky Power

Power is without a doubt the trickiest part of touring south of the border. Power (voltage and cycles) changes not only from country to country but also from city to city in some cases. But, the problem is not limited to voltage. If the cycles (also known as hertz) drop below 60, this can cause great damage to any device with a motor, and that includes fans in dimmer racks and consoles. In some cities, such

as Mexico City, you can monitor the input service and see a 20- or 30-volt change when the local factories shut down. During a sound check, the voltage might be close to 100 volts but by show time it could jump to as high as 135 volts. This overvoltage is potentially damaging to any and all electronics. At the other extreme, 20-volt computers are designed to automatically shut down for low voltages or cycles.

Latin Culture

The Latin world is not prone to discrimination if you do not speak Spanish or Portuguese (as in Brazil). There is some discrimination within the Latin world with regard to Mexican Spanish, and Castilian Spanish as spoken in Spain. I was quite shocked to see a sound engineer for a big Mexican star act superior around the rest of the Mexican crew. When I asked what the issue was he flat out told me that Spaniards thought their fellow Spanish-speaking people were inferior. I hope this was an isolated case, because I have always been treated extremely well in my travels in Mexico and other countries south of the border. Brazil is unique in South America because they speak Portuguese. They match or exceed their northern neighbors' colorful lives filled with multiethnic cultures and a gusto for dance and music.

TOURING IN JAPAN AND ASIA

The Asians may be late to enter the concert market as packagers of tour systems, but, following the lead of the Japanese, countries in Asia have become big markets for American and European artists. The extreme cultural and language differences would seem to be insurmountable, but to the Asians they are not. Many study and practice English all the time and welcome a chance to speak the language. We will focus our discussion here on Japan, because it is the largest Asian market.

The Japanese Theatre Culture

The cultural differences separating our two countries would require volumes to explain, even superficially. Ken Lammers, who was born in Japan of

American parents, has spent all but 3 years of his life there, and he worked his way up to become a manager in one of the largest Japanese lighting companies. Recently, he started his own company to supply bilingual backstage coordinators as well as consulting services. He would be the first to say that he will never be completely "Japanese." Inadequately stated, the Japanese view theatre as an art refined to its most simple common denominator. Light is there to make it possible to see and is not a dramatic part of traditional Japanese theatre. Therefore, the light should be bright, even, and clean so the audience can see the subtle lifting of an eyebrow or the twitch of a finger during a scene from a Kabuki or Noh play. Form and economy of movement are the keys to these ancient dramatic arts.

Theatres and Concert Production

All of the major theatres in Japan, some 2000 of them, were built after World War II. Most are government owned—by a city, school, prefecture, or federal authorities. These were built for community use, and each was designed to serve as an all-purpose theatre equipped to handle all of the performing arts. Most theatre systems have been standardized for efficiency. Because the theatre has to serve the whole community equally, no one person or group can book the hall for a long period. This means that most shows must start to set up on the morning of the show. Other than a few privately owned theatres, it is difficult to get more than 3 or 4 hours to set up and focus before rehearsal.

The use of color media and more complex control entered Japan in two ways—via Western-style theatre and American modern dance. Taking its cue from the likes of Martha Graham and others, the modern dance of Japan is very experimental and innovative and has sparked new uses for lighting.

In the early 1960s, Japan had no equipment rental houses, except those that catered to the movie industry. When television came out of the studio and began shooting in the theatres, a few lighting rental companies began to appear. This was when the 6-channel, 6-kW dimmer packs were first introduced to the market.

When rock & roll concerts came to Japan, the Japanese were surprised that people wanted to bring a full lighting system into a theatre that already had one. For a long time, artists had to use the house system unless they brought the whole thing with them overseas. The Budokan in Tokyo opened its doors to concerts and was the biggest influence on touring equipment. Trusses started to be used, although they were constructed of steel. Most of the rental houses were using the American-made 120-volt lamps because they were economical. The lumen output and color temperature were bad, though, because they were being used at the standard Japanese voltage of 100 volts. In the late 1970s, one of the lighting rental companies, Kyoritsu, made the first 100-volt PAR-64. Even with this lamp, the halls seldom have enough power for shows so they are forced to rent generators that can be set at any voltage. Most of the promoters have accepted the need for a generator as part of their production costs, so most tours from overseas can now have their own control consoles and dimming. A few of the big-name Japanese artists tour with full systems, but it is still common to use the theatre's equipment.

Japan has no full-service production company. The nearest thing is the Shimizu Group, but they do not own lighting or sound equipment. Shimizu is the biggest concert company, with seven stages and roofs; their staff totals 855 full-time employees and about 500 part-time ones, including ticket takers, security and trucking personnel, and crews, both freelance and staff. They have full-sized rehearsal facilities, more CM hoist motors than any other company in the world, and LED screens, cameras, and fabrication facilities. They also have branch offices in Beijing, New York, Dallas, and London.

Lighting Companies in Japan

In Tokyo alone there are many lighting companies plus some theatrical groups that have lighting departments registered with the Lighting Engineers Association of Japan. Only a handful provide equipment for concert tours for Western performers. Currently, these companies include Kyoritsu Ltd. (biggest lighting company); Kawamoto Stage Lighting; Big-1, Inc.; Lighting

Version Ltd.; Sogobutai, Inc.; and Tokyo Lighting, Inc. They normally supply their own crews.

Local Power

Japan is the only country in the world that uses 100-volt AC at 50/60 cycle power. It is also useful to know that northern Japan is on 50 cycles while southern Japan is on 60 cycles. Also, there is no law that requires each light to be grounded. Grounds are required for installations up to the main breaker box, but the individual outlets do not need to be grounded. Unlike Europe and Australia, the common outlets are the same as in the United States. If you bring equipment that needs to be grounded, make sure you do it yourself and do not depend on the house having a ground. Check before you connect anything.

Part of the reason for not simply putting a company disconnect in the theatre has to do with the electrical laws of Japan. Any temporary hook-up that is over 100 kW (about 300 amps) needs to have a qualified electrical supervisor on hand. It is easy to see that this would become expensive just for the six or seven shows a month that would need the service. This also applies to generators, but as long as there is an operator from the generator company on hand they seem to allow it.

Generators and Transformers

To deal with the 100-volt issue, generators that can be pushed to 120 volts are often used. Due to stringent noise pollution laws, generators must be extremely well sound insulated (blimped). The most common size is around 120 kVA (kilovolt–amperes). The largest available are around 300 kVA. Taps can be changed on all sizes so the voltage between the hot legs is either 440 or 220 volts and can then vary as much as ±50 volts. In this way, all needed voltages can be obtained. Also, the cycles can be varied between 50 cycles and 60 cycles. The other way to deal with the power conversion issue is to bring in a transformer that can be tapped at a higher voltage to match the imported equipment. Many consoles and moving luminaires do have switchable power supplies that will automatically change to the input

voltage, but be careful; in some cases, you may have to make the change manually.

Equipment Availability

- *Moving luminaires* are now common, especially Martin and Vari*Lite.
- *PAR-64 luminaires*—Even after the 100-volt PAR-64 was introduced, many companies continued to use the 120-volt lamps because only narrow-beam and medium flood lamps were available at the lower voltage. If you needed the very tight focus of a very narrow lamp and wanted it to be brighter than the 100-volt medium flood you had a problem. In 1984, a 500-watt very narrow lamp was produced. This light is so efficient that it is brighter than the 120-volt, 1000-watt lamp when used on their 100-volt systems. Now there is a whole family of very narrow, narrow, and medium floods.
- *Ellipsoidal luminaires*—The rash of Broadway musicals coming to Japan led to the Japanese companies purchasing a great number of ellipsoidal luminaires. The most common are 19- and 36-degree ellipsoidal reflector spotlights (ERSs) with 100-volt lamps. There is also a Japanese-made ellipsoidal with a zoom lens. Its beam spread is just about equal to a 19-degree ellipsoidal at the narrow end of the focus. If you need 10-degree ellipsoidals, it is best to bring them with you. Although the inventory has grown, only a few companies have them. Subrentals are common, but provide enough lead time to allow for negotiating between companies.
- *Projection equipment*—Even though RDS projectors (known in the United States as RDS Scene Machines) are made in Japan, the availability of the effects heads is limited, especially for projection of painted glass slides. If you need to use a remote system and change speed or slides during the show, it may be best to bring your own. You can get 35-mm slides and film projection equipment, and because most of it is imported the make of equipment you are used to in the United States is probably

available. Video projection equipment is also available. The new LED and media servers are now also available.

- *Follow spots*—Most of the theatres are equipped with 2-kW xenon units, usually four in a theatre. For performances in other types of venues, the rental companies will supply your needs. It is interesting to note that many Japanese operators disdain the color changers and place colors manually. Ken Lammers said that he gets quite a reaction when he tells foreign designers that their shows were just done this way. The smooth and accurate operation they achieve, even on fast rock shows, is undetectable to most people.

Other Lighting Equipment

In general, all sizes of Fresnels, striplights, and plano-convex spots are readily available. The Japanese-built equipment is excellent and compares very well with the U.S. and European equipment, but you need to work with it before you commit a full design to unfamiliar equipment.

- *Trusses and hoists*—CM Hoist equipment has been well accepted in Japan. Some trusses are imported, but some are made in Japan as copies of European designs. Hoists converted to Japanese voltage must be used because of differences in voltage from the United States and Europe.
- *Lasers*—There are no federal laws on laser use, but some local restrictions do apply. Lasers are available, and promoters can put you in contact with specialists.
- *Sets*—Because some venues have smaller stages than tours may be used to in the United States, the tour might be better off using cut-down versions built in Japan and save the shipping costs.
- *Pyrotechnics*—Pyrotechnic regulations are set by each hall, city, or prefecture fire marshal. Permission can take up to a month, and advance notice may be required in writing. On the day of the show, an inspector has the power to disapprove, even if you comply with the written order that was issued.

Rigging

If the lighting system can be arranged on straight pipes, it will make life much easier for everyone involved. Theatres do not like to see trusses hanging from their grid, even if they are lighter than the equipment they replaced. If you have to bring a truss, be prepared to ground support it. Fire curtain laws are strict, so no box truss configuration can be hung unless it is behind the curtain line. The only solution is to move the artist back away from the audience, which will not make the performer happy.

Gymnasium-type halls are gradually allowing rigging, but each has different weight restrictions and most seem to be ridiculously low compared to the U.S. limits. The Budokan, one of the best known and biggest venues in Japan, has gone through a few stages regarding rigging. The hall was built for the 1964 Olympics and has a concrete false ceiling. The rigging of one of the first foreign groups did some damage, so rigging was banned for some 6 years. After long consultation with the architects who designed the building, rigging points have now been installed. Each point has a dynamometer in it that gives an instant readout. If the weight restriction is exceeded, a warning bell rings and work stops until it can be fixed. Each point has a different weight limitation, depending on how the other points are being used. Only designated bridles and Japanese-approved cable hoists are allowed. A computer controls all the points, and there are ten patterns that it will allow. You must design the grid to fit one of these plans. The total weight allowed is 7.2 tons, but because of the weight distribution restrictions only about 5 tons can be over the stage area. If you ground support, there is no restriction.

Trucking

All trucks are 11-ton, straight-bed frames that are 36 feet long. Few tractor–trailer rigs are available in Japan, and no air-ride boxes are available. Luckily, all Japanese roads, even in the countryside, are excellently maintained.

Crew and House Staffs

There is no union of stagehands in Japan, although some of the privately owned theatres have in-house labor unions. The government-owned halls will have a two-person lighting staff to act as supervisors only. Between shows they maintain the facility in top form. The crew needed to load-in and run your show will be contracted from the outside. This is usually done by the promoter through the Shimizu Group or the lighting company, and the theatre has no say as to which firm is hired. If the show is going to travel outside of Tokyo, the key people will travel with the show and local crews will be picked up at each stop. This is arranged by the lighting companies through regional offices or agreements they have with local companies.

Crews get no set breaks or minimum guaranteed hours. They are paid a flat rate for the day with a bonus if the work continues through the night. The full crew will stay through the load-out, even if that is the next day. Most Japanese have the attitude, "Let's get the work over with and then rest." Most crews will want to work until the focus is finished before taking a break, even if that means no lunch. Japanese promoters are not in the habit of feeding the Japanese crews. If they do, make sure special meals are ordered for the American and European crew members. The typical Japanese worker's meals will not be palatable to most outsiders, even if they think they have eaten true Japanese foods in other countries. The diet is very different and takes some getting used to before you can eat whatever is available.

Payments and the Promoter

The promoter normally contracts for the lighting, sound, trucking, rigging, and staging for a tour rather than the artist's manager, as is done in the United States, South America, and Europe. There are no Japanese promoters who enjoy an exclusive at any facility. The lighting and sound companies have ties with different promoters, and there is not the bidding situation encountered in the United States. In the past, it was hard to convince the promoter to take a lighting system on tour. Now almost all tours will travel with a full system.

Just as in the United States, the production costs are mounting largely due to bigger and bigger designs. It is common to see a promoter write into the contract that the artist is limited to a fixed cost for production, and the artist who wants more must pay the promoter back for the additional cost. Lighting plans do get cut down during the production meeting between the Japanese promoter and the equipment company, so be prepared to convince the artist or manager to pay up or be willing to cut back the design. The cost for equipment and crew runs about one third more than in the United States.

The main Japanese promoters are Udo Artists, which primarily handles all the rock & roll concerts, and Positive Productions, which handles a lot of the rap and R & B artists.

Theatres and Other Halls

Most theatres are under 3000-seat auditoriums, and some venues have very wide stages by U.S. standards. 100-foot-wide proscenium openings can be found, but the working depth will be very shallow. At the other extreme, many stages are very narrow due to the needs of traditional Kabuki theatre, so do your homework carefully. The larger halls are usually gyms that hold 5000 to 7000. There are about four halls in all of Japan that hold 10,000. Recently, artists such as Madonna, Bon Jovi, and Christina Aguilera have used stadiums because of the greater demand for tickets.

As mentioned before, most halls are government run and thus there is often no discernible logic as to the rules they impose, even discounting the cultural differences. A bureaucrat is a bureaucrat the world over.

Booking a hall must be done well in advance, usually a year. This means that many times a promoter will book a hall even if he has no artist scheduled at the time. The halls, being government owned, have to give an equal opportunity to all who apply for dates, thus it is impossible to get more than a couple of days back to back. If an artist wants or ticket sales demand longer runs, the promoters will often get together and shift schedules when possible, but it is not a given that an agreement can be made, so do not count on an extended run.

Each hall has a different starting time, but the common rule is a starting time of 7:00 or 7:30 p.m. Halls will want the show to end by 9:00 p.m. This is not only because of the work schedule of the hall staff but also for the sake of the audience. Most people come to the theatre directly from work or school. Public transportation is used by the majority of people, who then face at least an hour's ride one way, so they need to be on their way by 10 p.m. It is common to see people leave a show before the last song to catch their train, not because the show was bad.

One other reason for the early stopping time is noise. The legal penalties for the promoter are very harsh for violations of noise pollution laws. Because homes are built on all available land, the loading doors probably look directly into someone's bedroom. Noise after 11 p.m. is not tolerated, and the theatre will get complaints immediately if this unwritten rule is disregarded.

Business Ethics

The way business is conducted in Japan can be a puzzle. We have all heard about the Japanese being interested in saving face above all other considerations. The formality of business entertainment is also strange to the Westerner. I believe that in Japan 4% of a firm's gross revenues are tax deductible for business entertainment. It is more than being friendly to ask a client out for dinner and drinks; it is part of the normal business day. I have often seen a businessman, briefcase still in hand, leave a bar at 9 p.m., bid goodbye to a group of similarly equipped men, and head for the subway train to finally go home for the day. It is an expected ritual for corporate executives, and it is a great offense to decline such an invitation.

Another thing to consider is the pecking order within a firm. If your boss (usually translated as superior) is to have dinner with an important client, you could be asked to attend. And, if you supervise a department, your assistant will also attend. All this is a show of status and position, not because they need the advice of their employees.

This brings up another rule of business in Japan. Often a question is asked directly of a technician and a clear answer is not given. The system in Japan requires that, when a superior is present, the question must be answered by the higher ranking person. Westerners might view this as an example of subservience, but you should realize that it all fits into the Japanese sense of order and respect for elders and those of higher position in business.

The Japanese pride themselves on being very Western in their entertainment business dealings, and their lighting market has grown to keep pace with tours coming into the country, so you should not have great difficulty in obtaining your list of lights, but make sure the request is made long in advance.

RESPECT FOR OTHER CULTURES

Touring in Japan or any foreign country is exciting and challenging. Do not be quick to judge the methods and work habits you encounter overseas; instead, look at the social and economic structure of the country's society. Americans have cultural links to Europe, but Asian travel can be a challenge. No matter what country you travel in, remember that the citizens are on their home ground, and you should respect their beliefs and ways of conducting business. Take time to prepare for your overseas tour, not only technically but emotionally and culturally, as well.

9

RISK ASSESSMENT AND SAFETY

Anyone who has been on a stage realizes very quickly how dangerous a place it can be if you do not watch your step. Scenery moving, pipes flying in and out, trap doors opening, or risers that are not stable—all are accidents waiting to happen. So many incidents are reported that many states have tried to enact hard-hat area laws. Even venerable Broadway experiences its share of accidents and deaths each year. Stars are not immune; I remember when Ann-Margret fell from a platform at Caesar's Palace in Las Vegas while rehearsing. A quick-thinking stagehand broke her fall and possibly saved her career, if not her life. He was injured in the attempt, as are many stagehands each year.

It is difficult to find a stagehand that has not suffered a personal mishap or injury or does not know of a friend who has. I recently lost a master electrician in a fall when a stage collapsed under the platform he was taking lights down from. Usually accidents result in only minor injuries, but when virtually everyone knows someone who has been injured it points to how often these accidents occur. A sizable portion of these injuries happen during set changes on a darkened stage. When you have only seconds to make a scenery move and one person is off the mark that night or there is a substitute who has not done it before, then there is the potential for error and possible injury. Working with local crews who have not seen the show and are doing in one day what theatrical shows are allowed to rehearse over and over again cannot help but cause a problem. That more accidents do not occur is reason for praising road crews and local stagehands.

The road offers enormously greater potential for accidents than do general theatrical productions. When you consider the long hours and the travel involved, fatigue becomes a major factor. The body just cannot stay at peak performance levels week after week for months at a time. The pressures on traveling road crews are tremendous. On the average, they must get the show loaded in, rigged, and sound checked and then go through a performance, strike, and load-out in under 12 hours. Now repeat that process 5 to 6 times a week for 6 to 8 weeks at a stretch and you have some very tired people.

One of the major concerns of the now defunct Professional Entertainment Production Society (PEPS), which the author helped to found in 1980, was the safety of road crews. Not only were we concerned about our own personal safety but we also wanted to promote the image of people who wanted to work as safely as possible. We felt that, as a whole, our fellow road technicians had an outstanding safety record, but there were no statistics to support or refute our beliefs. Compiling such figures is not easy for a small, new organization, because polling the membership would not be representative of a large enough percentage of the total industry. The cost of a properly supervised survey was beyond our means. Nevertheless, our informal surveys seem to back my original belief.

Our concern for safety was so great that as a primary requirement for membership a company had to have a $1 million public liability insurance policy. We wanted the hall managers, promoters, and producers to be aware of our concern not only for our own safety but also for the audience's well-being.

A safe working area is critical. Our work area is the stage. What goes into making it safe? Most minor accidents happen because people just did not think. We cut ourselves or bruise a leg and then declare, "What a dumb move," which is exactly right and implies that we really knew better than to do what caused the injury, so why do we let down our guard? A large percentage of the incidents can be attributed to trying to do things too quickly. We are obsessed with how quickly we can get a touring show set up or load out a touring show. This haste is a major factor in accidents.

The constant long hours of load-in and show after show in different cities without a break can be physically tiring and mentally boring. You begin to establish a routine and do things mechanically without thinking. There is a positive side to that, of course, but it can also be dangerous. The positive side is that, with so much to do and so little time in which to do it, things must flow and fit together without a lot of effort. In most cases, we do not even count items in a case; we just know they are all there, due in part to instinct, I guess, but also to repeating a process so often that it becomes an unconscious action. This is not meant to encourage beginners to disregard checklists and planned procedures; instead, it is meant to point out that disregarding these tools could be dangerous.

SPECIFIC AREAS FOR SAFETY CONCERN

What are the key areas to watch for potential accidents? The answer is anywhere accidents are not expected. The following areas are worth discussing in detail.

Truck Loading

People who are brought in to unload and load trucks have no idea of the weight of an item, what is in a box, or how delicate the contents are; they are simply trying to empty or fill the truck. A forklift driver cannot be excused for dropping a case, but the loader who is trying to get a top-heavy case down a narrow ramp is always in danger of its tipping over. Make it a practice to have one member of the road crew outside the truck to assist the loaders with information about case contents, weights, and so on and to direct the movement of the cases into the hall for organized placement. Generally, the truck driver

is inside the trailer supervising the load, but not necessarily. If the load has shifted in transit, an accident or at least severe damage to the equipment can occur. Make sure you know what they are contracted to do. Another crew member should be inside with the loaders, if the union rules allow.

Trusses

A daily visual inspection of welds is required. A fracture can occur in the best of units, and severe road conditions can do great structural damage. The author believes that yearly x-rays of trusses should be required. This would reveal any hairline fractures before stresses cause a serious structural failure. Second, and no less important, is to have one person responsible for rechecking all connectors—nuts and bolts, pins, and so on—that join truss sections together before the truss is raised into the air. The potential for truss failure is great when the truss is being raised or lowered. Additional stress, as much as 350% of the static load, can be placed on any one section when the motors come out of synchronization or a corner of the truss is caught on a pipe or set unit. The motor may have no load for only a second, but when the truss comes clear it can fall with such a force as to break welds or the anchorage point of the motor.

Rigging

When overhead rigging is required for the show, most often a professional theatrical rigger is traveling with the show. The importance of good, safe, competent riggers was highlighted when the Entertainment Services and Technology Association (ESTA) chose certifying riggers as one of their first certifications. This certification has already received wide acceptance from hall managers and the International Alliance of Theatrical Stage Employees (IATSE), as well.

The rigger, or *high man*, may also have a *ground man* with him, or this position can be filled by a locally qualified rigger. The ground man sends up the cable, bridles, shackles, chain, and other gear needed by the rigger. The road rigger is worth the added expense, especially during load-outs. Correct packing of cases can save hours on the next load-in. Eliminating the wasted time and energy of locating parts that were not packed in their proper place makes for a quicker, smoother, and safer operation.

If the show does not travel with a qualified rigger, it is wise to pay to have one go over the plans before the tour starts and suggest a rigging plot or, better yet, draw one out, giving insights into point placement, position, and weight configuration. This will greatly assist the local rigger by giving him a better idea of what must be lifted, thus promoting knowledgeable decisions based on load limitations at the specific facility. Even non-concert arena shows provide this information. The traveling show "Walking with Dinosaurs," a 2008–2009 arena spectacular, relied on the excellent rigging plot and truss/weight load chart shown in Figure 9.1 and Figure 9.2. A rigging plan for Def Leppard arena performance stage setup is shown in Figure 9.3.

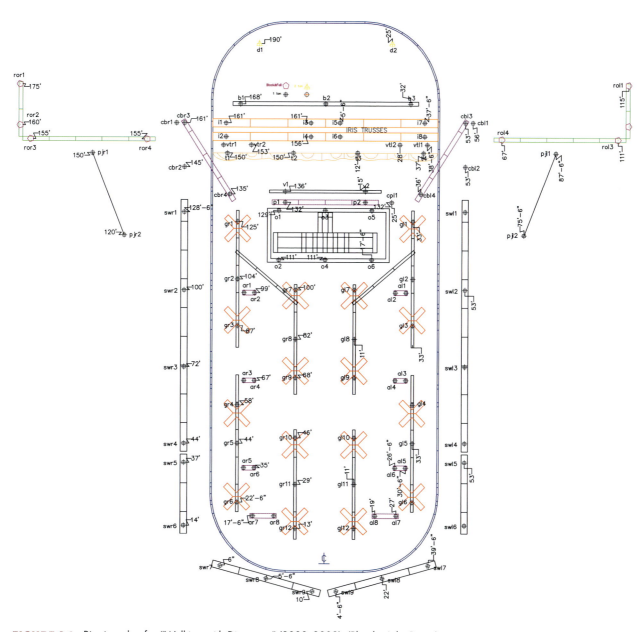

FIGURE 9.1 Rigging plan for "Walking with Dinosaurs" (2008–2009). (Plan by Jake Berry.)

WWD RIG DIMENSIONS 150' FLOOR 10/24/2008

Point		X	Y & –Y	Weight	Motor	
Discription		From Datum	From Center	Lbs	size	Hook Height
Left	Right			Each		
Interior Grid Truss				Point		
gl1	gr1	125'	33'	2000	1T	70'
gl2	gr2	104	33'	1200	1T	70'
gl3	gr3	87	33'	1700	1T	70'
gl4	gr4	58	33'	1400	1T	70'
gl5	gr5	44	33'	1400	1T	70'
gl6	gr6	22'6"	33'	1400	1T	70'
gl7	gr7	100'	11'	1300	1T	70'
gl8	gr8	82'	11'	1300	1T	70'
gl9	gr9	68'	11'	1300	1T	70'
gl10	gr10	46'	11'	1200	1T	70'
gl11	gr11	29'	11'	1200	1T	70'
gl12	gr12	13'	11'	1200	1T	70'
Perimeter Audience Truss						
swl1	swr1	128'6"	53'	1000	1T	70'
swl2	swr2	100'	53'	1500	1T	70'
swl3	swr3	72'	53'	1500	1T	70'
swl4	swr4	44'	53'	1000	1T	70'
swl5	swr5	35'	53'	800	1T	70'
swl6	swr6	14'	53'	800	1T	70'
swl7	swr7	2'	38'	1310	1T	70'
swl8	swr8	–3	21'6"	700	1T	70'
swl9	swr9	–8	4'	700	1T	70'
Light Towers and P4 Audio						
vtl2	vtr2	153'	28'	500	1T	70'
vtl1	vtr1	153'	38'6"	500	1T	70'
Upstage Lighting Vari Light Truss						
v1	v2	135'	15'	1500	1T	70'
Cable Bridges and Picks						
cbl1	cbr1	161'	56'	1500	1T	70'
cbl2	cbr2	145'	53'	900	1T	70'
cbl3	cbr3	161'	53'	1800	1T	70'
cbl4	cbr4	135'	36'	900	1T	70'
Masking and Orni Mini Grid						
o1	o5	129'	17'6"	1350	1T	70'
o2	o6	111'	17'6"	1600	1T	70'
o3 center		111'	0'	750	1T	70'
o4 center		129'	0'	1400	1T	70'
Video Projection and Audience Screens						
p1	p2	132'	15'	450	1T	70'
pjl1	pjr2	150'	87'6"	100	1/4T	40'
pjl2	pjr3	120'	75'6"	200	1/4T	40'
Teeth						
t4	t1	150'	37'	500	1T	70'
t3	t4	150'	12'	500	1T	70'

FIGURE 9.2 Rigging calculations for "Walking with Dinosaurs" (2008–2009). (Plan by Jake Berry.)

After the rig is flown, always have the crew place several *safeties*, or safety cables. These are non-load-bearing cables attached between the portable truss and the physical building structure. They are there in case a chain or cable slips or breaks. A few years ago, a truss did break as a safety was being released after a show. If the safety had not been in place the accident could have happened while the

WWD RIG DIMENSIONS 150' FLOOR 10/24/2008

Iris Truss						
i8	i2	156'	37'6'	1300	1T	70'
i7	i1	161'	37'6"	1300	1T	70'
i6	i4	156'	5'6"	1300	1T	70'
i5	i3	161'	5'6"	1300	1T	70'
Sound						
al2	ar2	99	26'6"	340	1T	70'
al1	ar1	99	30'6"	340	1T	70'
al4	ar4	67	26'6"	340	1T	70'
al3	ar3	67	30'6"	340	1T	70'
al6	ar6	35	26'6"	340	1T	70'
al5	ar5	35	30'6"	340	1T	70'
al8	ar8	17'6"	19	340	1T	70'
al7	ar7	17'6"	27	340	1T	70'
Run Off Drape 80' BnF						
rol1	ror1	155'	49'	100	BF	70'
rol2	ror2	155'	71'6"	100	BF	70'
rol3	ror3	155'	94'	150	BF	70'
rol4	ror4	155'	116'6"	200	BF	70'
Up Stage Blacks						
b1	b3	168'	32'	250	1T	70'
b2 center only		168'	0'	250	1T	70'
Dino Chain Hoists						
d1	d2	190'	25'		2T	43'

Show Weight Each Side	**49330**

Total Show weight For Both Sides	**98660**

FIGURE 9.2 (Continued)

show was in progress, causing injury to the artist as well as audience members.

[*Author's note:* The author is not a certified rigger, and this is just his own opinion. Always work with an experienced, certified rigger.]

Stage Support

Check the stage surface carefully. When the performance is to be on a portable stage, check for spongy, weak spots or uneven sections. Next, look under the stage. Are the sections secured tightly together? Is the deck secured to the legging, and are all the legs touching the ground? You do not have to be a structural engineer—just common sense will tell you if it looks unsafe. If it does not look safe, tell the promoter or facilities staff about it and make them change it. Even though stage construction is not a direct concern of the lighting technician, I always slide my foot over cracks and joints to check for leveling and open spaces that someone might trip over. A portable *dance floor* material is useful in eliminating this problem and at the same time it gives the stage floor a better appearance. Even carpet or throw rugs help.

Ground Support

The types of structures that are commonly used to support lighting are shown in Chapter 14. In this chapter, the proper, safe use of these structures is discussed. Some of the structures have safety devices built in, such as the Genie Superlift, which has a breaking device designed to prevent load forks from falling if the cable breaks. This lift is probably the best unit with regard to safety because of the cable-and-winch design it employs.

Few hydraulic lifts are available for this type of work, and they have problems related to leveling. Hydraulic lifts are so heavy that they can tip over

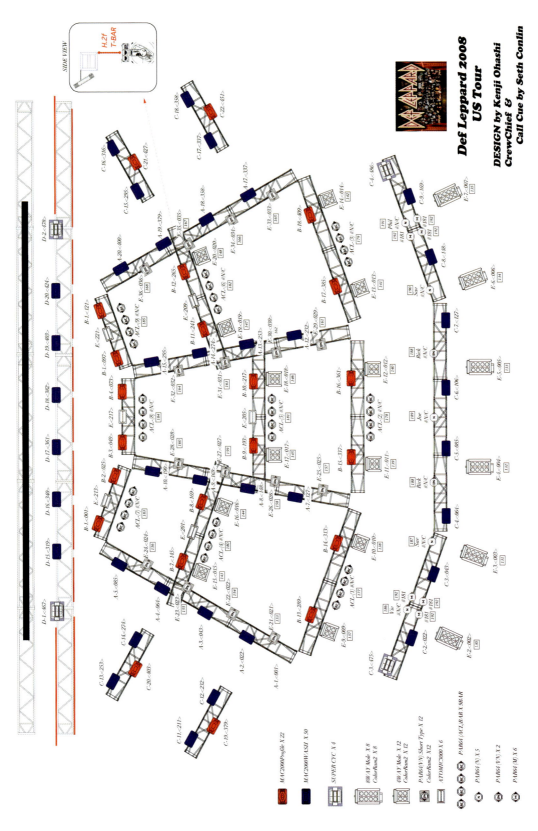

FIGURE 9.3 Truss layout for Def Leppard tour. (Plan by Kenji Ohashi.)

if not properly stabilized, or, if there is a weak point in the stage, they could break through the floor. In the latter case, though, the lift did not fail, the stage did. Nevertheless, it is a safety problem related to the device and the stage. Use outriggers or stabilizer bars on all lifts to establish the widest possible base.

Probably the two most common accidents that occur with lifts are pinched fingers (getting them between columns or joints as the device comes down faster than expected) and shin bruises when someone runs into a lift's outrigger leg. No matter which lift is used, recheck the leveling once the truss or load is raised to ensure that it remains level, and have a technician check all lifts again before the show to ensure that everyone on stage is as safe as possible.

FALL PROTECTION SYSTEMS

Fall protection became one of the entertainment industry's hot buttons in the mid-1990s. In 1970, the federal government passed a new series of laws administered by the Occupational Safety and Health Administration (OSHA). OSHA has stepped in to make it mandatory to wear safety equipment during rigging and any truss or overhead work. There have been several cases of stagehands falling, even with the new requirements, but, sadly, these stagehands had failed to employ the safety equipment properly or at all. Many falls occur while follow spot operators or crew members are climbing wire rope ladders and they lose their footing. One incident that had a happy ending occurred on the 1995 Billy Joel–Elton John tour. A stagehand was climbing a wire rope ladder to a follow spot position and missed a rung near the top. He was wearing a full body harness, and when he fell the retractable lifeline stopped him from falling all the way to the ground. He was able to regain his footing and complete the climb. The gear that saved him is referred to as *fall protection* equipment.

In a broad sense, the concept of fall protection refers to all of the effort involved in making sure workers are protected from accidental falls and, if a fall occurs, having a system in place to save the worker's life. All the rules for this are contained in the 1970 Occupational Safety and Health Act. The

Act created the *6-foot rule*; where there are no other restraints (such as a railing), anyone working more than 6 feet off the floor must wear a harness, not just when climbing on a truss or high steel.

Laws require employers to protect every employee from injury in the normal course of work. A specific law regarding this is CFR 1910.132 (Personal Protective Equipment). *Fall hazard* is a term used in the law to describe any situation where a worker could lose his balance, fall to a lower level, and be injured. OSHA uses the 6-foot rule to define fall hazard situations. If a worker is exposed to the possibility of falling 6 feet or more, some restraint such as a railing or other fall protection device must be provided and used. OSHA recognizes three basic methods of fall protection. The first is a guardrail system surrounding the walking or work surface, which is not practical on lighting trusses. The second is a safety net system installed below to catch a falling worker, but safety nets for lighting trusses are also a problem. Third, a personal fall arrest system can be provided to protect a worker in lieu of the foregoing methods (Figure 9.4). This is not simply a matter of putting on a rock climbing harness and attaching it to a rope tied around a steel beam in the ceiling. In fact, recreational climbing gear does not meet OSHA standards. The components that are used in these systems are very different from those a rock climber uses, and most of the rock climbing and rescue equipment used by riggers and firemen is not legal for fall protection. One obvious piece of gear is the *seat harness*, which is a full body harness that has an attachment point above the shoulders so the person falling cannot become inverted and fall out of the rig, as has happened to climbers. This is why OSHA has stated that the system must be designed as a *total system*. The agency does not leave it up to riggers to put their own systems together. OSHA's concern is that it is very easy to mismatch components, which weakens the standards. OSHA says the system must meet the following conditions:

1. Be continuous.
2. Be exclusive.
3. Reduce the arrest forces communicated to the fallen worker to safe levels.
4. Limit the free-fall distance before the arrest.
5. Be composed of components of safe design.

do not follow all the rules, and in 2008 one fell from the high steel, apparently because he did not hook up before unhooking. How can these tragic accidents be totally eliminated? More education is probably the only answer.

Frankly, there is another group of workers who are not skilled riggers but also need to take advantage of fall protection. For many years, truss-mounted follow spots have been used in arena shows. The stagehands responsible for operating them are not necessarily trained, proficient climbers, and many have a justifiable concern about climbing a 20-foot wire ladder up to the truss to maneuver into the follow spot chair (see Chapter 13). In addition, local stagehands who do the light focus are required to not only move along the truss but also to lean over to focus and put color media in lights from a very uncomfortable position. It is hot, sweaty, exhausting work. This group also deserves to be made safe, and a number of concert rigging suppliers have pioneered safety equipment for them.

FAILURE OF STRUCTURES

Despite all the emphasis on rigging and riggers, there are still factors that cannot be accounted for. The storm that caused the collapse of the Rocklahoma Festival roof structure is not the only example of such failure. Structural failure most often occurs in outdoor venues, where weather becomes an unwelcome and unpredictable factor. For a show in Australia, I made the decision to lower the roof only a half hour before the show because a wind storm had come up suddenly. After calls by the promoter to the weather service and consultation among all the concerned parties, we determined that the front had passed; we raised the roof, and the show proceeded a half hour late, but all agreed that the safety concern was legitimate.

In some cases, such decisions have not been made or adverse circumstances arise too quickly for an effective response. Roofs have collapsed all over the world, but I have not heard of one incident where there was a reckless disregard for the safety of the performer or audience by the roof supplier.

Accidents do happen, whether due to inadequate building designs, structure steel members weakened due to age or water or snow loading, inaccurate weight figures given to the rigger as to the trusses and luminaires, or miscalculations by the rigger.

Boardwalk Hall, Atlantic City

The most visible and highly reported accident involving an indoor rig occurred in the 75-year-old but newly refurbished Boardwalk Hall in Atlantic City, New Jersey. Some $90 million dollars was spent on upgrades in 2001, including a master truss grid over the stage. Before the renovations, riggers would attach chain motors to cables that were dropped out of a false ceiling. This was a difficult setup, because the points had to hang straight down or they would cut into the ceiling material (unfortunately not an uncommon situation in older venues). The new permanent grid that was installed had motors attached that allowed it to be lowered down to stage height; the tour motors would then be attached to the grid raised to the trim. After the grid was back in place, the rig would be attached to the motors, and the package would be raised to the appropriate height.

On August 8, 2003, the Justin Timberlake–Christina Aguilera tour was loading in and the road truss with speakers was being raised. Some 30 crew members and stagehands were on stage when the grid gave way and fell (Figure 9.7). Luckily, no one was seriously injured, but the rig was a total loss. What caused this massive failure? I am not an engineering inspector, and I can only rely on the reports that circulated after the accident, but it appears that what happened was not due to anything the road crew did wrong, with the possible exception that the weight of their rig had not been calculated correctly. I do not know whether or not the hall required a rigging plan to be submitted before the load-in to certify the anticipated weight-bearing points, so I can only speculate that somewhere along the chain of responsibility someone did not do their homework. Following are some questions raised by this incident:

1. Was the lighting and sound rig weight not calculated correctly?
2. Was there a damaged or broken weld or a tiny fracture in any of the many trusses?

3. Could a cable or chain have become kinked or overstressed during the raising process?
4. Could the motors have gone out of synchronization, thus putting excess stress on other lifting points or truss members?
5. Could power to some of the motors been cut off?
6. Could the refurbishing contactor skimped on the grade of steel added for the master grid?

I cannot assign blame, but we should all be on the lookout for problems; even when it is not your area of expertise, many eyes can help avoid potential accidents. I can assure you that you, as the lighting designer, as well as 100 other John Does, will be involved in the mountain of lawsuits that follow. How can you best protect yourself? Make sure you give accurate information to the parities that are building the system for the tour and that you get a receipt confirming that they have the particular plan in their possession.

Rocklahoma, "The Perfect Storm"

The engineering that goes into the structures used for touring and festivals is second to none. We know that we are working in exceptional conditions that have a lot of incalculable factors. An incident that occurred at a long-running festival in Oklahoma is a real exception, but the story serves to show that, even with all the right preparation and safeguards, things happen. Always be on guard, especially when things are going right!

According to weather service, a microburst that occurred on July 12, 2008, close to Pryor, Oklahoma, where the Rocklahoma festival was set up, caused the collapse of the portable roof (Figure 9.8). The weather service had predicted 30-mph winds with rain that day. The report after the incident stated that the winds were approximately 75 mph at the time of the collapse. The roof system was rated only for 35-mph winds.

The roof system had withstood 30-mph winds and rain for the previous 5 days. Oklahoma is known for its wind, and the festival people constantly monitored

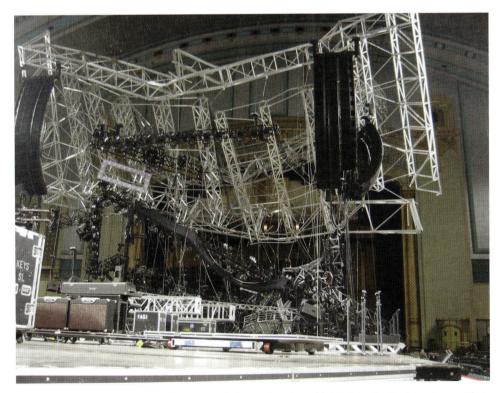

FIGURE 9.7 Justine Timberlake–Christina Aguilara tour rig failure at Broadway Hall in Atlantic City, New Jersey. (Photograph by Peter Roberts.)

FIGURE 9.8 Truss collapse at Rocklahoma caused by storm in Summer 2008. (Photograph by Robin Alvis.)

the wind speeds and weather forecasts throughout the event, but the powerful microburst could not have been predicted. There was an onsite weatherman, and the festival was given the all clear to start 5 minutes before the high winds hit.

In the past, the festival had safely deployed roof systems during winds up to the rated 35-mph limit, but the organizers had never encountered high winds that arose so quickly that nobody could respond in time to bring the system down to the ground. Right after the incident, the weatherman said that in his 25 years of experience he had never seen weather like that before. A National Weather Service satellite photograph of the storm showed that it came from all four directions and the winds actually formed a cross.

Nobody was hurt, because everyone had run to the tents for safety. The promoters and technicians felt that the system had actually held up quite well compared to what it was up against. An engineer from Thomas Engineering (the manufacturer of the roof trussing) who was flown in after the event went over each piece of truss and took out of commission every piece damaged from the event.

SEISMIC CONCERNS

Portable, outdoor structures are subject to more of nature's forces than just wind, rain, hail, and snow. We must also be concerned about seismic activity. Before anyone says, "Well, my show isn't in California," let me clue you in. Large earthquakes have been reported in the United States in areas other than California—for example, the New Madrid, Missouri, quake in 1811, which recorded 8.0 on the Richter scale; the Valdez, Alaska, quake of 1964 (which I experienced), which registered 8.6 on the Richter scale; and the Northridge, California, quake of 1994 (which I also experienced), which was a mere 6.4.

Although you, as a lighting designer, may not be concerned about seismic testing, it is good to know that some people are, and they have set standards. The American National Standards Institute (ANSI) standard E1.21–2006, Entertainment Technology—Temporary Ground-Supported Overhead Structures Used to Cover the Stage Areas and Support Equipment in the Production of Outdoor Events, has quite a long title, to be sure, but it addresses *wind loading*, which is relevant to the Rocklahoma incident. California has become the latest state to adopt a building code based on the International Building Code (IBC). This has challenged roof and support structure designers because this code represents a fundamental change from the Uniform Building Code (UBC), which has been the basis for the design of all structures, both permanent and portable, throughout the history of U.S. building codes.

It is important to ask the promoter of your next outdoor event whether or not the stage conforms to these codes. If the answer does not satisfy you, go to your production manager or road manager; after all, if your client is hurt you lose business.

LUMINAIRES

Some safety devices for luminaire accessories are not common. Most PAR-64 luminaires have a spring-loaded clip to hold in the gel frame. I like to take a tip from television and film lighting technicians, for whom it is an accepted practice to safety-chain any barndoor or snoot to the luminaire. Usually a commercial safety cable made of wire rope and a carabineer or dog clip is best. If color changers or other effects are added to the luminaire, a similar device should be used without question. The last thing you need is a color frame or color changer falling from the truss onto a band member during a performance.

Even when attaching luminaires to a truss, add a safety cable to all luminaire yokes and the truss or pipe. If you are using lamp bars with four to six luminaires attached to a unistrut track, a safety should be added between the bar and the truss.

Another major danger is lamp failure. Lamp manufacturers refer to this as a *catastrophic failure*, not an explosion. It can happen with an ellipsoidal or Fresnel lamp, but because these are designed with a lens between the lamp and the outside, it would be rare for glass to fall out of them. The PAR lamp, however, has little or no real protection for this eventuality. The most extreme measure is to place a wire screen in the color holder. Other than this, there is little that can be done, although hot glass showering down can cause serious burns and cuts. Lamp manufacturers have made great strides in eliminating this problem, and I have not heard of a case of this

happening during a performance in many years, but a word to the wise is always good. If catastrophic failure is going to occur, it most likely will happen soon after the luminaires have been unloaded and power is put to the lamp circuit too quickly.

The problem is intensified by the fact that quartz lamps are sealed under pressure; therefore, a crack in the quartz glass acts just like a puncture in a balloon—it explodes! The PAR lamps have a glass lens sealed onto the face of the reflector-coated glass backing. If the cement has not completely sealed the gap, the heat created by the lamp can expand the glass, allowing pressure to escape and causing further fracturing of the seal, which could lead to an explosion … violent failure! This is a layman's view and should not be considered a technical evaluation by an expert in lamp design, but it summarizes popular understanding of the issue.

Lamp manufacturers have taken great pains to ensure that the lamps are airtight before shipment; however, bouncing around in a truck can cause damage or cracks to occur. I have seen lamps come out of a truck neatly separated at the seal and the lens lying against the gel frame. It is not practical to check each lamp every time they are moved because visual inspection does not reveal fractures. This is one area of safety that continually plays the odds. I always try to do a *preheat*, where I have the board operator bring all of the lamps up to 20% while the crew does a lamp burn check while the trusses are still near the ground. This procedure allows the filament to warm up slowly which usually prevents the sudden rupture of the glass face from the envelope.

Another problem with lamps in the PAR family is that they are focused by manually rotating the porcelain connector cap at the back of the luminaire. These caps are constructed of two pieces of porcelain held together with nuts and bolts. In transport, they can come apart, thus exposing the metal contacts. Without looking inside, people can reach in to rotate the lamp and come in contact with these electrically hot leads. Always look before reaching inside or at least wear gloves as an insulator. It is wise to use gloves in any case, because the lamp housing is very hot.

Your daily inspection of PAR lamps on the rig should also include looking for socket caps that are off the lamps or that have come apart. One method used to keep these caps from separating is to use a high-temperature glue in the hole that contains the hex nut. I have also seen wire wraps tied around the two parts, although I do not recommend that practice, as it increases the potential for fire. Some manufacturers have begun to install caps that go over the porcelain. This is a great idea, and all PARs should be retrofitted in this way.

FOCUSING

This is the most dangerous time of the day. When someone is high on a ladder or walking a truss, wrenches can slip out of hands and color frames can fail when the rigger is inserting or changing color in the frame holder, so use extreme caution on stage. Warn people of the danger, and do not allow others to work in the area. This is difficult to enforce, and you will probably have people mad at you, but it is for their own safety and if you explain it as such the muttering usually stops. The falling wrench problem can be minimized if the rigger attaches a line to the handle and places it around his wrist. When something falls, generally the person hurt is the stagehand holding the ladder. He or she should have been looking up anyway, but that is not an excuse for the accident.

It is a constant problem to get ladder movers (holders) or man-lift pushers to pay attention. If you are the person on the ladder, always make clear, before you climb up, how you will direct the movements. No one else should call for the man-lift to be moved, only the person up high. Try to get at least one person to keep looking up, both to hear you better when you call an order and to watch for falling objects. If no ladder is used and a person is up high on the truss, station a stagehand in the area to watch for falling objects and to keep people away.

POWER HOOKUP

The potential for accidents with a temporary power hookup are tremendous. Because we are dealing with temporary hookups, it is not always possible to be sure the correct connection has been made. Another problem is that many situations require that the road electrician give the feeder connections (pigtails) to a house electrician to make the actual connection. Take the time to clearly show the color code markings on your pigtails. Do not assume that everyone knows green is ground (earth), at least it is in the United States. What kind of ground is available: equipment ground, grounding rod, or water pipe? Is the service single-phase or three-phase power? Check all of these things carefully. Then recheck everything at the source after the hookup is complete before energizing the lighting system. I strongly recommend volt meters on each leg of the portable dimmer system you are taking on the road to double-check the incoming power before the system is energized. At least carry a good-quality volt/ohm meter.

Grounding is the number one cause of power problems. The two main problems are potential shock to someone working or touching the lamps, truss, or lifts and a buzz in the sound system created by problems between lighting and sound grounds. A good ground is essential, especially outdoors. Many road electricians take the added precaution of grounding the lifts used to raise the trusses via a ground stake at outdoor shows. It is an excellent practice.

No matter what country you are working in, even when you are not the electrician, I feel it is prudent to be familiar with the power system and grounding procedures. Be especially careful around portable power.

PYROTECHNICS

This is an area best handled by a licensed professional. Too many cases of accidents, even if only superficial burns, have been recorded. One example of an unfortunate outdoor incident occurred at the California Jam at the Ontario Motor Speedway in the summer of 1974. The English group Deep Purple used flash pots in their act. The effect was to happen at the end of a guitar solo when the musician destroyed his guitar and (fake) amplifier, with the result being an explosion. After the licensed pyrotechnician had loaded the device with the charge per California regulations, it is believed (but to my knowledge never proven) that someone with the group decided the charge was not powerful enough and added more powder without informing anyone. The result was seen by a quarter of a million fans at the concert and by millions of people watching the television special. The stage caught on fire, and men with fire extinguishers could be seen running on stage to put out the fire, and they were not part of the act! If the concert had been indoors, the potential for disaster would have been much greater.

Another tragic example is the fire that broke out during a Great White concert in 2003 at The Station, a club in West Warwick, Rhode Island; 100 people lost their lives because of a combination of factors, including a poorly placed pyro-flame effect and flammable material placed on the walls by the club owners. It is wise not to get involved in making any decisions if you are not a skilled, licensed pyrotechnician. Luckily, this incident is extremely rare because most road people have immense respect for what pyrotechnics are capable of doing if not handled correctly.

SMOKE

Chemicals are used to make theatrical smoke or haze (see Chapter 19), so beams of light can be seen better. The Actors' Equity Association (AEA) and the League of American Theatres and Producers (LATP) jointly funded a 2000 report on the effects of these chemicals on dancers and other performers. The 180-page report, *Health Effects Evaluation of Theatrical Smoke, Haze, and Pyrotechnics* (prepared by Environ International Corporation; environcorp. com), can be found on the AEA website (www. actorsequity.org) in their public access library, along with three other later studies and recommendations. The results were not particularly kind to the manufacturers of the smoke-producing devices; for example, the report opposes the use of glycol fog on stage. Even before the report, though, considerable effort had been exerted by manufacturers of these devices to correct many of the problems. Chapter 19 discusses the materials and methods used to make haze and smoke for theatrical purposes. It is wise to consider that some artists have respiratory problems, and these people should be warned about remaining too long in areas where heavy concentrations of some forms of fog and haze are to be used. A Material Safety Data Sheet (MSDS) must be attached to *any* material or chemical that poses a hazard to personal safety such as solvents, dyes, or glycols, which can be ingested by inhalation or absorption by unprotected skin.

In recent years, manufacturers have reformulated the base materials. The way in which these effects work is that particulates hang in the air so light can bounce off and be seen by the audience. Pure pharmaceutical chemicals were tried initially and are still used in some applications, but most manufacturers now use a pure water base to create these effects. Still, people complain, and as the lighting designer you will have to fight for these effects if the artist wants to put on a real light show.

The Fog and Smoke Working Group of the ESTA Technical Standards Program (TSP) has contributed to several ANSI standards and recommended practices that have been published. These papers and standards establish use and testing procedures, in addition to being a source for manufacturers of these products.

These effects may be the second most discussed issue today, right behind fall protection, but remember that no one issue makes or breaks a show. Work with all those involved to arrive at a consensus of what level of fog or smoke or haze the show can tolerate.

MURPHY'S LAW

Murphy's Law—"If anything can go wrong, it will!"—is certainly true on the road. We place terrible strains on equipment and personnel. Fatigue can bring Murphy's Law into play without warning. The biggest deterrent to accidents is to be aware at all times that they occur when least expected. So, expect an accident or you may become a statistic. This is why I place so much importance on addressing the stress and fatigue that occur on the road (see Chapter 7). You may have done the setup a thousand times and can do it in your sleep, but if you are asleep you are an accident looking for a place to happen. Insist that everyone on the crew be alert, because someone else's carelessness can be the reason you get hurt. I have at times told people, "You might not care if you hurt yourself, but *I* want to be around to do this job for a long time."

SAFETY PROBLEMS: CORRECTIONS AND SOLUTIONS

There are four areas of concern in maintaining safety. First, during preproduction, check the manufacturer's certification for the load capacity of the truss, and check the safety record of the company or equipment to be used. Has the equipment been inspected recently? Even if you are "only the designer," you do have a moral as well as legal obligation to ensure that what is used on the production is safe. Do not try to put this responsibility on the head electrician. Everyone must be concerned with safety. As the designer, you must know that the design you have submitted can be safely executed by the equipment supplier. Second, as the designer or crew chief, you have a continuing obligation once the tour is on the road to make sure the equipment is maintained and the crew is working in a safe manner. The third area relates to the safety of the work environment—specifically, the stage upon which the production is to be performed. Even if no lighting is placed on the

stage, as you walk around you should be on the lookout for wobbly sections or other signs of an unsafe stage. Report any problems to the stage manager or the promoter's representative immediately.

Finally, make sure the equipment supplier has proper insurance on the equipment and the people working for them. Legally, this is an area that is not entirely clear. If you enter a facility and see or suspect an unsafe condition and report it, you do not relinquish your responsibility if the house manager, promoter, or stage manager tells you it is okay and to mind your own business. If the safety problem falls within your area of expertise, you could be held criminally negligent if an accident happens subsequent to your making others aware of it. The "good Samaritan law" does not apply here. If you are going to be an independent designer, I suggest very strongly that you obtain a public liability insurance policy. You owe it to your coworkers and to the public to protect them. I do not know of anyone ever being sued over this point, but our PEPS legal counsel advised the members of their ultimate obligation in this matter: *You are responsible for the safety of others.* Insurance companies pay big settlements on accidents arising from people tripping on cables or having things fall on them during a show. Do not put yourself in the middle; avoid such a possibility by making your production as safe as possible.

10

FINDING SOLUTIONS

The human brain has selective memory; we tend to forget, or at least push into the far recesses of our minds, the disasters we've all faced—you know, the important lamp that blows just as the curtain goes up or the color that looked great in rehearsal but now seems washed out. These things happen to every lighting designer and are unavoidable. The important thing for a designer or technician to know is not how to place the blame or make an excuse but how to deal with the problem as quickly and effectively as possible.

PROBLEM SOLVING, STRESS MANAGEMENT, AND INTERPERSONAL COMMUNICATION

The best designers are skilled in the art of problem solving. They communicate well and possess the ability to make on-the-spot decisions. They do not procrastinate. On a day-to-day basis, I choose to work with other designers who possess communicative and decision-making abilities over more creative persons who cannot verbalize their thoughts and ideas. Concerts are a collaborative venture; when one member of a creative team is unable to deal effectively with the others, that person drains everyone's energies.

Let me tell you a personal story. When I was in graduate school, I was assigned to stage manage a show. The lighting designer assigned to the show was very highly regarded by his professors, who thought he had a bright future. On this particular production, the director was a visiting professional. He took a nontraditional approach to the play, forcing us to look at it in a new way. In the early meetings, we all heard his explanation, and in subsequent creative conferences the design team contributed ideas. When it came time for the lighting designer to explain his thoughts, the director listened carefully. After hearing him out, he commented on his approach, saying everything was fine except for one key scene, which he asked the designer to rethink. In the next meeting, the lighting designer reiterated the exact same concept. Again, the director said, no, rethink it. Technical rehearsals of the play began, and guess what? The look the director had rejected showed up on stage. Again, the director was patient (an uncharacteristic quality among his peers), but he said, "Okay, I've seen it, and it still does not fit the imagery I wish this scene to create. Show me something different tomorrow."

As the stage manager, I called the crew together for tech cleanup the next day before the dress rehearsal. The electrical crew was on time, but no lighting designer. We waited 2 hours, and still he did not appear. The board operator said, "What do we do? The director will hate it if I bring up the same preset again tonight." Figuring the designer had an emergency, we decided to cover for him and programmed a different look before the rehearsal began. At rehearsal time, the designer walked in—no emergency. He simply said, "I went to the beach to think."

We need not carry this story any further, except to say that this highly promising designer never made it past small community theatre. His talent was wasted, to be sure, but even more tragic was that his teachers did not educate him in the art and science of problem solving, stress management, and interpersonal communication.

DECISION MAKING

Part of the author's life was spent as an air traffic controller. How that happened was simple: I joined the Air Force, and they said, "You're an air traffic controller." Life often plays tricks on us; we may see no reason for certain events as they're happening, but they turn out to be pivotal influences later on. That was the case here, because air traffic control is essentially a game of juggling schedules and making commitments. The greatest lesson learned was:

> *Do not be afraid to make decisions.*

As an air traffic controller, even the slightest hesitation to commit yourself can be fatal to the people on a plane; therefore, a lot of time was spent *teaching* us to make decisions. The ability to make decisions is a learned skill, and being able to work under great pressure is equally critical.

A failure of our educational system is the lack of instruction in the psychology of taking action or making decisions. Theatre, film, and television are action-oriented professions. We all must work to a production timetable that, it is hoped, brings all the technical and acting elements together at the same time. My earlier story is a classic case of a creative mind being unable to deal with the realities of group-created art.

The best designers seem to be those who work well under pressure. Sadly, many who have the creative and even the communication skills necessary to be good lighting designers cannot deal with pressure. The best lighting designers know how to handle disasters, both physically and psychologically.

What do you do when you encounter a problem—for example, discovering you've used the wrong color? The solution is simple: Admit your mistake and change the color. For most people, the problem is not in changing the color, but in admitting that they could have made a better choice, and that is what it is—a choice. There is no black and white here, no life and death decision to make. It is no different when an actor cannot seem to find a special. Though it may be the actor's fault, do not even think about arguing the point with the director. Change the special. And don't think the crew is standing behind you laughing; they'll only laugh if you are too foolish not to change it.

DESIGN, CREW, AND EQUIPMENT FAILURE

An important thing to realize is that we fail as often as we succeed. Every success contains elements of failure, and *vice versa*. The types of failure that lighting designers commonly experience can be put into three categories: design, crew, and equipment.

Design Failure

There is no such thing as a perfect design. Based on that premise, all of our designs fail in some way. Artists, critics, audiences, managers, producers, directors, and teachers all decide in their own terms whether a production succeeds or not, and their evaluations are important, especially in the professional world. But, honest self-evaluation will do you the most good. Learn to step back and review your own checklist, asking if the show worked for you—as the director or artist defined the problem, as the physical limitations of the production required, and as you conceived it on paper. Ask yourself what you learned from this design experience. What could you do differently next time? Each production requires you to rethink ideas or solutions that may have worked perfectly in other situations. To think you will succeed on the basis of your past glories will be the death of you as a creative person.

Crew Failure

Regarding the crew, ask yourself if they failed or if you failed them because of your bad instructions, your supervision, or your lack of communication skills. When crew members fail, look to yourself first and then *do not look any further!* Did you give them all they needed to succeed? If they failed, it was probably because you did not communicate your needs adequately. You would be wise to study the many good books available about motivating coworkers; I particularly recommend *People Skills*, by Robert Bolton (1986, Touchstone Books).

Television pilots most often have a hectic pace. On some productions, usually underbudgeted ones, we often encounter problems that would be easy to blame on the crew if we chose to do so. It's not uncommon to be short an adapter or cable or to forget a lamp, especially when you're working on a limited budget. You can't bring the whole warehouse. Most likely, the problem is that you've put more effort into the complicated production and did not spend a lot of time working out the nuts and bolts of this "simple" project.

Why waste time fixing blame? Do something! Maybe you can gang circuits, or look for a place where a luminaire can be moved and used to solve the problem. Maybe you did give the correct list to your gaffer or the rental shop, but that does not help you now. Before the next show, say a word to the person who you believe forgot to check the equipment. The person will appreciate that you did not reprimand him or her in front of their co-workers and will be more careful the next time.

Equipment Failure

No one is necessarily to blame for equipment failure, but you are responsible for finding a quick solution. While working on a series for the USA Cable Network, we shot more than 120 half-hour wraparounds. Those are the in-studio segments that lead into and out of a field report such as you see on "60 Minutes" or "20/20." These had been shot in groups of 20, with a month off between sessions. When we came back for a fourth session, the lighting was put back but we couldn't get the same intensity on one part of the set dressings, even though we had the same studio, same dimmers, same luminaires, and same set. Why? We never figured it out. Under pressure to get tape rolling, the solution was to add a small luminaire that was available, and to achieve the lamp intensity that was needed I put it at spot focus. About an hour into the taping, there was a violent failure (see Chapter 9). Luckily, the Fresnel lens kept the hot filament fragments and glass inside the instrument. We quickly changed the lamp and went on taping. An hour later, it happened again. We then exchanged luminaires, and the problem was eliminated.

When the producer asked, "What the hell is going on?" I could have blamed the lamp, the luminaire, or the equipment rental house. What I said was, "I screwed up." Finding a scapegoat or going

into a long explanation on focus and lamp failure would only have extended the problem and wasted everyone's time.

DIMMER PROBLEMS

Here is a common problem: dimmers that develop minds of their own. It doesn't happen often in permanent installations, but rental gear is subject to invisible damage. Portable control cables get run over by forklifts, and constant patching and bouncing around in trucks can take a toll on dimmers.

One of the live award shows I have worked on, "The Golden Globe Awards," had such a problem. The show was syndicated at that time and therefore did not have a big budget. I pride myself on being able to work with a limited equipment budget, but this time it caught up with me. The stage was lit adequately, but I did not have a lot of toys to create different looks. I was depending on creating most of the livelier looks in one area where singers were to perform and otherwise keeping the lighting simple. In addition, because backstage space was limited, the fire marshal ordered us to put the dimmers outside the building.

About halfway into the live broadcast, one of the dimmer packs overheated and started flashing individual dimmers on and off. Naturally, it involved a critical lamp: the key light on the master of ceremonies' podium. The first time it went off, I thought we had a blown lamp, so I quickly got a follow spot onto the speaker. Only a semi-disaster, I thought, and the show was saved by my quick action … or was it? When we went back to the presenter, I put the follow spot on the podium. The lamp then came up. What was the problem? Was the problem a short in the luminaire? Or was it a bad twofer or cable? Frankly, the last thing I suspected was a dimmer problem, as the equipment was from a top rental company and had worked perfectly all during rehearsal.

The next time we used the luminaire it came on as it should but then started flashing on and off. Then I knew it was a dimmer problem; however, figuring out what the problem was took several minutes, and all the while the flashing continued on camera. Now *that's* a disaster. I asked the master electrician to shut the dimmer down and we finished the show using a follow spot to cover the area.

This was not the time to remind the producer that he had cut the budget so tight that there was no money for backup luminaires, which normally would have been built into the design to cover such a failure. I, after all, had ordered the equipment. On live broadcasts, it is considered standard operating procedure (SOP) to double hang areas such as the emcee's podium in case there is a lamp failure or to use only follow spots; I could have saved a couple of lamps from other areas and used them to back up this critical area. I did not.

COMPUTER PROBLEMS

The author prides himself on being an early advocate of computer lighting control. I believe I was the first to use a Century–Strand Light Palette on a concert tour with John Denver in 1974. But, failures do happen. My most recent failure came at a corporate show for a major auto manufacturer. We had rehearsed for a week, two shows were behind us, and we had one performance to go. After a quick run-through in the afternoon, we were ready for the next audience.

I was sitting next to the console reading when my board operator returned 20 minutes before the curtain was to go up. As I looked up to greet him, I saw both of the console's video monitors go blank. My first thought was that we had lost power. We checked and the power was okay. I reset the switch, the screens came back on, and then we watched the computer go through its internal diagnostic program. The show program came up and immediately went blank again. We went through the check and diagnostic program once more, and again the screens went blank, only this time smoke came out of the back of the console.

Now we *knew* there was a serious problem. While the operator opened the board, I ran for the telephone to get a backup console on the way. Then I went looking for the producer to tell her we were dead in the water, so to speak. After the color came back to her face, she asked the obvious question: "What can we do now?" I calmly replied, "I'm not sure there is anything we can do."

By this time, the cover was off the console. We found burn marks near the power supply, and a burned wire lead. We replaced the wire and the computer was back on line—with no show program! The power failure had caused an electronic spike that destroyed the disc drive module. The backup console arrived, but it wasn't a match, so at this point I had to make a decision. Should I go with the new console and reprogram the show from scratch, or stay with the one I had? I decided to use the backup board. By reprogramming the soft patch and reassigning channels to the 24 submasters (luckily the board had that many submasters), I felt I could at least do the basic show looks. To reenter all 137 cues would have taken too long, so while the board operator entered the dimmer-to-control channel patching information, I went to work laying out the submasters so we could run the show manually. We brought the curtain up 40 minutes late, but that final show turned out to be better than the first two.

However, the problem never should have happened. I should have required a computer with dual power supplies or a standby duplicate console. It's easy to say that in all the times I had used this particular board nothing had happened. I failed by falling into the *it never happened before* syndrome. There is always a first time.

BEING PREPARED

Early in my training as a designer, I learned a wonderful lesson from Dr. Sam Selden, author and for many years chairman of the Theatre Department at UCLA, who came to Southern Illinois University as a visiting professor. I was assigned to be his stage manager on a production of "Peter Pan." Things had been going pretty well, and I was feeling cocky when he came up to me and asked what I intended doing if a particular hydraulic lift did not come up on cue. I hesitated, and he told me to come to him later with three solutions to the problem. His point should be taken: We should constantly be considering *what ifs*.

Several of my colleagues brought up this point. A good production manager or lighting director always has Plan B and Plan C, so when a truck goes missing with half the lighting we can move to Plan B. When the house can't hang a truss where you want it, use Plan C. It's like being prepared with an excuse when you think someone is angry with you. Your mind races through all manner of possible excuses. Well, this is no different; you will be damned by someone no matter what you decide, so just do it.

I'm not saying we should or could build redundancy into every part of every system, because that isn't economically feasible. But, we should always be prepared for the worst and learn from our misfortunes. The greatest problem is being unprepared.

II

EQUIPMENT DESIGNED TO TRAVEL

11

LIGHTING CONSOLES

The tremendous advances in console design, especially computer engines, cannot be overlooked, but it would be unfair to say that they came about purely because of rock & roll demands on the established manufacturers of the 1970s. It is true that the most complex manual boards of the 1980s, with flash buttons (instant-on buttons), chasers (grouping to auto-timed changes), and pin matrices (shunt boards allowing one fader handle to operate random dimmers), were not available as standard items in the theatre consoles of the 1970s. Up until that time, most lighting consoles were pretty much the same, except possibly those that had a manually set group of presets that could be attached to the console. There were some very ingenious concepts, including a drum system designed by legendary educator and designer George Izenour at Yale Drama School. I had the privilege and sometimes pain of operating one he built at the University of California, Los Angeles, as a graduate student and later at Harrah's Casino in Lake Tahoe, California. The standard of the day was to have a grand master to control the total output power of whatever low-voltage system the manufacturer used (there was no standard). Two scene masters would have master control of all the channels, usually no more than 48 individual channels (dimmers). A group of submasters, a way of grouping channels together in a discrete number, could be added. Nothing else except the "wing" idea of George's was generally available.

Later, British concert companies started making their own *desks*, a term they liked to use instead of *console* or *boards*. An early manual preset board by Electrosonics, a British company (which still exists, although today they only develop video gear such as servers), was based on a layout concept I brought to them (Figure 11.1) and was probably the first mass-produced board to use flash buttons and pin matrices. It was aimed specifically at the concert market. Showlites, the London rental company that Eric Pearce started in the early 1970s, began manufacturing Alderham lighting console products to accompany dimming systems in 1976, and they stayed in use until well into the 1980s. The rock & roll desks, as they were called, had two presets, each with 60 faders, 20 submasters, and a 20×60 pin matrix. The Alderham 602 model did not have a microprocessor and was all hand-wired. Ian "Avo" Whalley, a touring technician at the time, continued its development and started Avolites a couple of years later. Figure 11.2 shows his Avolite QM-500.

FIGURE 11.1 Electronics console. (Photograph by Sundance Lighting Corp.)

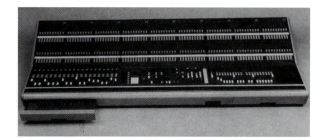

FIGURE 11.2 Avolite QM-500. (Photograph by Avolite Production Company, Ltd.)

At about this same time, the major manufacturers were gearing up to bring out computer-based consoles. They were convinced that there was no market for a more complex manual console, but then they did not see rock & roll as being a financially profitable market either—too bad. Not that the introduction of computers was wrong, but rock & roll lighting designers still thought they needed to play the console to the music and that a computer spitting out a string of preprogrammed cues couldn't meet their needs. When I took the first computer out on tour (the #2 Light Palette by Strand), the company delivered it without a manual. I used it for the first time on a European tour for John Denver in 1978 without ever having used it before on a live show, just playing with it in my shop.

All rock & roll tours have replaced the earlier models with manual and even early computer control consoles. What is needed today, not just for rock & roll touring, but for most lighting that uses moving luminaires, media servers, or LED walls, are systems to assist in handling the larger number of channels required and to improve programming ease.

The main reason why most concert lighting designers took time to adjust to the preprogrammed consoles was that they wanted that hands-on feel that allowed them to play the music, to be able to activate flash buttons and faders in time to the music. Designers said they needed the manual control to feel a part of the music, and they did not seem to get that feeling when they touched only one button to execute a complex command. What I liked about the computer, though, was that I could be sure that I would get the same look show after show, which is what my client wanted. None of us wants surprises, at least not on tour. The same could be said about Las Vegas showrooms, where many artists eventually started to perform. Some highly paid lighting designers discovered that they did not have to stay for the full run; with the computer, the artists and designer could be assured that the house operator would run the cues just as programmed.

Yet, there are times when that live feel is best, such as at one-off shows when there has not been a full rehearsal or at a festival of artists not used to a full lighting rig. In these cases, the lighting designer can contribute to the show's mood and tempo on the fly. The LD can set up a series of palettes, instead of preprogrammed scenes, that offer the option of picking various combinations—for example, assigning colors, gobos, and different sweep movements to change moving luminaires "live." The spontaneity can be exciting for both the audiences and the designer.

There are many reasons why a designer might choose a particular console; their preference can be likened to a musician who has a favorite maker of piano. Partly this is due to familiarity and a certain comfort level, to be sure. In some cases, designers may be in such demand that they simply do not have time to get comfortable with another unit, even when a newer, different console might work better for their projects. And, because there are so many options available, it has gotten harder and harder to keep up. The following discussion provides examples of several of the models currently being offered.

Another twist is the addition on many high-end consoles of visualization programs that allow the designer and programmer to see what the cue will look like as they are building it. This is extremely handy, because most allow you to print out the looks to show them to the manager or artist so they get a better idea of what the designer has planned. Some of these programs can be used live during the performance to allow editing on the fly.

HANDS-ON CONSOLES

This group of consoles is often favored for live events such as festivals or one-off shows. They are designed on a simpler platform that does not have as complex a program methodology or does not handle as many cues. These consoles may be used for television

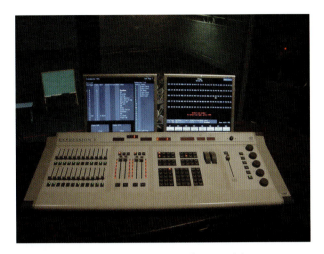

FIGURE 11.3 ETC's Expression 3. (Photograph by James Moody.)

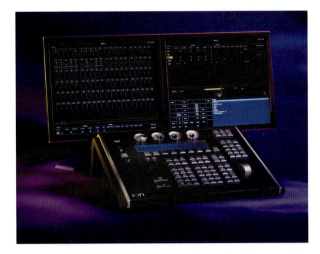

FIGURE 11.4 ETC Ion lighting control console. (Photograph by Electronic Theatre Controls, Inc.)

shows or other special events that also do not require the expense of a more complex console or do not use many moving luminaires or effects. Based on the old manual, two-scene concept with submasters and a grand master, they have been expanded to include the concert idea of *flash buttons*, which allow individual channels or groups of channels assigned to a submaster to be instantly energized without sliding a fader handle up and down. It gives the operator the ability to *play* the console. This idea has its roots in a concept called the *light organ*. In today's world, a light organ is a device that controls the brightness of a light according to sound levels—that is, the louder the sound, the brighter the light; the softer the sound, the dimmer the light. A *color organ* is a slightly more sophisticated device with several channels, each one sensitive to a different part of the audio spectrum. So, low notes might flash a blue light, middle notes might light up yellow, and high notes might flash a red light. For each channel, the louder the sound produced in the specified pitch, the brighter the light.

Expression 3

Probably the most popular console has been the Electronic Theatre Controls (ETC) Expression 3 (Figure 11.3). This console has been around for a while and just recently has been archived (it will no longer be built); however, there are so many of them out there in regional theatres, television stations, and educational

facilities that they have to be included in this chapter. According to ETC, they shipped 6085 of the consoles, and their research suggests that there are still 90% of them in use. The line originated in 1982. It has 24 submasters on 10 pages, 2 sets of cross faders, and 5 attribute encoders for the controls needed to program moving luminaires with 1536-DMX-512 dimmers/attributes on 400/800/1200 channels and 600 cues. It also has a feature first introduced in the original Light Palette (discussed later in this chapter)—the wheel fader/timer. This is an endless rotating wheel that has no 0 or 100% fixed point. The designer does not have to relate to a number, only the visual response, and the computer can remember it. It also works to modify time fades up and down using a "take control" instant touch method that can speed up or slow down a retimed fade, thus allowing the designer to match a change in speeds of the band's tempo or fades.

Ion

Ion is ETC's replacement for the Expression 3. Its design may have evolved from the old Strand 300 series console in that it takes a systems approach of components that can be joined in several add-on configurations. The basic Ion unit (Figure 11.4) is 19 inches wide and controls conventional and moving luminaires equally well. It includes split faders, a grand master, four function wheels, key pad, and no submasters on the main unit; submasters/channel

FIGURE 11.5   Leprecon LP-1600. (Photograph by Leprecon.)

FIGURE 11.6   Avolites Diamond 4 Vision. (Photograph by Avolites, Ltd.)

control can be added via up to 6 "fader wings" of 1 × 20, 2 × 10, and 2 × 20 for a total of 240 additional submasters with paging control. The system can support one or two external high-resolution digital video interface (DVI) monitors (which can also be touch screen). Configuration of 1024 to 2048 outputs, with channel numbering from 1 to 5000, is available. An additional feature is that the Ion can be used as a remote programming station for the Eos system. ETC says that the operational style is the same as Eos, which can make it much easier for an operator to move up to the more powerful console.

Leprecon LP-1600

This console is in the lower cost range of quality consoles. Leprecon's LP-1600 series is similar to the Ion but does not offer the moving luminaire capability of dedicated encoders (Figure 11.5). It is a true manual console with a memory. Sizes range from 24/48 to a maximum of 48/96 channels. A limited memory for cues allows for only 288 but point cues can be added on top of this number. A disc drive and VGA monitor outputs are optional. To be fair, Leprecon does make two consoles with moving luminaire controls and offers a luminaire library for their LP-X24 (512 channels) and LP-X48 (192 channels) models.

Avolites Diamond

In the ever-evolving, hands-on category because they have retained channel faders on the console logic are the Avolites Diamond Series and smaller counterpart Pearl Series consoles. The Diamond 4 Vision (Figure 11.6) was featured in the second edition of this book over

14 years ago and is still a highly regarded console. It is designed for large-scale shows, festivals, and concerts. Features on the consoles that many designers appreciate have resulted from engineering advances as well as from input by touring designers. It is loaded with 28 cue playback masters of over 200 pages and with 128 preset faders, which can now control a chase, cue list, or luminaire intensity of over 1000 luminaires. Also available are multiple times on attributes and luminaires. An interesting feature is what Avolite calls "dedicated buttons for theatre plotting" with dual theatre configuring of palettes, groups, luminaires, and cues. The unit has eight DMX-512 outputs, as well as Ethernet capabilities. This is a rock & roll lighting console legend that is highly praised by leading programmers and lighting designers. I see many more years of valued service for this series of consoles from England.

CONSOLES DEDICATED TO MOVING LUMINAIRES

History of the Hog

Behind the development and naming of the Wholehog was a console designed by Nick Archdale with Andy Neal. They used the DLD6502 desk they had built to run dimmers, scrollers, and Clay Paky Goldenscans from 1988 to 1991. In 1988 and 1989, they moved in with SpotCo, and the DLD desk was thus brought to the attention of Peter Miles and Tim Bayliss, who saw the potential for such a console in the blossoming DMX wiggle-mirror market. SpotCo in London provided a portion of the seed money and the encouragement for the development of a new console, the blueprint of which was largely formed in Nick's

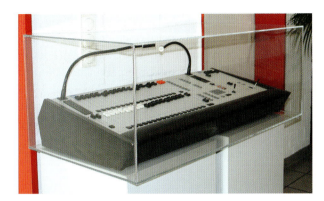

FIGURE 11.7 First Wholehog DLD Serial No. 001. (Photograph by Ralph-Jörg Wezorke.)

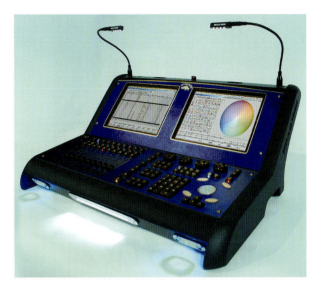

FIGURE 11.8 Road Hog Full Boar. (Photograph by High End Systems.)

head, as he had spent hours behind the DLD6502, Vari*Lite Series 100, and Artisan consoles. He flew to the United States in 1991 to find Tom Thorne. They drove across the country that summer and conceived the names Flying Pig Systems and the Wholehog. They then worked on the system for less than 10 months before the first prototype was used for the Irish band Simply Red. Figure 11.7 shows the original console, DLD Serial No. 001, under glass in the lobby of Lightpower GmbH in Paderborn, Germany, today.

Flying Pig Systems was founded in 1991, and the first Wholehog console was shown at a Professional Lighting and Sound Association (PLASA) show in 1992. They felt there was a void in the market that they could fill with a desk that served equally as well for conventional as well as automated lights. Nick explained, "Hence the idea of 'going whole hog' about controlling large numbers of all kinds of fixtures."

The Wholehog was a departure in console logic. The idea of parametric, high-priority programming using *highlight*, *blind*, and *clear*, as well as a library of fixture personalities, first appeared with the Wholehog. A visual two-dimensional rig schematic was visible to view the placement of all the luminaries. But, maybe the most radical thing was that … *you could buy one*! Up until then, Vari*Lite had refused to sell their luminaires or consoles; they would only lease them.

The Wholehog II (conceived in 1993 and released in 1995), fondly called the "Son of Hog," lost the rig schematic and a few thousand channels but gained a real-time effects engine, three-dimensional space (XYZ position programming and updating), and "what you see is what you get" (WYSIWYG) integration. Flying

Pig seed-funded Cast's WYSIWYG visualizer, manufactured the DMX cards, chose the name (see Chapter 21), and finally added the use of touch screens.

Unlike the original Wholehog, which was specifically designed as a touring board, the Wholehog II was created with a much wider market in mind (e.g., theatre, television, trade shows) and certainly seemed to hit the spot, as thousands were sold.

The Wholehog III is the latest version. The general idea was to produce a scalable system, so all DMX and IO processing was moved off the board to networked (Ethernet) components that could be added as required. This was also the point where the syntax was changed to bring it in line with more established computer techniques, but Nick said he feels the best part is the abstracted fixture library model that uses real word units to control any fixture in exactly the same way, thus allowing for the Holy Grail of touring consoles—one with the ability to swap out or clone fixture types and preserve all programming.

High End Systems bought Flying Pig Systems in 1999 and took over product development in 2003. Subsequently, High End Systems has been sold to Barco, as noted in Chapter 1.

Road Hog Full Boar

The console shown in Figure 11.8 with the Wholehog III includes many of the same features. They are

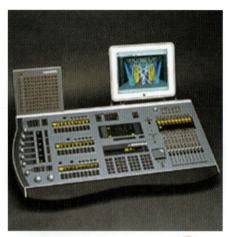

FIGURE 11.9 Virtuoso DX2 console. (Photograph by PRG.)

equipped with Flying Pig Systems software and are able to create, manage, edit, and play back moving luminaires at some of the fastest speeds possible today. They feature a backlit trackball with four configurable buttons for cursor and position control and ten playback faders with familiar playback controls: choose, go, pause, and flash—with dedicated intensity and rate wheels and 36 soft-keys for quick toolbar selections. Most often the Wholehog will be used with a second console dedicated to conventional luminaires. It is not unusual to see this combination on a tour or television special.

Virtuoso DX2

The Virtuoso DX2 console (Figure 11.9) from PRG was designed to handle moving luminaires right from the start. Vari*Lite Series 300 luminaires were its prime target. The console was originally built by Vari*Lite but has been updated and is now manufactured by PRG; however, digital luminaires and conventional luminaires are at home here, too. In addition to having 8 DMX-512 outputs and supporting over 600 universes, it has an Ethernet communications protocol. The system has a fully integrated three-dimensional programming environment that provides real-time status display and offline editing. This is a Mac platform, unlike most others that are PC based. To prove that this console fits right in

with media servers, it has a media window for graphics control and preview of the numerous media files required for digital lighting. For the moving luminaires, it provides 2000 multiple parameter, 2000 to 10,000 cues per luminaire, 1000 presets, and 1000 effects. Add to this 1000 board cues, 1000 groups, and 30 submasters and you have a very powerful moving luminaire and digital lighting system control.

V676

The V676 is a recent entrant into the market by PRG. In fact, it was not ready to ship at the time of this writing. There was talk that it will eventually replace the Virtuoso DX2 but this was not confirmed by PRG. The V676 console (Figure 11.10) is designed to be intuitive for the programmer. The company claims extraordinary speed, both in programming and responsiveness, as well as innovative interface advancements in hardware and software design. In an unusual move, the processor is mounted outboard of the console, an Apple Quad-Core Mac Pro, for better graphics and faster processing. The console is marketed for moving luminaires and digital media.

THEATRE CONSOLES

It may no longer be fair to make this distinction because so many theatres now plan highly complicated lighting cues for their plays and musicals, and you will often find moving luminaires and other effects and media. It is fair to say that all of the traditional lighting console manufacturers have awakened to the need to make their consoles more technically appropriate to today's diverse lighting uses, such as corporate shows, theme parks, architectural, dance, and opera, as well as concerts. In all of these cases, there is a good chance that up to several hundred moving luminaires will be included in the lighting plan; therefore, the old 48-channel, two-scene console with split-faders and a grand master just won't work anymore, even for houses of worship.

Expression 3

The ETC Expression 3, as noted earlier in this chapter, is very common and will remain on the market for years to come.

FIGURE 11.10 V676 console. (Photograph by PRG.)

FIGURE 11.11 Strand Light Pallette VL. (Photograph by Strand Lighting, Inc.)

Strand Light Palette VL

Strand Lighting still holds sway with their Light Palette series. The Light Palette VL (Figure 11.11) is the newest in the series and is based on the Classic Palette, which dates back to the 1970s; it has been a prominent feature of Broadway shows lately. Light Palette VL added 24 submasters and a 100-key direct action keypad. It delivers power in a traditional console with 2 timed split cross faders and 12 combination submaster/playback faders with 4 additional rotary encoders and a built-in touchpad. It has core processors for optimum performance, a color picker for direct color selection, 2 timed playbacks, 12 direct-access playback

faders, 4 rotary encoders, and 2 grand masters, but touch screens are optional. The capacity for memory is up to 8000 control channels/attributes. The console comes with four DMX-512 outputs with optional touch-screen monitors.

Congo v5

Congo was developed when Electronic Theatre Controls (ETC) acquired Avab's console line. When ETC made the announcement that they had acquired the Avab line, they touted the fact that both companies had been designing consoles for 30 years. The press release said, "Melding together the engineering brilliance and experience of ETC and Avab systems, Congo maintains the simplicity of classic systems with the feature-rich functionality of a dedicated moving luminaire console." The manufacturer tried to maintain the functions desirable for everyday theatre work with conventional luminaire but added the advanced control of hundreds of moving luminaires. The Congo console (Figure 11.12) allows the user to move independent channels such as work lights, conductor lights, follow spot, and smoke machine control to a special section that isn't affected by the rest of the system. The unit comes complete with a theatrical-style main playback fader pair, 40 multipurpose masters for group submasters, a submaster, effects and additional sequence controls, 40 direct selects, and a dedicated moving luminaire control section. In 2008, the company released Congo v5, which is essentially a hardware upgrade but well worth the trouble.

Hybrid Consoles

Hybrid consoles do not follow the old channel methodology. Many of these consoles no longer even call a controller a "fader." These consoles cost tens of thousands of dollars and are more than worth it, *if* your needs justify the sophisticated and complexity

FIGURE 11.12 ETC Congo family. (Photograph by Electronic Theatre Controls, Inc.)

they offer. I have spoken to several operators who admit they feel that have not scratched the surface of the capability of units in this group but are more than anxious to try as shows become more and more sophisticated. There are many times, such as during Super Bowl halftime shows, when many of these consoles must be linked together to operate the show. Tours often operate with more than one console. In some cases, these might be two different units determined to best handle particular sections of the lighting system, such as conventional luminaires or moving luminaires.

HOG iPC

The Hog iPC console is a small, mid-range controller that offers a lot of flexibility for programmers (Figure 11.13). It is designed to use the Wholehog 3 control system, so you get advanced features and functions, but it is backward compatible with the earlier Wholehog 2 software. It's the industry's first hybrid console.

The Hog iPC uses four universes of DMX output direct from the console, expandable to eight universes using USB DMX Widgets or a USB DMX Super Widget. An unlimited number of DMX universes is possible when using networked DMX processors (DPs). The unit has two 12-inch, high-brightness color touch screens with adjustable viewing angles, and a third optional monitor or touch screen can be

FIGURE 11.13 Hog iPC. (Photograph by High End Systems.)

connected to the DVI port of the console, allowing additional windows to be viewed and adjusted simultaneously across multiple screens. A backlit trackball with four configurable buttons provides cursor and position control. Wholehog 3 playback controls include choose, go, pause, and flash. The Hog iPC has an internal hard disk drive, a rewriteable CD-ROM drive, and five USB ports for touch screens, external drives, printers, and Wholehog accessories (e.g., playback wing, expansion wing, and additional USB DMX Widgets). An Ethernet connector allows communication with Hog 3PC computers, Road Hog Full Boar consoles, another Hog iPC console, and Wholehog 3 console systems and devices. Remote focus functionality is available when networked with a computer running Hog 3PC software.

Eos

The Eos line by ETC (Figure 11.14) is trademarked as Complete Control™. It won several awards for innovation after its release in 2007. ETC's brochure says it provides simple, approachable control in a nuanced programming environment, with unmatched depth and power. The idea of fast and easy access to moving luminaire control without sacrificing what was needed for conventional control was the watermark they wanted to achieve. This console, along with the grandMA, was the first console to rethink how common functions could be handled in different, even simpler ways. By spending a lot of time with lighting designers, programmers, and electricians, the company was able to develop intuitive solutions while reducing keystrokes. The operator can issue a command via a keypad or the touch-screen controls. Some people have trouble with touch screens, feeling that if they aren't looking at the point they are touching they can't be sure they have initiated the command. Eos says they have found a way to provide touch-screen flexibility with the tactile response of buttons, so operators can keep their eyes on the stage, not the console. Plus, the mouse has been eliminated. The operator accesses commonly used programming functions via buttons that are within easy reach. Each device on the system has a discrete workspace; designers don't have to sacrifice their needs for those of programmers. Partitioned control provides a safe and effective method for multiple programmers to build content into a show file. All this comes with 10,000 channels (devices); 4000, 8000, 12,000, and 16,000 outputs/parameters; a dedicated master playback fader pair; 10 definable motorized faders with 30 pages of control; 999 cue lists; 200 active playbacks; 300 submasters; 3 programmable grand masters; dedicated pan/tilt or XYZ encoders; 4 page of encoders; 2 built-in 15-inch LCD touch screens; and electronic magic sheets. The console's output can use the new Net3 (ACN based), as discussed in Chapter 20, and ETCNet2 native devices, as well as multiple MIDI and SMPTE inputs.

Maxxyz

Maxxyz (Figure 11.15) is Martin Professional's top-of-the-line lighting console, and it has some original and unique features. The console has extremely fast processing power, direct luminaire access, an effects generator, USB connections, motorized faders, touch screens, and SMPTE/VITC/LANC/MIDI time coding. Maxxyz was one of the first of its kind to incorporate digital LCD buttons designed for tailored, flexible programming. This feature lets the programmer group functions directly to an LCD button and customize the console layout to ease navigation and provide fast recognition. Maxxyz's three-dimensional visualizer is fully integrated into the console software. The visualizer allows the designer to view the light show in highly realistic three dimensions. The

FIGURE 11.14 ETC Eos. (Photograph by Electronic Theatre Controls, Inc.)

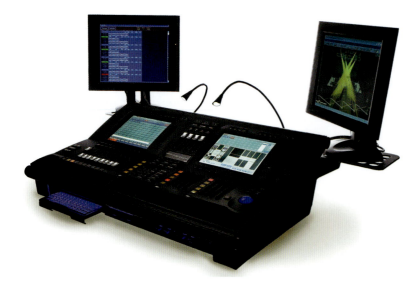

FIGURE 11.15 Martin Maxxyz.
(Photograph by Martin Professional.)

full real-time preview allows designers to create and refine designs without interrupting the running show. This console covers all lighting needs but also integrates the multimedia that today's tours require. The console has a rewriteable DVD/CD drives so you can synchronize your light shows with audio and video. The combination drives also let you save your shows to CD. Maxxyz also has 39 GB of space available for backup purposes. In addition to the 8 DMX universes available directly from the console, Maxxyz is also fully Ethernet enabled. Its amazing 32 DMX universes, 8 of which are available directly from the console, make this an ideal console for large LED screen applications. Dual processing power (two Pentium CPUs) controls up to 16,384 luminaires; 1000 cue lists with motorized fader control and 1000 cue lists with playback control buttons round out the impressive features of the EOS.

grandMA

The long-anticipated announcement has arrived regarding the release of the German-built grandMA2 (Figure 11.16) by MA Lighting and marketed in the United States by A.C.T. Lighting. The MA in the name stands for one of the coinventors, Michael Adenau. The grandMA is the product of the two inventors teaming with Lightpower to build a series of consoles that soon were being talked about throughout the concert touring community. Many of the features now seen on the Congo and Eos were first built into the radical design of the original grandMA. The grandMA features extensive integration of conventional luminaires, moving luminaires, LEDs, and media server control in one console. It was the first console to have built-in touch screens and the first to drop the fader and submaster concept for *executors*. These are 20 motorized controllers that can run a single channel, a single cue, a whole sequence, or the entire show. Each executor has 128 pages, and there are also an additional 40 executor buttons. For moving luminaires, there are 4 encoder wheels that accept 2048 or 4096 parameters and a unique ball that can be energized to pan and tilt luminaires. Among its other features is that there is no restriction of the amount of cues that can be recorded; the number is virtually unlimited. Presets are set in 10 groups of 999, and an additional 999 sequences can be stored. In typical German manner, it is heavy and rigged, built for touring. The hinged feature allows the programmer to set the touch screens at an angle that allows the best viewing in various lighting conditions.

I was first introduced to this console over 12 years ago when I was the director of photography on "Wheel of Fortune." Up until this time, when we shot on locations, the board operator would bring two consoles, one to run the moving luminaire and one for the conventional luminaires. When we

FIGURE 11.16 grandMA2, full size. (Photograph by MA Lighting.)

switched to the grandMA, he only needed the one console because he could write cues, program moving luminaires, and bring up individual channels during tech, all on the same board in real time. Now there are other consoles that can do the same thing, but the grandMA and the upgrade to the grandMA2 are the leaders in this field. Most consoles have a short lifespan in this highly competitive market of touring, but the grandMA has been in service for longer than most and is still a highly sought after console.

SPECIALIZED CONSOLES

Several PC-based consoles do not require anything more than your laptop to run the show. See Figure 11.17 for the names of companies that provide this option. There are also specialized consoles such as ones that are designed to run large LED displays. One, called Madrix, comes from a German company and runs in Windows XP and Vista with a minimum system requirement of a 2-GHz processor, a video card capable of at least DirectX 9.0c, and a USB port for the dongle. This is a software solution that outputs 16 universes of DMX-512 or ArtnetII. The best use seems to be as a pixel-mapping control interface. The stored programs you create can be played back

in several ways, including being triggered by audio sources, but during playback they can be manipulated live. One drawback is that only about a third of the effects work without an audio signal. This is part of the new era of expanding use of LEDs. While this console may not control the entire concert, there are many uses for this type of outboard control.

SUMMARY

The market for high-end, highly customizable, cross-platform (media servers to moving luminaires) consoles is very competitive, and the units are expensive (upwards of $35,000 to $89,000 or more for a top-of-the-line console). This is not an item to be purchased on a whim. It may seem unusual, in some ways, that these consoles have enjoyed such longevity until you remember that programmers and designers are creatures of habit who are highly paid for being able to do their job fast, efficiently, and accurately. Doing so requires an enormous amount of knowledge, practice, and real-time experience on how to make the console work for you. Yes, there are programmers who can move between several of these consoles without breaking a sweat, but at some point the market will have to settle

ABD Lighting Technologies	www.adblighting.com
A.C.T. Lighting	www.actlighting.com
Avolites America, Inc.	www.avolites.us
Celco	www.celco.co.uk
ChamSys	www.chamsys.co.uk
Coemar	www.coemar.com
Compulite	www.compulite.com
Electronic Theatre Control (ETC)	www.etcconnect.com
High End Systems/Flying Pig Systems	www.flyingpig.com
Jands	www.1.jands.com.su
Leprecon	www.leprecon.com
Leviton	www.leviton.com
Martin Professional	www.martin.com
Production Resource Group (PRG)	www.prg.com
Pulsar	www.g1limited.com
SGM Technology	www.sgmtechnologyforlighting.com
Strand Lighting	www.strandlighting.com
Zero 88	www.zero88.com

FIGURE 11.17 Lighting consoles manufacturers.

out. You can see by the long list of manufacturers provided in Figure 11.17 that this is not a game for the faint of heart. Get as much time on as many consoles as you can. Most dealers and rental houses have a demo area where you can arrange to spend time with the console. ETC recently set up several suites in their Los Angeles offices for programmers and designers to practice or program their shows. Most rental houses have several consoles set up in a demo room for designers to spend time learning on. Most of the high-end consoles offer free online downloads that provide a console image and allow users to practice the functions and write cues. These downloads are great for learning new consoles that you may not have physical access to, even if you cannot use the cues for a show unless you purchase the program. Make your choices with all due consideration.

12

PORTABLE DIMMING AND DISTRIBUTION SYSTEMS

L ate in the 1960s, when I started working for a lighting manufacturer, I was told, "We sell what we have. We don't encourage the customer to ask for specials." That remained true into the 1980s. Like legitimate theatre, concert lighting borrowed from any market or service that had something that could be adopted, modified, or used to its benefit.

By the mid-1980s, though, things were changing. New manufacturers were trying to meet the needs of concert and touring lighting. Products appeared that were not adaptations or reworkings of old ideas but instead were exciting new products cut from the whole cloth of a new vision.

Dimming is the area where packaging, not electronic innovation, made the first real impact in the early years of touring. In the 1970s, the established manufacturers were just beginning to market compact groups of 6 to 12 dimmers in a portable unit, usually a 19-inch rack-mount configuration. What the manufacturers did not think of was mounting input connections as well as output strips directly to the dimmer packs so there would be much less to assemble each time the dimmers were set up.

Distribution systems with low voltage control patching were a true touring innovation initiated largely by British manufacturers. This design allowed designers to assign one or more dimmers to a single control channel on the console. Later, electronic or *soft patch* systems appeared that used the lighting console's computer to assign these control circuits. Some British and U.S. manufacturers put a "mini-pin" cord-type patch bay into their dimmer racks (Figure 12.1).

Another way to assign luminaires to a dimmer was to make panels with more than one female connector. The electrician could simply place the appropriate cable's male connector into the dimmer rack's female connector and, assuming here is control of the dimmer, one or more luminaires could be energized at different locations by the same dimmer. If there is an internal low voltage patching system in the dimmer rack or a soft patch in the console, it is possible to make the connection between the luminaires and the dimmers assignable while using Socapex 6 circuit connectors, without the need to connect each individual pin connector.

The placing of several of these *package dimmers* (Figure 12.2) in a single, castered, road case meant that 24 to 72 1-kW or 2.4-kW dimmers could

FIGURE 12.1 ETC 48-channel road rack with mini-pin patch bay. (Photograph by Electronic Theatre Controls, Inc.)

be connected and ready to operate in a fraction of the time it would take to stack and wire each pack individually. Figure 12.1 shows the front view of an Electronic Theatre Controls (ETC) rack equipped with casters and 24 double modules (48 2.4-kW dimmers), with 3 spare modules (lower right). The electronic brains are at the top, and a mini-pin patch cord panel that facilitates output assignment is on the right. Shipping damage is considerably reduced by the use of slide-in wooden covers to protect the front and back during transit. In the early dimmers, it was necessary to take the packs apart to add lock washers and epoxy to keep the components from rattling to pieces

FIGURE 12.2   Strand touring rack. (Photograph by Strand.)

FIGURE 12.3   Avolites Art dimmer rack. (Photograph by Avolite.)

in the truck. But, here again, the wise manufacturer listened and quickly made the components more secure. The most popular dimmer racks used on the road were those that used a modular construction of one or two dimmers in a slide-out track for quick and easy replacement. A concert setup goes too fast for a technician to shut down an entire rack to disassemble it to get at one dimmer, and then it is very difficult to find the problem and replace the component. With modules, a spare can be shoved in and when time allows the bad unit tested and repaired. Recently, new dimmer packs have been introduced by several manufacturers that are only semimodular, because the dimmers have become so reliable that the added cost and space required for a modular design have been eliminated (Figure 12.3).

In the early days, before touring gained a foothold, U.S. dimmer modules were offered in a wide range of capacities: 2 kW, 2.4 kW, 3.6 kW, 4 kW, 6 kW, and even 12 kW. However, these were bulky and very heavy, as they were designed for installations where weight was not an important criterion. A big advance came when British manufacturers utilized small 1-kW dimmers. While these were not true modules that could be easily removed from the rack, the road crews liked them because they were more compact, and the lighting designers liked them because they gave them individual control of each luminaire. This design allowed each PAR-64 (99% of all the luminaires on the road were 1000 watts at that time) to be patched without twofers or multiple outlets. Now the low-voltage patch could take over and assign the dimmer and associated luminaire to any one or more control channels. The 1-kW dimmer also meant that more circuits could be housed in a single, movable rack with the main circuit breakers fully enclosed in a protected unit. American manufacturers such as ETC went with 2.4-kW modules. Today, you will find people that are fans of either for their own reasons, but both clearly have balanced shares of the market.

The evolution to true road status meant that a number of things had to become integrated into the same case. The first was the power distribution system, although there are still reasons to carry separate main breakers designed to distribute power to multiple dimmers and power racks, and they are certainly available for special needs. After trying out a number of connectors, the U.S. touring industry settled on the Cam-Lok[1] "J" Power series, which is used exclusively on the industry's standard feeder sizes: #2/0 and #4/0 single-line feeders. These single feeder cables are color coded to a universal standard so neutral, ground, and hot legs are readily apparent to electricians making the connections (Figure 12.4). These feeder sizes were chosen because most house disconnects are either 200 or 400 amp (per leg), which matched very well. They are rated for 225 amps and 400 amps, respectively. To work the simple math, that means a 200-amp service can safely run 42 1.2-kW dimmers; however, a PAR-64 is most often used with a 1000-watt lamp, which translates into 8.3 amps, so we can safely run 48 1-kW dimmers on this feeder. The same goes for the #4/0 and 96 1-kW dimmers. Some of the racks have a power feed-through system, so a 48-channel 1-kW rack can feed onto another without establishing a "home run" for the second rack.

The next addition was digital multiplexed (DMX) capacity. The Engineering Technical Committee of the United States Institute for Theatre Technology (USITT) drew up the standard we still use today, DMX-512. One of the features of DMX compared to analog is that 512 channels can be controlled with one cable instead of running a cable to each dimmer rack. Low-voltage analog requires a control wire for each dimmer, resulting in a large bundle of wires. As the number of dimmers increased, it became more difficult to build control cables that were flexible

FIGURE 12.4 Cam-Lok "J" series connector color coded for U.S. hot, neutral, and ground. (Photograph by CBI Cables.)

---

[1]Cam-Lok, by Cooper Crouse-Hinds, is a trademark for single-pole locking connectors used widely in film, television, and concert power feeders.

enough for portable situations. Now all that is about to change (see Chapter 20), and the new system will give us the capacity for many more channels to handle moving luminaires, scrollers, and LEDs.

Another standardized item that speeds assembly of dimming and luminaires is multicable. A 6-circuit version is the most popular; however there is also a 12-channel version. The Socapex beat the competition, and now all rental and road houses are standardized to this connection. It provides a quick, secure connection between the dimmer rack and the raceway or "break-out." Most dimmer racks provide some indication of power connection and a main breaker to protect the unit. Some even have digital meters that show voltage and the amperage being consumed.

Another push has been into *high-density dimming*. Because the big manufacturers were behind in the total packaging of dimmers with ancillary gear, they began to focus on the research and development of miniaturized components; however, the concert field is still dominated by newer start-up companies that were not around in the 1970s. Dimmers that provide fully integrated systems in a road case, complete with main breaker and multicable connections, are available from some old-line companies, but ETC still holds a commanding lead in this area. The initial dominance by British manufacturers was mainly due to their direct involvement in touring and seeing firsthand what worked for the road. Eventually, they got a major piece of the installation business which probably has led to the demise of most of the old-line companies.

The newest dimmers today use *sine-wave* technology, which effectively eliminates hum in the lamp filament. This is generally not an issue for concerts, but in sales to theatres and television studios this is an important consideration. Naturally, these dimmers cost more.

The most popular racks in the United States are ETC Sensor racks configured with 48 2.4-kW dimmers. There are several strong contenders; for example, Avolite, Strand Lighting, LightRack, and Celco's Fusion series use 10-amp dimmers. (*Note*: The Fusion racks use groups of 12 dimmers in each module.) The reliability of the ETC Sensor dimmers has been tested over many years, and touring folk tend to stay with what they know is reliable (see Figure 12.5).

You need to be able to not only recognize the power box but also decipher what it can provide. Most installed company switches in theatre are on the wall opposite the fly rail. Some are on the back wall (not ideal, as it takes more cable to get to them—note to self: bring more feeder cable). Now for the important part: What power will this panel deliver? In the United States, we speak of 3Ø/5 wire power; translated, that means that on the face of the box a label should show a number (i.e., 100, 200, 400, 600, or 800 amps). That number is multiplied by the number of phases (Ø). Therefore, a 400-amp, 3-phase system will deliver 1200 amps. Figure 12.10 shows a Union Connector company switch. Note the color-coded, panel-mounted Cam-Lok output connectors at the bottom of the panel with protective covers.

The features include no exposed screws or holes and a positive connection made via the exclusive Lektralink principle, which makes a secure connection easy. In vulcanized versions, the connection is also waterproof, a good thing when working outdoor festivals with the possibility of rain showers. As best we can determine, no rental house uses anything other than Cam-Lok connectors on #2/0 and #4/0 single power feeders, so you should be able to travel anywhere in the United States and be able to rent additional feeder cable for situations that require longer runs.

If you do not see any connectors, then the system is using either a buss bar or a bear-end into lug connection.

For a buss bar connection, you need sister lugs on the end of your cables. For a bear-end connection, you should have no connector on the end of your power cable. It is standard practice in the United States to carry 5 runs of 100 feet of #4/0 feeder cable with Cam-Lok connectors, plus a set of 5- to 10-foot cables with Cam-Lok single connectors on one end and bear ends on the other (with a set of sister lugs in your drawer, just in case). Understanding this doesn't make you a master electrician, but it may help you look not too foolish to the survey team.

TRANSFORMERS

It is somewhat unusual to find U.S. tours carrying transformers, although we once needed them on an aircraft carrier to do a television show, and they are often found on temporary festival sites where the local power company has made a high-voltage (say 480vac) power source available and that power must be reduced to a useful level, usually 208/120vac or 240vac. The need for transformers becomes a more important issue while touring overseas. As was already mentioned in Chapter 8 in our discussion on touring in Japan and South America, the need to match power for moving luminaires can sometimes require that a transformer be placed in service.

A *transformer* (Figure 12.11) is a device that transfers *electrical energy* from one *circuit* to another through *inductively coupling*. Two conductors are referred to as being inductively coupled when they

FIGURE 12.10 Company switch. (Photograph by Union Connector.)

FIGURE 12.11 Transformer. (Photograph by Paul Dexter.)

are configured such that a change in current flow through one wire *induces* a voltage across the ends of the other wire. A changing *current* in the first circuit (the *primary*) creates a changing *magnetic field*. This changing magnetic field *induces* a changing voltage in the second circuit (the *secondary*). This effect is called *mutual induction*. If a *load* is connected to the secondary circuit, *electric charge* will flow in the secondary winding of the transformer and transfer energy from the primary circuit to the load connected in the secondary circuit.

Put simply, a transformer can either raise (*boost*) or reduce (*buck*) the output voltage when voltage is applied to the device. Therefore, whatever voltage is available onsite can be, in a sense, modified to the needs of the equipment you are trying to power. The transformer also offers the advantage of isolating the power from other sources that may be creating hum or buzz in electronic equipment.

PORTABLE GENERATORS

Many sites, especially festivals, require more power than what is available. Temporary power can be brought in by the local power company, but it may be more cost effective for a one-day show to simply rent portable generators to provide the additional energy necessary. Movie companies have been doing this for years. Figure 12.12 shows Cat generators; the housing is called a *blimp*, and it reduces the noise produced by the diesel engine. A generator is essentially a self-contained transformer that can produce either A/C or D/C power (although there are some D/C-only portable generators). In this case, the generator provides its own mechanical energy via a motor to convert that energy into electricity to serve as a power source for other machines. Electrical generators, for our purposes, are normally diesel powered. Depending on the situation, they may be blimped, or made quiet, by adding a covering and exhaust system that reduce the noise factor. This blimping does raise the rental price, so determining if it is needed can impact the costs you will incur.

There is one other issue with generators. The ones used to power electronic equipment must have electronic control, a feedback system that electronically maintains the selected output voltage and frequency. The cycle or frequency for any device must be maintained with a motor or fan, such as via computer control or cooling fans placed in the dimmer racks or luminaires so they receive a steady 60 Hz. Not all generators come with this feature.

CABLE REELS

A weak link in the lighting system is the 100- to 200-foot control cables to the lighting console that are placed out in the house. Normally, a show, the cable is disconnected from the console, and a local stage

FIGURE 12.12 Portable generators. (Photograph by Paul Dexter.)

FIGURE 12.13 Cable reel. (Photograph by Marcaddy.)

FIGURE 12.14 Cable crossover. (Photograph by Yellow Jacket.)

hand starts pulling it to the stage to coil it up and dump it in a road case. Along the way, the connectors gets caught under chairs and dragged through spilled soda. Save your cable by asking for a cable reel (Figure 12.13).

CABLE CROSSOVERS

If your main feeder cables get cut or pulled out, or the Socapex cable that you ran across the backstage area or your control console cable that is taped down with gaffer tape across the aisle in the audience gets damaged, you are not likely to be putting on a show. Look into cable crossovers (Figure 12.14). They can save expensive cable and prevent a law suit when someone trips over the cable. Don't expect the hall to have them; carry them as part of your touring package.

13

LIGHTING TRUSSES

The use of found space for a theatrical production is not new. Barns, grassy fields, arenas, convention centers, and all manner of multipurpose rooms have been used for performances. The elaborately equipped buildings designed for large symphonies and opera, as well as those built specifically for drama in the early 20th century, are in the minority when it comes to what is being used for concert entertainment venues today. Because nearly all of the large theatres were built with stage houses of a similar size and design (except for thrusts), it used to be relatively easy to mount a touring opera or theatrical production. Most of the buildings had permanent lighting pipes in neat rows to provide lighting luminaire support. But what about buildings that do not have accommodating theatrical lighting as their primary mission?

TRUSS DESIGN

There are as many variations on the theme of truss design as there are companies designing and building them. Each company feels that its design offers the best solution to a particular problem, such as simplicity and speed of setup, strength, luminaire capacity, or method in which the sections are interconnected. For flexibility in design applications, remember that, whether you mount luminaires internally or not, the four sides of the box and even the ends are available for mounting, so bottom corners can be used as well as top corners. A major advantage of truss use is the flexibility afforded by corner blocks and angle blocks to connect truss sections into shapes. The many

possibilities for truss layouts are discussed toward the end of this chapter.

THE FIRST TRUSSES

The early rock & roll concerts were considered not artistic enough for many city and college theatres, and they were generally relegated to school gymnasiums. Where do you hang luminaires in a gym? The answer was to bring along your own structures to mount them, to create a performance area in a found space. Finding or constructing *portable* units that could be trucked easily from show to show became the Holy Grail for early designers.

Portable units also had to have *time-saving* features. It is not unusual for recording artists to travel and perform in a different city almost every night, whereas plays usually have a run of a week or longer with possibly a day or two of rehearsal and time to make adjustments to the new space. For concerts, speed of setup is extremely important. At first, solutions to this problem were hit and miss, but eventually new design ideas emerged. Creative people took on the challenge, and a whole new concept was born: a structure, somewhat like a bridge, that allows designers to place luminaires overhead instead of on standing poles or trees.

The structures themselves have no historical precedent in theatrical design. The first touring truss was designed by Chip Monck and Peter Feller, with Bernie Wise, for the 1972 Rolling Stones tour. During a recent conversation with Chip from his home in Australia, we reminisced about that truss. It might have been 37 years ago, but Chip had his facts down.

FIGURE 13.1 First Rolling Stones truss with short-nose PARs. (Photograph courtesy of Chip Monck.)

It was built of 6061T aluminum alloy built into a 4 × 4 box square of 5- to 10-foot sections. In the photograph shown in Figure 13.1, the truss shows sections stacked for loading with short-nose PAR-64s that fit in the truss. After the bars holding them were lowered below the truss, the snoots were added to each luminaire to extend color life. These bars were staggered in height so the front row did not interfere with the backlight beams (Figure 13.2). There were five sections to complete a 50-foot span. The truss was supported by a single hydrologic ram on each end. These were made by Gallaway Company (Azusa, CA) and could go much higher than Chip needed, but they were raised to 32 feet for optimal light angles. I asked him how they stopped the ram from fully extending. The answer was that they attached a string and weight, and when the string was fully extended they stopped the ram manually! Each truss section could hold up to 9 PARs in each of two rows; however, the final 5 feet were needed for the lifting ram, so a total of 72 PARs could be mounted. After that tour, the system went out with Rod Stewart and Faces. I became involved when I left my first tour and joined Bob See in Los Angeles for the rest of that tour. Seeing that structure over our heads each night was quite a sight, and the audience loved it.

In 1973, a box truss with luminaires mounted inside during travel was introduced by Bill McManus, with the assistance of Peter Feller and Bernie Wise. It was the first truss grid and measured 50 × 28 feet.

FIGURE 13.2 Rolling Stones truss stacked for shipping. (Photograph courtesy of Chip Monck.)

It was flown, not ground supported, using Columbus McKinnon (CM) LodeStar hoists. It was designed for the Jethro Tull tour of that year, known as "The Passion Play." Other young rock & roll companies such as Showco, Tom Fields Associates (TFA), Sundance Lighting, and See-Factor were not far behind in developing individualistic designs that with each tour brought new ideas to the road. It was also a time of sharing; Tom would try a new idea and, if it worked, would call Bob or me and let us in on the idea, and we would reciprocate. Maybe this sounds Utopian, but the fact was we were not getting clients willing to take us clear across the country, so we often all participated in the same tour regionally, and the goal was for the client to have the same system when he crossed over into another company territory.

TRIANGLE TRUSS

Triangle trusses fall into two groups: commercial towers and specialty trusses. Trusses built commercially as antenna towers are available in several widths and tubing sizes. *Do not use them*. They are not designed

FIGURE 13.3 Triangular truss. (Photograph by Prolyte Products.)

to be placed in the horizontal position. Yes, they were used in the very early days of touring, but there are many commercially built units that are specifically engineered for horizontal use now. I only mention antenna towers here for historical reasons.

If you want the look of a triangle, several commercially make versions available in different sizes that offer special features and are engineered to withstand the horizontal stress and loads placed on them by concert lighting (Figure 13.3). Initially, some were constructed of a heavy chromium molybdenum (chrome-moly) material, but most today are made of a lightweight aluminum. Again, be warned to use only commercially made products that are specifically designed for horizontal stress.

Two types of design are used in triangle trusses. The first is a solid triangle with each side ranging from 12 to 30 inches. The second design is constructed with a hinged joint at the top and a removable or hinged spreader bar attached on the horizontal side (Figure 13.4). Removing the spreader bar allows the sides to close for compact storage. In both cases, luminaires must be attached once the truss is supported in position. For a one-time production, this may not be a problem, but the following disadvantages should be considered before using these trusses:

1. The luminaires should only be focused safely from the ground using a ladder.
2. The luminaires must be attached and plugged into power each time the truss is set up, a time and labor disadvantage.
3. Because luminaires must be attached to the triangle, the usual method is individually via

FIGURE 13.4 Folding triangular truss. (Photograph by Prolyte Products.)

C-clamps or hanger straps on either end of a 6-lamp bar. Adding 60 to 200 C-clamps, at about 2 pounds each, is considerable additional weight. Also, the C-clamp is very prone to denting the aluminum structure and making it unsafe. Specialty clamps are available to solve this problem.

SQUARE OR BOX TRUSS

Square or box trusses come in many configurations. Some, even though constructed as rigid units, still use hanger straps to mount the luminaires to the structures onsite. An example of a box truss is shown in Figure 13.5. They can be used not only to mount lighting luminaires but also to rig drapery, follow spot chairs, and sound systems. The sections are easily connected, but if a corner or an angle must connect sections, something like the 6-way corner block shown in Figure 13.6 can be used to join trusses in several configurations.

Other box trusses are large enough to semipermanently mount the luminaires internally to T-connectors that slide on a 1/2-inch solid pipe mounted in the truss. Some have a system that allows the bar to raise and lower to expose the luminaires below the truss for a better focus angle. The prerigged truss sections shown in Figure 13.7 demonstrate both single- and double-lamp bar designs in the transport

For location television lighting, an issue may be the size of the luminaires. More and more we see such situations, especially award shows, music specials, and other event programming that use a combination of PAR-64, moving luminaires, ellipsoidal reflector spotlights (ERSs), and cyc lights, all of which would fit comfortably within these trusses. However, if a 5-kW Fresnel is desired, then perhaps a very small box truss or triangle truss would be better so the luminaires can be underhung. These are generally one-time events and are not toured, so storage does not enter into the picture.

Speaking of moving luminaires, yes, they are often bigger than the PAR-64, so more room is needed in the truss, and a size different than the standard 18 inches may be required and must be factored into the design. Truss manufacturers have come up with a number of designs that deal specifically with this problem. Some are designed to hold only a specific number of moving luminaires. Others are flexible and can accommodate both conventional luminaires as well as moving luminaires. Some are specifically designed for a moving light manufacturer and can accommodate only their products. It should be noted, however, that because of their delicate electronics many moving luminaires are boxed separately and attached to the truss onsite. Before designing a truss structure, the designer should be aware of the recommendations of specific moving light manufacturers with regard to transportation, which can affect the load-in scheduling time. For example, removing 100 moving luminaires from boxes and attaching them to a truss is a two-person job, and it is easy to see how the load-in time could be extended.

Engineering and Construction

Most trusses are built of 6063-T5 aluminum tube with a 1- or 2-inch outer diameter (OD) or HE30 aluminum alloy with a fairly heavy inside wall thickness. Chrome-moly can be used because it is less expensive and easier to weld; however, the added weight (about twice that of aluminum) makes the use of chrome-moly undesirable for touring purposes. If the trusses are for semipermanent installations, chrome-moly could be considered for its cost savings, but the additional weight must be factored in. Its other advantage is that it can be arc welded, whereas aluminum requires the use of the more difficult heliarc method. The welder does not have to be as highly skilled to arc weld as he does to use the heliarc method.

Although steel welding is less expensive with regard to materials and equipment and labor is more readily available, steel is seldom used in the United States, largely due to its weight and to the controversy surrounding the employment of amateur or semiqualified welders, which can create a tremendous liability problem and should be avoided. The potential for a wrongful death or injury suit is substantial.

Touring trusses should be built by companies that specialize in entertainment structure design. A list of some of these companies is provided in Figure 13.12. Unless your design requires custom fabrication, most rental lighting companies will have a complete collection of different style and size trusses in stock. Make sure you are provided with certified mechanical stress test information (Figure 13.13) or x-ray records if you are simply renting gear yourself and are going to assemble the lighting. Because the rental house is not the person attaching luminaires or other things on the trusses, they cannot be held responsible if you misuse the truss. Many manufacturers now provide stress data information in their brochures (Figure 13.14).

The engineering of trusses is critical and is probably the biggest reason why one-off shows should not try to contract for special built trusses. Figure 13.15 shows the basic procedure used to test the trusses for load. Most rental companies I have surveyed say they do this procedure about once a year, although if the trusses come back from a long tour they will sometimes test them before sending them back out. The trusses can be sent to a lab that does stress analysis, but some rental houses keep the equipment in-house to check for the manufacturer's recommended deflection. A few admit to only doing a close physical inspection, looking for stress lines in the welds. Even fewer actually have the trusses x-rayed for very fine cracks. There is no industry standard here. Maybe the Entertainment Services and Technology Association (ESTA) will get involved and set American National Standards Institute (ANSI) standards.

All Access	www.allaccessinc.com
Global Truss America	www.globaltruss.com
James Thomas Engineering	www.jthomaseng.com
LiteStructures	www.litestructures.co.uk
Milos Structure Systems	www.milos.cz
Prolyte Products	www.prolyte.com
Tomcat (The Vitec Group)	www.tomcatglobal.com
Total Solutions	www.totalsolutions-group.co.uk
Tri-Lite Truss	www.opti-kinetics.com
Tyler Truss	www.tylertruss.com
Xtreme Structures	www.rigging.xsftruss.com

FIGURE 13.12   Truss manufacturers.

Tests

Test 4: 16' Span of DT2525 Drop Truss

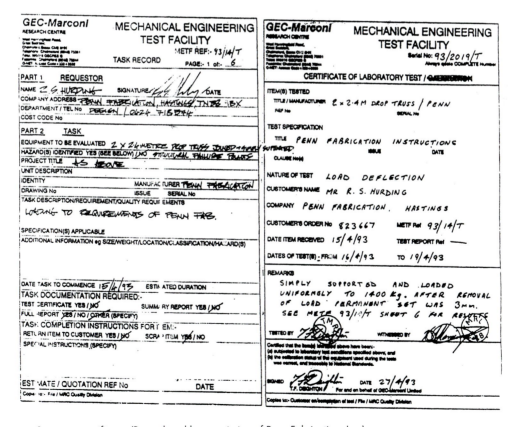

FIGURE 13.13   Stress test certificate. (Reproduced by permission of Penn Fabrication, Inc.)

PROLYTE X30D CIRCULAR TRUSS - ALLOWABLE LOADING																					
DIAMETER		3 SUSPENSION POINTS				4 SUSPENSION POINTS				6 SUSPENSION POINTS				8 SUSPENSION POINTS				10 SUSPENSION POINTS			
		UDL		CPL		UDL		CPL		UDL		CPL		UDL		CPL		UDL		CPL	
m	ft	kg/m	LBS/ft	kg	LBS	kg/m	LBS/ft	kg	LBS	kg/m	LBS/ft	kg	LBS	kg/m	LBS/ft	kg	LBS	kg/m	LBS/ft	kg	LBS
4,00	13.1	110	73.76	302	667	195	131.02	434	958	389	262.13	652	1440	590	397.37	798	1762	789	530.79	892	1968
6,00	19.7	55	37.30	220	486	105	70.52	331	731	227	153.00	538	1187	361	242.76	695	1534	495	333.00	806	1779
8,00	26.2	33	22.22	173	382	66	44.08	267	590	151	101.88	457	1009	249	167.90	616	1359	350	235.76	735	1623
10,00	32.8	22	14.51	142	314	45	30.04	224	494	107	72.08	374	825	185	124.52	503	1110	265	178.54	631	1394
12,00	39.4	15	10.04	121	267	32	21.21	193	425	74	49.60	309	681	133	89.34	417	921	209	140.45	524	1158
14,00	45.9	11	7.22	105	231	23	15.18	165	365	54	36.04	262	578	97	65.24	355	784	153	102.79	448	989

FIGURE 13.14 Example of an allowable loads data sheet. (Photograph by Pyolyte Products.)

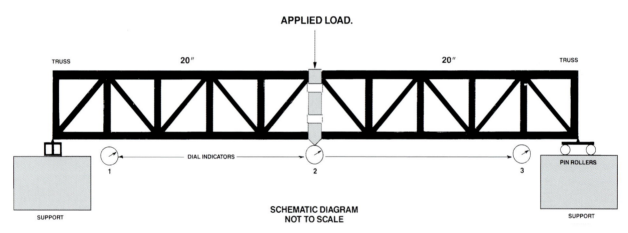

FIGURE 13.15 Stress procedure. (Provided by Rigstar Rigging, Inc.)

Prolyte Products Group, a Netherlands-based company, has published a good booklet called *Prolyte Black Book*, which discusses rigging methods, safe spans, stress, and types and use of hardware. Remember, though, that this is only a guide, and an experienced rigger should always be in charge of any rigging.

It is best to lease trusses from an established concert rental company. If you are considering constructing your own, I recommend you use only certified welders. Actual construction time could be as much as 5 days for a 40-foot truss, but it is essential to take the time to have a certified structural engineer design or check your plans. The added cost and time are other reasons to lease if your project is short term. Be sure to ask the company for certification of the structural stress and load capacity. This can be done by specialized engineering firms and should cost less than $1000, depending on what procedures they use and how far you carry the tests.

SPANS

As there is a large difference between the strongest and weakest truss, a certificate of load capacity is very important. Moreover, the clear spanning capability of the truss must be determined. Some trusses can only be supported up to a clear span of 40 feet. Not only length but also the size of the tubing and design of the truss are important considerations. You must also add in the weight of the luminaires, cable, someone focusing the luminaires, and maybe truss-mounted follow spots and operators. Generally, truss sections come

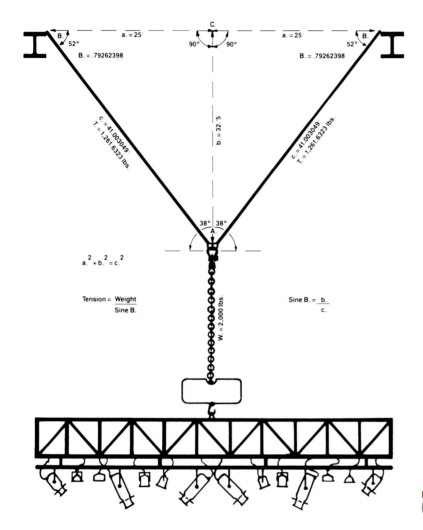

FIGURE 13.16 Rigging bridle calculations.
(Provided by Rigstar Rigging, Inc.)

in 8- and 10-foot lengths. The rental companies will help you fit your design into the lengths they have in their inventory. Even better, contact the company for a breakdown of what is available before you begin a design, then have an Entertainment Technician Certification Program (ETCP) Certified Arena Rigger create a rigging plot.

Now that most trusses are flown, the full 40 feet of a standard portable stage are usable (ground support lifts reduced the usable width to 32 feet). That is why 40 feet is considered the average truss length. Larger productions, however, are now calling for 50- and 60-foot lengths.

It is rare to see lengths over 40 feet being floor supported. If your design requires a truss over 40 feet wide, it is time to consider a flown system. Why can a longer truss be flown when it cannot be ground supported? The solution is found in placement of the rigging pickup points. Bridles are assemblies that enable the suspension of an object at a particular location; they are composed of two overhead load points of lesser load-bearing capacity that are joined to lift a heavier load. They help to distribute the weight evenly. It is common for a 40-foot truss to have two motor or winch pickup points. These, in turn, will usually be bridled about 4 to 6 feet apart. The proper bridle configurations for a given load must be determined by a qualified rigger. Figure 13.16 shows some of the calculations needed to determine the stresses; however, the method for making these calculations is much more complicated and should only be done by an experienced rigger.

Ground-supported trusses are generally not considered as safe as flown because they are subject to ground movement. No, I am not talking about a California earthquake. Out of necessity, the ground-supported trusses are at the mercy of many factors, such as portable stages with uncertified construction that could collapse under the weight. In several reported cases, lighting companies have refused to set up a ground-supported truss because they felt the stage was unsafe. If the towers supporting the truss are on the cement off the portable stage, there is a better chance that the system will be safe, but do not let down your guard; make sure the structure is level and securely attached to the truss. If the event is outdoors and the support tower is on grass, use extra caution; for example, thick plywood squares to spread the load could be placed under the tower (see Chapter 14 on lifts and towers).

INTEGRATION OF ELECTRICAL CONNECTIONS

The electrical raceways and cables attached to the truss affect the structures. There are several ways to get power to the luminaires; the simplest method is for luminaires to be wired onsite. Another method is to use a six-circuit Socapex multicable assembly. Some lamp bars are designed with internal wiring to a male Socapex connector so the technicians only have to make the one connection to energize six luminaires at a time. This speeds up the load-in.

The standard electrical raceway takes this method a step further. It is either placed on the truss onsite or attached semipermanently to the truss. The connection of the luminaires can then be easily accomplished if the luminaires are to be mounted each time or patched and left for the run of the production if the raceway is an integrated part of the truss. The luminaires can be permanently wired to a raceway, as well. This last method, however, inhibits design changes and luminaire replacement and is generally considered inefficient for any use other than a straight all-PAR-64 production design. DMX-512 low-voltage control cables and AC power for moving luminaires can also be integrated into trusses to speed assembly.

LIGHTING GRIDS

A total lighting grid can be formed with one type of truss or with a combination of sizes and designs. It is not uncommon to see a single square truss in the front and on the sides and a double-row truss in the back for the more important backlight. These trusses are joined together in several ways. End blocks with bolts, aircraft fasteners with a ball-lock capture pin, cheese-boros such as are used on scaffolding, and other highly specialized connectors have been designed especially for this use. Some trusses have been designed in such a way that a single-row truss will attach to another to create three or four rows of luminaires.

The way in which designers lay out the configuration of trusses to accomplish their lighting needs is limited only by the load limitations of the truss, lifts, winches, and motors used to place these structures in the air. A current trend that many sports arenas have adopted is to lower a master or semipermanent grid for entertainment events. One advantage is the time savings for the tour rigger in establishing high steel attachment points. Also, riggers are given a plan with the grid size and weight-bearing points to make the installation quicker. The tour may still use their motors to keep their trusses well below this grid when raised into the air. A big advantage is that facility managers can be assured of the structural loads on their motors and lift points. They also don't have to worry about riggers falling from the high steel. But, *caution:* the added weight of the truss, luminaires, etc., that will be suspended from the structure must be calculated.

The simple tour truss design in Figure 13.17 consists of two 32-foot trusses supported by Genie Superlifts (see Chapter 14). Note that the backdrop is also hung from the truss; this should only be done when wind conditions are not a factor. Better yet, don't do it at all. Grids can also be dressed up such as in Figure 13.18, which is a photograph of the 2008 papal visit. Figure 13.19 is a photograph of the 2008 Def Leppard tour that shows the grid and luminaire before being flown.

A designer, with the guidance of a qualified rigger, should also consider the capability of the roof structures of the halls before using a flown truss layout. While designers are not expected to be qualified riggers, they should have some understanding of types

FIGURE 13.17 Two 30 × 2-foot trusses. (Photograph by Sundance Lighting Corp.)

FIGURE 13.18 Mountain Production's staging of the 2008 papal tour. (Photograph by Mountain Productions.)

FIGURE 13.19 Def Leppard truss before raising. (Photograph by Paul Dexter.)

of venues and what shows have played in them previously. Consulting a tour rigger before you are committed could save you a lot of problems.

ADVANTAGES OF PORTABLE TRUSSES

A truss that has been loaded with luminaires and cables and has been checked in the shop prior to going to the location will offer savings in onsite setup time. The efficiency of the structure means that less labor is used in the field. This does not mean that the main reason to use a truss is to put people out of work. Trusses still require labor to prepare them in the shop and touring crew to guide the local stagehands in their assembly. Use of prerigged trussing can even mean savings of a day or more of site setup time, which translates into possible rental savings. More important, it may be the only way to get the production to fit into the very tight schedule of the venue.

Trusses provide a very convenient, adaptable lighting system, but the safety element must be stressed again. Getting luminaires up to a fixed pipe 20 feet in the air creates a very real hazard. I have ducked out of the way of many falling items during many a theatre load-ins. The problem of working at this height rather than at ground level should be obvious. Trusses have been in use since the 1970s and have proven a reliable and efficient method of supporting luminaires of all types as well as drapery, projection screens, and scenery. Their adaptability for use in films, television, and theatre rigging has been a great boon to flexible and safe mounting of luminaires under a wide variety of conditions.

One other thing is attached to trusses—people. A very popular design element is to place the follow spot on the flown truss to give an angle not possible from the house positions. These chairs can be placed either on top of the truss or below. The follow spot has a post already attached, so the follow spot and person need only be added (Figure 13.20). Two notes of caution: First, never place someone on a ground-supported truss until you can be assured that any movement has been secured via a safety system appropriate for the task. Second, place stagehands on a flown truss only after the supplier and rigger have approved the trussing and its rigging for such use. Anyone who climbs the truss to focus luminaires, to man follow spots, or for any other reason must wear appropriate fall protection gear.

FIGURE 13.20 Spot chair setup during the 2008 Metal Masters tour. (Photograph by Paul Dexter.)

BUILDING A GRID

A major advantage of truss use is the flexibility afforded by corner blocks and angle blocks to connect truss sections into shapes.

The basic structure of a triangle does not allow much room for creativity in internal design. The box truss, however, has spawned some highly creative modifications to its basic form. With all the moving luminaires now available, special trusses had to be constructed to hold these larger units and protect their more delicate mechanisms. Some units are purpose built to hold only moving luminaires, while others can be adapted to support both conventional and moving luminaires.

14

LIFTS, HOISTS, AND ROOFS

W hen trusses were first used to solve the problem of luminaire support they were ground supported. When trusses began to be tied to the building's structure, the process, called *rigging* (borrowed from the circus and construction), came with its own set of considerations. First, many of the buildings did not have primary and secondary beams with adequate weight-bearing margins to hold the added weight. Second, the cost of hiring riggers was prohibitive. As the lighting systems and production grew in complexity in later years, it became necessary to use only qualified rigging specialists. This meant that the tour had to restrict itself, in some cases, to larger facilities such as basketball and hockey arenas.

But, the market for artists playing in smaller settings was still thriving. How could their shows be expanded but avoid rigging? The logical solution was improved ground support. Devices used in the construction trades could be adapted for touring use. Units were already on the market that allowed workers to change light bulbs and mount materials overhead in buildings. Since those early days of borrowing, structures have been devised by truss manufacturers to more safely lift bigger loads.

This chapter provides a summary of some of the types of lifts adapted for concert use and some examples of the finished products, including trusses and complete portable roof structures. Not all manufacturers are represented here, but we have tried to present an example of every generic type of lift, hoist, and roof, paying special attention to the ones that have received the most use through the years. These items represent savings of thousands of hours in labor.

They often make the difference between performing with the lighting available or not taking the performing into the venue which limits bookings to theatres with existing structures. These devices are key to bringing theatre, dance, opera, and other entertainment to portions of the population that do not have equipped theatres in their communities and for opening up some exciting possibilities in found spaces.

AIR AND CABLE CRANK-UP LIFTS

We have moved on from the first generation of lifts—the Genie Tower and the Vermette Lift. These were lifts pulled right off construction sites. One lift from that era is still used, the Genie Super Tower.

Genie Super Tower

The Genie Super Tower operates using a 3/8-inch wire rope over a series of pulleys and offers the unique advantage of having a safety braking system approved by the Occupational Safety and Health Administration (OSHA). Another advantage is that the columns nest inside one another and are pulled out as you crank the forks up (Figure 14.1). The telescoping sections allow it to be used at less than the maximum extension. Two models are available—18- and 24-foot versions, each with a load capacity of 300 pounds. These units have two types of base configurations, both with excellent leveling jacks. It must be noted that Genie Industries (now owned by Terex Corporation) was one of the first outside companies to take a real interest in concert touring needs.

FIGURE 14.1 Genie Super Tower. (Photograph by Genie Industries.)

Eventer Stage Lift

Modeled on the Genie Super Tower, Sumner's Eventer Stage Lift (Figure 14.2) has updated 20- and 25-foot versions of the crank-up lift. Designed to fit compactly onto a truck tailgate or through a standard-height door, their lifting capacities are 640 and 800 pounds, respectively. It comes with wide outboard stabilizing legs that have end-mounted cranks for leveling. Note that the load-carrying capacity is double or more than that of the Genie Super Tower.

Air Deck

The Air Deck is a compressed-air-operated lift that is an adaptation of the old Genie Tower. Essentially, it ties together three air columns with a basket on top to create a lift for a person (Figure 14.3). If the load is not centered over the columns, the air seal is compromised and air leaks out, thereby preventing the lift from going higher or, if extended, it may come down slowly. Must not used in lifting or supporting trusses, the Air Deck is very popular as a focusing platform and sometimes as a follow spot platform. It has operating heights of 24 and 36 feet. The unit weighs in at 351 pounds and has a traveling minimum height of 7 feet, 5 inches. It is marketed by Upright Scaffolding, Inc.

FIGURE 14.2 Sumner Eventer Stage Lift. (Photograph by Sumner Manufacturing Co.)

RAM LIFTS

Several oil-operated ram lifts were used in the early days and may still be used in other countries, but in the United States they are no longer popular. Although some models can go very high (55 feet or more), the weight and size of these lifts have usually prohibited their use for touring. Load capacities of 500 to 1600 pounds can definitely be an advantage, but they have the same disadvantages as a Genie Tower: If the load is not directly over the column, the seals are broken and the unit could leak oil or slip.

SCISSOR LIFTS

Oil and fluid systems have given way to scissor-type units. They have the advantage of stability, reach, drivability, and battery operation. There are many lift manufacturers due to the large construction

FIGURE 14.3 Air Deck. (Photograph by Sundance Lighting Corp.)

market. Figure 14.4 lists quite a few of these manufacturers, but for concerts the leaders seem to be Genie, UpRight, and Skyjack. Virtually all of these companies also make boom-type units, but they are not normally used for tours, although I have been in arenas where large units have been utilized to take riggers to high steel or other attachment points so they do not have to walk the steel. It is an obviously better and safe way to work.

Many theatres also have smaller units that they can use on their stages, but be sure to check with the building engineer or architect as to the weight loading of your stage floor, especially if there is a basement below.

A conventional scissor lift generally has a platform height of from 20 to 38 feet and a load capacity between 500 and 1200 pounds (Figure 14.5). More compact units extend to 21 and 25 feet but with reduced load capacities of 550 to 600 pounds. On the other end of the scale are units that have a working platform at 50 feet, are drivable over rough terrain, and hold 2400 pounds.

At times, these larger units are employed as front follow spot platforms because they can be moved into place quickly, they require no preconstruction time or safety wires, and the working height (beam angle of the spots) can be set by the designer onsite.

All Access, Inc.	www.allaccessinc.com
Applied Electronics	www.appliednn.com
Doughty	www.doughty-usa.com
Genie Lifts	www.genieindustries.com
Global Truss America	www.globaltruss.com
James Thomas Engineering	www.jthomaseng.com
LiteStructures	www.litestructures.co.uk
Milos Structural Systems	www.milos.cz
Prolyte Products	www.prolyte.com
Sumner	www.sumner.com
TMB Products	www.tmb.com
Tomcat (The Vitec Group)	www.tomcat.com
Total Solutions	www.totalsolutions-group.co.uk
Tri-Lite Truss	www.opti-kinetics.com
Tyler Truss	www.tylertruss.com

FIGURE 14.4 Lifts and towers manufacturers.

FIGURE 14.5 Scissor lift. (Photograph by Skyjack.)

TRUSS TOWER LIFTS

Because of the unique problems associated with touring, several truss companies have developed their own lifts. Most market their towers as support for standalone speaker clusters or support for trusses or portable roof systems. The units discussed below represent only a few of the companies because there is a great duplication of systems and ideas. It should be noted that the towers manufactured for vertical use (i.e., legs) are generally different than a truss span or horizontal unit. Check with the manufacturer and do not mix the two.

Thomas Tower

James Thomas Engineering offers towers with either manual cable cranks or chain motor hoists to raise the trusses or speakers; the hoists ride up with the truss rather than being secured at the base. The lifting capacity of can be up to 4 tons. Towers can be built out of Thomas' 15 × 15-inch and 20 × 20.5-inch trusses that extend to 40 feet. The smaller 12 × 12-inch tower can be raised to 33 feet with a 2-ton load. If the towers are used to raise a truss, they are erected after the truss is assembled (Figure 14.6) and are placed inside the box of trusses (Figure 14.7). Note that in the figure six towers are

FIGURE 14.6 Thomas tower being raised. (Photograph by James Thomas Engineering.)

FIGURE 14.7 Thomas tower with box truss. (Photograph by James Thomas Engineering.)

lifting the total grid. The motors are not synchronous, but they travel at a close enough speed to keep the grid reasonably level. Control units allow for a single motor or any combination of motors to work together. There is an optional base with outriggers that allows a tower to stand alone (Figure 14.8). The photograph shows the base detail with a LodeStar motor attached. The light frame holds 36 PAR-64 luminaires. Leveling and bracing are critical to the stability of any ground-supported truss system.

Versa Tower

The Versa tower is a self-climbing version that uses a motorized cable winch and can be up to 40 feet tall (Figure 14.9). It is supported by six legs that extend out from 6 to 7 feet in each direction. It has a 40-ton lifting capacity, which makes it ideal for full lighting grids or outdoor roofs. It is made by All Access, Inc.

Prolyte Products

Prolyte, a Netherlands-based company, has provided for either hand winch crank-up or motorized hoist

FIGURE 14.8 Tower with outriggers. (Photograph by Prolyte Products.)

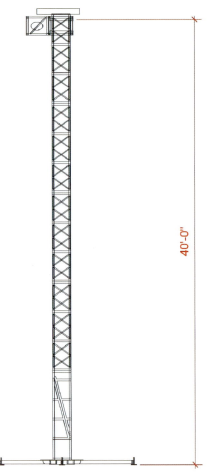

FIGURE 14.9 Versa tower fully raised. (Graphic by Versa Stage, Inc.)

power. The hoist travels with the truss rather than remaining in the tower base. The base is a four-leg outrigger assembly. Several different truss sizes and heights are available. One interesting thing is that they publish a free book called the *Prolyte Black Book* (see Chapter 13), which covers safe working practices, basic truss knowledge, and the latest developments in regulations and standardization. There are many great companies who manufacture both trusses and towers.

HOISTS AND RIGGING

Rigging trusses or portable roofs can be accomplished in only a limited number of ways. The devices made for this purpose are very specialized. Some methods of rigging have been developed by

FIGURE 14.10 CM LodeStar hoist. (Photograph by Columbus McKinnon.)

individual companies that have not gained wide acceptance. For large systems, such as in arenas or large convention centers, the circus approach to rigging is generally employed. By securing cables to the ceiling support beams (often referred to as the *high steel*) of the building, a truss or grid can be lifted and temporally suspended.

CHAIN HOISTS

CM LodeStar

Probably the most widely used hoists are LodeStar (Figure 14.10) motorized chain climbers, which were electric chain pulleys that originally had to be modified to operate in an inverted mode. The original motors had a gravity switch that had to be field modified. Since then, Columbus McKinnon (CM) learned what the concert market was doing with their motors and developed a new switch that would work in either condition with no liability, no matter how the motor was used, inverted or upright. It should go without saying that any modification

FIGURE 14.11 Truss raised with eight motors. (Photograph by Sundance Lighting Corp.)

should only be done by a factory-authorized agent. The company is very aware of the touring applications and works closely with road technicians and rental houses to ensure proper maintenance. A certification program is also in effect through the manufacturer.

The load rating of the hoists in this series is from 1/4 ton up to 5 tons. The biggest drawback is that the chain can fall out of the collector bag, and there have been reports of slippage or brake release when the motor is disengaged. Many of these problems are due to poor operator handling and should not be interpreted in any way as being the fault of an unsafe product. The person using the device has a great responsibility to make sure the device is checked frequently and used properly at all times.

How the hoists are attached to the trusses is up to the user. The placement and size of each hoist are determined by the rigger. Figure 14.11 is a photograph of a flown lighting system with at least eight chain motors visible. Note the bridles on the two upstage points. In Figure 14.12, the motor to the left is bridled to the lighting truss to distribute the lifting load to two points instead of one. Note the "horse

bucket" that catches the chain as the motor climbs up the suspended chain.

Stagemaker

Stagemaker hoists are made by R&M Materials Handling, Inc. Some hoists have built-in overload sensors and double brake systems, and they are available in sizes from 1/16 to 5 tons, with a 10:1 safety factor. The company also manufactures both fixed and variable speed hoists.

ChainMaster

Another contender in the field is ChainMaster (Figure 14.13), particularly its VarioLift series, which offers maximum precision and reproducibility in the positioning of chain hoists. The company claims a positioning accuracy of 0.2 millimeters using their integrated vector-controlled frequency converter, which is supplied by a high-impulse incremental encoder. Also available is the Jumbo Lift (Figure 14.14), which is designed to lift weights up to 6000 kilograms (13,200 lbs.) for heavy video cubes.

FIGURE 14.12   Hoist and bridle. (Photograph by Sundance Lighting Corp.)

FIGURE 14.13   VarioLift. (Photograph by ChainMaster.)

FIGURE 14.14   ChainMaster Jumbo Lift. (Photograph by ChainMaster.)

DIGITAL REMOTE CONTROL

All hoists commonly used in touring have their own multichannel control systems of from four to eight motors; however, some digital systems can control hoists in both directions and at different speeds at the same time. The StageMaster has been designed and tested to integrate with Nisco motion controllers Configuration E-Raynok (Figure 14.15). It has

FIGURE 14.15 Motion controller, 240 channels. (Photograph by Prolyte Products.)

FIGURE 14.16 Skjonberg eight-channel controller CS-800 LX. (Photograph by Skjonberg Controls, Inc.)

a computer-based program that can accommodate 12 controllers and 96 hoists in a preprogrammed set of cues. Skjonberg Controls, Inc., is an independent manufacturer of controllers that has worked in touring since the 1980s. Four- and eight-channel controllers and a 40-channel design with a wired load cell capacity are available (Figure 14.16). ChainMaster not only makes hoists but can also provide a console that controls hoist movement along all three axes, called the XYZ Stage Controller.

LOAD CELLS

The Ron StageMaster line not only offers multiple channel hoist controls but also features a wireless system to monitor weight overloads (Figure 14.17). This system is built by Eilgon in Israel. It includes a rather new safety device that is worth attention. A wireless device, called a *load cell*, can be attached between the hoist and a load point; it sends information via wireless signals to a laptop, which can have a computer-aided design (CAD) truss plan overlay that displays over- or underloads placed on points on the rig. No more human calculation mistakes—the device reads out the load in digital numbers. The systems can be set for 5:1 or 10:1 safety factors. The systems show real-time loads on a computer load map and can handle up to 256 load cells. As stated earlier, Skjonberg Controls makes a similar unit, but it is a wired system of up to 40 channels. Another company is Kinesys, which also uses a wired system.

FIGURE 14.17 Ron StageMaster computer and wireless load cell. (Photograph by Eilon Engineering.)

Figure 14.18 provides a list of hoists and accessories.

SAFETY FIRST

The ground lift and hoist methods discussed here are representative of what is now in use in the field. None of these items is foolproof. Accidents occur mostly because we sometimes use these devices for purposes other than what they were designed for by the manufacturer. Safety should be the watchword for anyone using these methods. Consult the manufacturers. Most of them have become interested in the concert use of their products and can offer suggestions on the proper procedures for these applications. They want to protect you and themselves from law suits. Always use a skilled rigger whenever hoists are involved to protect the building as well as the audience, the artists, and the production company. This is not an area where the manager can skimp.

Chainmaster	www.chainmaster.com
Columbus McKinnon Chain Hoists	www.cmindustrial.com
Eilon Engineering Weighing Systems Ltd.	www.ron-crane-scales.com
J.R. Clancy	www.jrclancy.com
Motion Laboratories	www.motionlabs.com
Niscon, Inc.	www.niscon.com
Rigging Innovators, LLC	www.rigginginnovators.com
R & M Materials Handling, Inc.	www.stagemaker.com
Skjonberg Controls, Inc.	www.skjonberg.com
Upstaging, Inc.	www.upstaging.com
Verlinde	www.verlinde.com

FIGURE 14.18 Hoists and accessories.

FIGURE 14.19 Arched roof. (Photograph by Prolyte Products.)

ROOFS

Roofs are generally trusses designed for portable concert rigs combined with their towers or lifting devices or hoists; however, there are exceptions. Specialized trusses have been built for this specific purpose, including domed designs, such as the one by Prolyte shown in Figure 14.19. The purpose of this section, however, is more about discussing companies that may not be the actual manufacturers but instead provide the services of constructing roof systems, staging, and even grandstands, in addition to providing the crews to assemble and transport the equipment.

Brown United

One such company, Brown United takes a unique approach to roof support construction that involves a base of I-beams that are set under the stage with the 4- to 30-inch round steel columns that are sometimes as tall as 55 feet and raised via motors in the roof to give the most unobstructed view possible (Figure 14.20). Sometimes four additional poles are used to build outboard sound wings. Figure 14.21 shows a side view of lighting and speaker clusters all supported by four poles. I have worked with this system several times on the beach in Hawaii with the I-beam buried in the sand; people can hardly believe such a large roof is supported by these

simple posts. The system also allows for one of the tallest freestanding roof structures at 80 × 100 feet (Figure 14.22). Wings can be attached at the sides for sound or projection screen placement using

FIGURE 14.21 Four-pole roof raised with speaker clusters and lighting trusses supported. (Photograph by Brown United.)

FIGURE 14.20 Four-pole freestanding roof being assembled. (Photograph by Brown United.)

FIGURE 14.22 Detail of single pole support side view. (Photograph by Brown United.)

FIGURE 14.23 Mountain Productions outdoor stage. (Photograph by Mountain Productions.)

FIGURE 14.24 Stageline mobile stage opening up. (Photograph by Stageline Group.)

additional poles. Naturally, the size of the columns and the number vary depending on the total roof structure desired. Each is engineered based on the purpose and the additional weight the roof must support.

Mountain Productions

Mountain Productions, a leader in staging and outdoor roofs for 27 years, is headquartered in Pennsylvania. They manage over 200 events a year and provide staging, grandstands, towers, scaffolding, trucking, and staff for a wide variety of events besides rock & roll concerts (Figure 14.23).

Traveling Stages

Several companies have designed stages that are affixed to a flatbed trailer body and can be hauled into position and set up quickly. While they may not be doing the next Rolling Stones tour, they are a vital part of touring for many up-and-coming bands and can be seen at many outdoor club-size tours. They are also very active in the county fair circuit. One such company is Stageline Group; Figure 14.24 shows one of their mobile stages as it is opening up. Figure 14.25 shows a portable stage in use with supporting sound wings and video screens.

For a list of roofing and staging companies, see Figure 14.26.

FIGURE 14.25 Stageline mobile stage in use, Elton John. (Photograph by Stageline Group.)

Accurate Staging	www.accuratestaging.com
All Access	www.allaccessinc.com
Brown United	www.brownunited.com
Milos Structural Systems	www.milos.cz
Mountain Productions	www.mountainproductions.com
Prolyte Products	www.prolyte.com
Stageline Mobile Stage Inc.	www.stageline.com
Staging Concepts	www.stagingconcepts.com
Staging Dimensions	www.stagingdimensionsinc.com
Tait Towers, Inc.	www.taittowers.com
Tyler Truss	www.tylertruss.com
Xtreme Structures	www.rigging.xsftruss.com

FIGURE 14.26 Roof and staging companies.

15

MOVING LUMINAIRES

T he most radical innovation in lighting to come along in more than four decades has been moving luminaires—remote-controlled lights, computer-controlled lights, intelligent lights, wiggle lights, movers, moving lights, whatever you choose to call them. The introduction of the Vari*Lite in September of 1981 by Showco, Inc., a concert lighting company from Dallas, stunned the concert community, not only because of the innovative quality of the product but also because they were not a manufacturer looking to sell a product.

How could the giants of the lighting world miss this idea? Actually, remote control of luminaires had been around for a long time. From my trips throughout the world—specifically, what I saw in Japan's NKH television studios and at the BBC in London in the early 1970s, as well as reports out of Germany—I was aware of earlier usage of motorized luminaires. However, these were only pan and tilt devises added to the fixture, not a total new design with RGB, CYN, or dichroic color changes. So what made this such an historic introduction? One explanation would be that the established lighting community saw it as a rock & roll effect *only* and didn't visualize its potential in theatre and television. This is not the first time the mainstream has been slower than rock & roll to pick up on a concept. (The list is long, including the PAR-64, trusses, multicable, and portable packaging of dimming). Again, tour lighting ingenuity had taken an existing technology and improved it tenfold.

THE HISTORY OF MOVING LUMINAIRES

To fill in some historical background, I was fortunate to be able to have a conversation with Louis Erhart, a Yale graduate who had assisted the legendary Stanley McCandles from 1932 to 1934. He joined Century Lighting in 1937. In 1941, he helped establish their West Coast factory operations, retiring as vice president in 1972. I called him to find the historical facts on foreign installations, only to be told that American ingenuity had not been lacking. He produced a copy of a data sheet on a luminaire Century marketed as the Featherlight. It was the outcome of a joint venture with Paramount Studios to develop a commercially saleable remote-controlled luminaire.

The story of that luminaire began toward the end of 1949 when production began on Cecil B. DeMille's cinematic spectacular *The Greatest Show on Earth*, subsequently released in 1952. A unit was desired that could be mounted high in the Big Top without an operator. A Mr. Hissorich is credited by Erhart as being the developer for Paramount. The joint venture ultimately produced a fully automated television studio in New York in the mid-1950s (NBC's Studio H). Unfortunately, some people thought it represented a potential loss of jobs, so they reportedly did everything possible to sabotage the concept. It is too bad they did not realize that automation increases productivity, thereby increasing usable production time and the need for even more crew. This has been proved in studios in Japan and London. Sadly, the Century/Paramount project was dropped soon after. The designers cited technical shortcomings in motor design, which could have been resolved if they had been willing to stick with the concept. Several of these units exist and were in working order at Los Angeles Stage Lighting until that company closed in the 1980s.

A British firm also has offered information about an early product. Dennis Eynon founded Malham Photographic Equipment, Ltd., in the early 1950s. After meeting William Cremer in Paris in the 1960s,

FIGURE 15.1 Photograph circa 1960 of William Cremer. (Photograph courtesy of Malham Lighting Design, London.)

they collaborated on many projects, including the Top Rank Bristol Suite (Figure 15.1). They installed Cremer's Mixlight, which projected patterns onto balcony walls and used 45-degree mirrors to project patterns onto the dance floor, according to James Eynon, Dennis' son and currently codirector of a company called Malham Lighting Design, Ltd., in London. To quote a letter written by James Eynon on August 7, 1995:

> The Mixlight was a development of a 1000w spotlight which contained 2 colour hanging systems. One was a conventional colour wheel with gel colours in an aluminum segmented wheel. The other was a four-blade paddle wheel with perforated gel panels, which rotated axially in a plane at 90 degrees to the conventional colour wheel.

These projectors could also take gobos (made from perforated metal) or glass patterns made from standard pattern glass. They were not remote controlled but changed continuously and used 240-watt AC synchronous motors supplied by Crouzet of France. The projectors had been formerly used at an exhibition, "Formes et Lumieres," in Liege (commissioned by Phillips).

Cremer was somewhat eccentric, it seems. After a few years he left the company in the hands of his

secretary, bought a yacht, sailed the Caribbean, and wrote novels. He returned to Paris and ran a restaurant until committing suicide in 1980. Dennis Eynon now lives in Ireland.

James Eynon went on to recount that, after a falling out with the Rank Organisation, Cremer concentrated on television studio designs for his remote-controlled devices. He developed a 1000-watt spotlight that had a motorized pan and tilt and could be focused remotely. The Rank Organisation, persuading Cremer to sell them components, continued supplying Top Rank, Bristol Suite, and at least two other ballrooms.

The luminaires developed in Croydon, England, had six functions: pan, tilt, focus, iris, blackout shutter, and color wheel. It is thought that the iris mechanism was supplied by Strand Electric. The motors were from Crouzel of France and as far as can be determined were 240-V AC. Control was by means of hardwiring through spiral cables to an operator panel, which had momentary switches.

Michael Callihan, in an article in the August 2006 issue of *PLSN*, said that he believes he has uncovered even more history. Famed New York lighting designer Jules Fisher apparently patented a pin spot with remote pan and tile in 1965. But, an even bigger discovery was that a patent was issued to Edmund Sohlberg from Kansas City in 1906 for a balcony-mounted follow spot; cables ran backstage so a stagehand could control the beam size and direction. While hardly an electronic solution to remote control, it did have an electrically controlled color wheel, according to the article.

The discovery of a patent granted to Dr. Fritz von Ballmoos in Switzerland adds further to the automated luminaire saga. A physicist by trade, with no theatrical experience, von Ballmoos entered a design competition with a friend. He studied what was considered to be the state of the art at the time and devised an automated 200-luminaire system for the theatre. Their design won and was installed in the early 1970s. The luminaires had pan, tilt, and intensity control and could change colors using two color wheels. The system apparently remained in use for some 20 years. He also applied for and received patents in a half-dozen countries, and the initial patents have survived two reexaminations. Callihan believes that these documents provide the best documentation for the history of luminaires, and Dr. von Ballmoos' design was perhaps the most influential for the next phase of development of modern moving luminaires.

In the meantime, in the United States, Showco, a successful concert lighting and sound equipment supplier in Dallas, was working to develop a color changer for the PAR-64 luminaire. In 1980, a team of designers drew upon a couple of emerging technologies—dichroic coated glass and metal halide lamps. The team was headed by Jim Bornhorst, then an audio engineer and console designer who worked for Showco, and it included Brooks Taylor, a software designer; Tom Walsh, a digital hardware designer; and John Covington, an analog systems designer.

Rusty Brutsché, president of Showco at the time, guided the project from start to finish. He was considering dropping the add-on color changer idea in favor of a completely new luminaire. In the fall of 1980, Brutsché and his partner, Jack Maxson, along with Bornhorst and other Showco employees, were having lunch at a local barbecue restaurant. During a discussion of the feasibility of building a new dichroic filter-based luminaire, Maxson remarked, "Two more motors and the light moves …." An all-out building effort ensued, and by December of 1980 a prototype and a controller existed. Brutsché decided that the group Genesis would be a likely candidate to preview the system, now known as VL Zero (Figure 15.2).

Brutsché and Bornhorst flew to London to present the idea to the band members. In the hayloft of a 500-year-old barn in the English countryside, they executed two simple cues. A band member exclaimed, "I expected the color, but by Jove, I didn't know it was going to move." With that, an active personal and business relationship began between the two groups.

On September 25, 1981, in a bull ring in Barcelona, Spain, Genesis opened their Abacab tour with 40 Vari*Lite prototype luminaires. The audience reacted immediately the first time the luminaires came on and moved into the crowd. Each time it happened, the crowd's reaction grew. "I remember thinking that we had seen our little system work its magic with the music and knowing that things somehow had changed," said Bornhorst on the occasion of the 15th-year anniversary of that night in Spain.

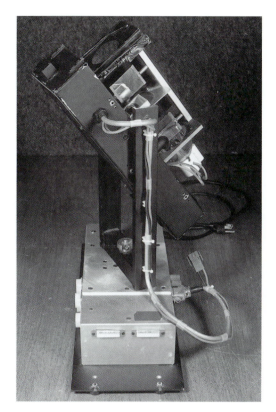

FIGURE 15.2   VL Zero. (Photograph by Showco, Inc.)

According to Tom Littrell, also with Showco in those early days:

Vari*Lite, Inc., engineers began the entire project which was to become the Vari*Lite by thinking in terms of a wholly unified system of automated luminaires and control. The GE Marc 350 bulb, designed initially for slide projectors, had made its appearance in the entertainment world in truss-mounted follow spots and came to the attention of Jim Bornhorst and his team in the late 1970s. They also began to investigate dichroic glass as a permanent color medium not susceptible to heat. The Vari*Lite, Inc., engineers created a practical fading color changer built around the properties of dichoric glass and packaged it, along with the new arc source, into a realistically sized stage lighting instrument.

A photograph of the bench model (Figure 15.3) was finally revealed after years of concealment. For many years, Vari*Lite was very quiet about how everything worked because they feared patent infringement.

FIGURE 15.3 Vari*Lite bench model. (Photograph by Showco, Inc.)

Their fears came true, and for a number of years they vigorously defended their patents in court.

VARI*LITE HISTORY

Vari*Lite was formed to handle the manufacturing and marketing of Showco's new product. The Vari* Lite was described as "a self-contained, computer-controlled lighting luminaire." The luminaire consists of an upper box assembly that houses the lamp power supply, pan mechanism, and other electronics. The lamp housing, or head, contains the lamp, color mechanism, mechanical dimming system, and tilt control. The upper box is also where all the mounting hardware attaches. All of the luminaires are controlled by a multiplexed digital signal distribution system. This means that a single three-wire microphone cable from the computer provided all of the control data for all of the luminaires.

The original Vari*Lite used a GE Marc 350-T16 lamp that could produce 140 fc at 40 feet with a color temperature of 5600 Kelvin. It took 2 seconds to rotate the unit 180 degrees, and the position was accurate to within 1 degree on either axis. It had a mechanical douser that went from full on to full off in under a half second. The beam spread could be varied by choosing any of eight available aperture openings.

Probably its most unique feature, and not copied until later, was the color system. The luminaire produced 60 colors using dichroic filter prisms rather than standard color media. It could change color in one tenth of a second. In addition to the 60 preselected colors, it was possible to dial in a custom mix of colors at the computer console.

Originally, the luminaires were all controlled via a custom computer console with proprietary protocol processing. The processors and cue memories were in the luminaires themselves, and a high-speed bidirectional data link allowed them to achieve a sophistication level that a central processing system couldn't provide in large rigs, according to Rusty Brutsché. Because of this, Vari*Lite products could not be run from any other console, except those designed by Vari-Lite. The original console could store 250 cues from 96 luminaires. It had no tape or disc drive storage; the unit used integrated-circuit storage. Cues could be written for each luminaire or for groups of luminaires and could be retrieved at will or in sequence. The board operator could manually manipulate any feature of any luminaire during a cue.

I am not privy to the details of the next turn of events in this company's long history, but in 2003 Genlyte Thomas Group acquired all assets of Vari* Lite's manufacturing and sales division including the Vari*Lite name, brand, and all patents associated with the Vari*Lite products. There is no doubt that the company has continued to build on its solid reputation and has many loyal designers, even after a further sale to the Philips Group took place.

CURRENT DEVELOPMENTS

This look at the early development of motorized, remote-controlled luminaires should not be taken in any way as minimizing the efforts of the highly inventive work done in the late 1970s and beyond. I am in awe of such creativity and the desire to advance our media.

Further progress in the design of moving luminaires (or motorized yokes or moving mirror luminaires operated via computer control—all names for the same general product) can be attributed to two nontraditional manufacturers: Showco, the touring lighting and sound company from Dallas, and Morpheus Lights, Inc., a San Jose, California-based touring lighting company. The development of their products took similar paths but had interesting variations on the theme. More than 18 other companies are now in the field, with many more offering new entries every month. Some are close copies, but many have added features that have enriched this innovation. So, the rock & roll computer luminaire has obviously disproved expectations that it was only a gimmick or a short-lived effect.

If the stigma of their being introduced by rock & roll companies caused the moving luminaires to initially be shunned by theatre and television, that prejudice has long since been overcome. Television, especially, has embraced them, awarding Emmys as far back as 1983 and 1984 to shows that featured them. Vari*Lite received the 1991 Academy of Television Arts and Science Emmy Award for the development of the Vari*Lite Series 200 automated luminaire system (Figure 15.4). The company was honored again in 1994 for its silent, compact VL5 Wash Luminaire with the innovative DICHRO*TUNE color changer for smooth, full color spectrum crossfades.

Broadway has embraced moving luminaires (see Chapter 24) and made them a staple in their lighting packages, including "Starlight Express," "The Will Roger Follies," "The Who's Tommy," "Miss Saigon,"

FIGURE 15.4 Vari*Lite Series 200 automated luminaire system. (Photograph by Vari*Lite, Inc.)

and most currently "A Tale of Two Cities." Films such as *Streets of Fire* and *Batman* have also used them extensively.

Rock & roll has proven that the luminaire, with its ability to be repositioned, change color, add patterns automatically, and even change focus on some units, is a very valuable tool. Even the least expensive luminaires can give a big ballyhoo effect for your money. The range of functions is as wide as the companies that develop them. Whereas it was a very small group that had the money and ability to use these luminaires in the 1980s, today and beyond the market is wide open on every level. Moving luminaires are as likely to be seen in a display window as in a Broadway show or your 8-year-old's dance recital. The uses for and variety of tools available under the generic name *moving luminaire* have not seen their zenith. I could never hope to keep up with this quickly changing segment of the industry.

The luminaire manufacturers listed in Figure 15.5 are among the leaders today, but I can't say what will happen tomorrow. Look at the proliferation of companies manufacturing luminaires that have been pared down to include only those units rugged enough to be viable in the touring industry. Luminaires designed for installation in clubs are not listed. This long list of manufacturers can only mean there is a thriving market not only for concerts but theatre, corporate shows, theme parks, Broadway, store displays, museums, and even architecture.

In 1999, a complete about face from its policy of not selling luminaires occurred when Vari*Lite

ADB Lighting Technologies	www.adblighting.com
Clay Paky	www.claypaky.it
Coemar	www.coemar.com
D.T.S. Illuminazione	www.dts-lighting.com
Elation Professional	www.elationlighting.com
G. Lites	www.g-lites.com
High End, Inc. (The Barco Group)	www.highend.com
Irridant	www.irridantthg.com
Martin Professional	www.martin.com
Morpheus Lights	www.morpheuslights.com
OmniSistems	www.omnisistem.com
PR Lighting	www.pr-lighting.com
Production Resource Group (PRG)	www.prg.com
ROBE	www.robelighting.com
SGM Technology	www.sgmtechnologyforlighting.com
Tei Electronics, Inc.	www.teilighting.com
Vari*Lite (A Philips company)	www.vari-lite.com

FIGURE 15.5 Moving luminaire manufacturers.

announced plans to offer a new line of products for sale. Up until then, shows leased units directly from the company or from one of their licensed distributors. Lighting professionals now had the choice to rent or own Vari*Lite automated lighting systems. Another transitional year for Vari*Lite was 2000, when the Vari*Lite Virtuoso DX console was introduced at Live Design International (LDI) along with the VL2000 Spot luminaire and the VL2000 Wash luminaire. The VL2000 Wash luminaire went on to receive the EDDY Designer Award for Lighting Product of the Year. The next year, they received their third Prime Time Emmy Award for outstanding achievement in Engineering for the Virtuoso console. In 2002, the Vari*Lite Series 3000 product line was introduced, and a color-mixing spotlight luminaire began shipping the next year.

The Vari*Lite Series 3500 Wash (Figure 15.6) and Series 2500 Profile (Figure 15.7) product lines were introduced next, followed by the Vari*Lite Series 500

FIGURE 15.6 Vari*Lite Series 3500 Wash luminaire. (Photograph by Vari*Lite, Inc.)

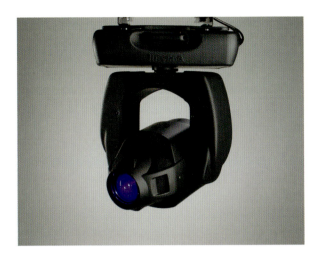

FIGURE 15.7 Vari*Lite Series 2500 Profile. (Photograph by Vari*Lite, Inc.)

(Figure 15.8) product line, which was touted as being the next generation of the VL5 Wash luminaire. Most recently, the Vari*Lite VL3500 Wash luminaire was introduced which boasted lumen output and beam control.

This innovative company brings more than 25 years of imagination and skilled research and development to the entertainment industry. Although the business plan may change, the reader should expect many more innovative products from this skilled team.

COST AND AVAILABILITY

Even with increased competition, costs have not lowered appreciably because the features, longevity, and accuracy have been increased. The increased demand for quieter, smoother, more powerful light output and special features does not help to lower the market price, even with a huge influx of products from manufacturers all around the world. Whether costs are coming more in line with theatre or the theatre market has decided they need these luminaires because of the desire of designers to use them, I cannot be sure. So many new products have been introduced into this market that there was bound to be a breakthrough, and it has happened. The introduction of light-emitting diode (LED) sources and digital moving luminaires (see Chapters 16 and 17) has sped the development of newer and better luminaires at a dizzying pace. Certainly we have not seen an end to the

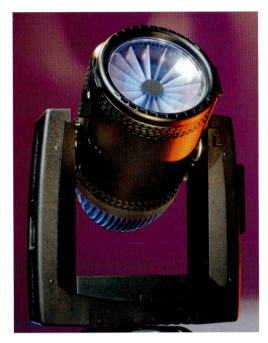

FIGURE 15.8 Vari*Lite Series 500. (Photograph by Vari*Lite, Inc.)

creativity in this area. It is not possible to describe all the moving luminaires because the list would be out of date by the time you read this. What follows is a description of a sample of luminaires offering a variety of features, as well as the units that started it all.

MORPHEUS LIGHTS

When Morpheus Lights entered the moving luminaire market shortly after Showco, they went a step further and made two different luminaires available. The short description that follows focuses on their first technology. The Panaspot was a unit much like the Vari*Lite, with a single housing for all control functions and motors. The GE Marc 350-T16 was also their lamp of choice. A mechanical douser was used to dim the light, but they did have a fully functioning iris instead of the template idea Showco used. There was a slot for a mini-ellipse-sized pattern. The beam size was altered by a magnifying iris. The beam varied from 2 to 25 degrees. Color was via a boomerang setup of seven user-selected colors.

The other luminaire introduced by Morpheus was the PanaB. While similar to the Panaspot in size and somewhat lighter, the light source was a standard

FIGURE 15.9 Morpheus PanaB TRX2. (Photograph by Morpheus Lights.)

PAR-64 lamp. Any beam width could be used, including an ACL lamp. No dimmer was built in, so intensity control had to be provided within the regular dimming system. This unit used a scrolling color changer with six colors and clear.

Right from the beginning, Morpheus took a different road by not designing a custom console. Rather, they chose to use a stock Kliegl Bros. Performer 2 computer console, assigning each luminaire a position via the soft patch. It allowed for 125 lamps to be controlled and 225 cues to be stored. Their thinking was that, because the console was mass manufactured by a mainstream lighting company, it could control the moving luminaires plus any standard theatrical luminaires at the same time. Repair and replacement was easy via Kliegl dealers around the world, so service was simplified. Kliegl Bros., though, has gone the way of many old mainstream lighting companies who did not see the future in compact, portable, reliable equipment.

The second generation from Morpheus includes many advances. The PC–Spot has a new lamp source, an Osram 400-watt HTI. A significant factor is that the complete luminaire weighs 35 pounds, considerably lighter than the VL2 and VL3. The hanging center has also been reduced to 20 inches.

The tilt coverage is not so good, coming in at 240 degrees. Proven zoom optics are once again able to provide a 2- to 25-degree beam spread.

This was the first luminaire other than Vari*Lite to use a dichroic color system; however, they added a ten-frame scrolling system to provide color correction, diffusion filters, or special colors the designer may require. Six pattern holders are built in, and three are capable of rotating with programmable speed and direction, which makes for some very nice added movement in the luminaire besides pan and tilt. Patterns are combined with the zoom lens to control the actual pattern size. Remote focus of the beam from hard to soft is still possible.

The other luminaire is the PC-Beam. The lamp is a 1000-watt FEL, and the luminaire has an internal dimmer. A parabolic reflector produces variable beam spread from ACL-type to wide-floor PAR fields. Color is via an 11-frame scrolling changer. Weight is a lean 20 pounds, which includes dimmer, luminaire head, and electronics. This luminaire also has preprogrammed looks built in; 100 are stored. They can be used as programmed or modified for user needs.

Recently, the company introduced the PanaB XR2 (Figure 15.9). This moving wash luminaire uses a 1200-watt short-arc source at a color temperature of 6500 Kelvin with CYM color mixing technology. It also has the ability to pan and tilt 360 degrees and has a beam control from 10 to 22 degrees. This company has chosen the lease philosophy of business, leasing rather than selling for most of the products.

MARTIN PROFESSIONAL

Martin Professional certainly ranks high on the list of manufacturers of durable, efficient, and user-friendly moving luminaires. The full family of luminaires totals over 15, including the MAC TW1 (Figure 15.10), which uses a 1200-watt tungsten source and a set of interchangeable zoom lenses—14.5 to 27, 20 to 41, and an exceptionally wide 97 to 105 degrees. This luminaire is also based on the demands of the television and film entertainment needs for accurate color rendering of 3200 to 7500 Kelvin to match the needs of those segments of the industry while providing

FIGURE 15.10   MAC TW1. (Photograph by Martin Professional.)

a full CMY (cyan, magenta, yellow) color mixing system. The luminaire is also very quiet, producing only 40 dBA. Power is automatically switched internally so the unit can be used globally without repowering.

The other lead product is the MAC III Profile (Figure 15.11), which uses a 1500-watt short-arc discharge lamp. It incorporates CMY plus variable CTO (color temperature orange) color mixing and has zoom tracking so gobo patterns remain in focus while zooming, if desired. This luminaire is also Remote Device Management (RDM) ready (see Chapter 20) and Art-Net II. Martin makes products other than moving luminaires and consoles; some examples of other such companies will appear in other sections of this book.

HIGH END SYSTEMS

High End Systems has made a business change recently because of its acquisition by Barco. This company also produces LED source moving luminaires (see Chapter 17). Their full range of luminaires also contains a long-standing moving mirror luminaire called the Cyberlight Turbo (Figure 15.12), which is the successor to their original unit, the Intellabeam. It has a unique place in lighting design, and I hope the company retains or improves on it. Generally, we are seeing a lot of club-level luminaires being produced in China in this style

FIGURE 15.11   Martin MAC III Profile. (Photograph by Martin Professional.)

FIGURE 15.12   High End Systems Cyberlight Turbo. (Photograph by High End Systems.)

but not by the major players here in the United States. The use of a moving mirror source is limited, as the mirror cannot rotate nearly as far as a yoke–type luminaire (pan, 170 degrees; tilt 100 degrees), and they tend to take up more room than their brothers, but for the right application on tour they are invaluable. Color, patterns, and the other basic functions are all there.

FIGURE 15.13   Studio Spot. (Photograph by High End Systems.)

FIGURE 15.14   Studio Color. (Photograph by High End Systems.)

FIGURE 15.15   SHOWGUN 2.5. (Photograph by High End Systems.)

I have used them very successfully in play and musical productions, as well as television.

High End Systems has enjoyed another niche market in television and theatre with their Studio Spot (Figure 15.13), designed specifically for beam effects of projection patterns. It has two rotating gobo wheels and an 18- to 30-degree stepless zoom lens. Studio Color (Figure 15.14) offers beam shaping that is selectable from 8 and 22 degrees, in addition to an infinite color selection. It is equipped with variable frosting, strobe, and shutter. Both use conventional light sources. At 575 watts with CMY color mixing, they are a total package for many applications. These luminaries have proved extremely popular with television and film people, as well as theatre because of their virtually silent operation.

Richard Belliveau, chief technology officer, has introduced several products that would have to be considered flagship products. The big gun, so to speak, is the distinctively designed SHOWGUN 2.5 (see Figure 15.15) which is a very large and powerful automated luminaire that can project images, change and mix dichroic colors, and switch from hard edge to soft edge all within a compact system. The SHOWGUN 2.5 offers 130,000 lumens of light.

SHOWGUN provides lighting in three dimensions: first, in the air with high-energy focused or soft edge beams; second, with images projected on the stage surface; and, third, at the SHOWGUN luminaire itself by using an innovative LED tracking system. The optical design incorporates a proprietary, optically accurate polymer micro-Fresnel lens that allows SHOWGUN to produce a true focused hard-edge or a brilliant soft-edge combination, without the need for two separate luminaire types. In the hard-edge focus, it can project HES LithoPattern images. SHOWGUN's main output uses a 2500-watt short-arc metal halide lamp designed as a joint venture between Philips and High End Systems and listed as the MSR 2500/2. Another unique feature is the LED tracking system, which mixes an RGB LED circular array that allows the user to either match the color of the main output beam or project a complementary color by mixing RGB values. According to High End Systems, "The LED tracking system alone produces more than 5000 lumens of output, and works in tandem with SHOWGUN's other features to redefine the rules of lighting design."

Standard luminaires from this company include the Studio Command series (Figure 15.16) in both MRS 700- and 1200-watt lamp versions, as well as a halogen version for 3200 Kelvin needs. This luminaire group has all of the preferred features and uses CMY dichroic color mixing. The wash version is called the Studio Beam, which also uses CMY color mixing with an 18- to 30-degree zoom range.

The company also produces two unique luminaires called the SHOWPIX and the StudioPix (see Chapter 16). SHOWPIX, a combination high-power LED wash luminaire and programmable high-intensity, graphic image-displaying luminaire, represents the first in High End Systems' new product line, Pixelation Luminaires. SHOWPIX is more than an LED wash luminaire on a moving yoke. Its 18-inch-diameter head features a circular array of 127 homogeneous 3-watt LEDs with an output of 24,000 RGB lumens. "These pixels offer unlimited visual possibilities by projecting not just washes of color but also displaying images and other eye-catching effects that transform the luminaire head into a display device," according to the company brochure. SHOWPIX is equipped with 411 stock content animations and patterns for lighting design. Users can easily upload their own files and images through the new High End System Echo application, a software program offering content

FIGURE 15.16 Studio Command series. (Photograph by High End Systems.)

visualization as well as RDM management features. Instantaneous conversion of many file formats automatically optimizes or "supercharges" the images and maps the files to the 127 LED pixels. The mere size of the unit is also impressive; with a face area that is 17.7 inches wide, the total luminaire is 23.4 inches wide and 33 inches tall. Although big, it is still about the size of a number of new luminaires, such as the Bad Boy by PRG and the VL3500. The company claims that SHOWPIX is economical because there are no lamps to replace and maintenance costs are lower because of limited moving parts. The luminaire in use is certainly impressive and stands as a watermark for other LED-based moving luminaires.

ROBE LIGHTING

A company with headquarters in the Czech Republic, ROBE Lighting offers a solid line and wide variety, including a dozen moving yoke luminaires, several moving mirrors (scanners), and digital and LED movers, as well. Their ColorWash 750 AT Tungsten

FIGURE 15.17 ROBE Lighting ColorWash 750 AT Tungsten. (Photograph by ROBE Lighting.)

FIGURE 15.18 Bad Boy. (Photograph by PRG.)

(Figure 15.17) uses a Phillips Hi-Bright 750 FastFit halogen lamp and is available in both European and U.S. models for power service considerations. The color system is via CMY with CTO mixing, and it has the ability to place five custom colors in the unit. The luminaire comes with a 12- to 34-degree or optional 24- to 60-degree beam spread and a rotatable beam shaper. The largest unit is the ColorSpot 2500E AT. This is a luminaire with the standard features found on most of the units already discussed. ROBE Lighting is only one of a long list of companies such as Clay Paky, Coemar, and OmniSystems that you will find listed in Figure 15.5.

PRODUCTION RESOURCE GROUP (PRG)

Although PRG only offers one luminaire (for rental only), and it is tied to its own console (see Chapter 11), it will serve as another example of the next generation of moving luminaires. PRG is more than a manufacturer of moving luminaires; in fact, their

name says it all. They are a large conglomerate of companies woven together by Jere Harris; however, their decision to develop their own light and console is a departure because they are tied to Vari*Lite in a marketing arrangement. They are the undisputed preeminent source worldwide for lighting and set designers, production managers, and video and audio engineers—so says their website, at least—but it is true that it is difficult to avoid the company on some level during a tour.

The Bad Boy (Figure 15.18) is a hybrid luminaire that combines the qualities of a traditional automated luminaire with a large venue luminaire. Powerful and brilliant, road ready and reliable, the Bad Boy could be your choice when a big-beam look is required. It produces a powerful 48,000 lumens and couples its high-definition optics with a Philips MSR Gold 1200-watt SA/SE FastFit lamp that can be set at any level between 700 and 1400 watts. Its brochure boasts of an optical efficiency of over 40%.

Optical clarity and smooth fluid control of focus, zoom, dimming, and imaging are obtained by bringing together high-quality lenses and high-speed servomotors, perfect for high-resolution gobos. Bad Boy incorporates a zoom lens system with exceptional optical imagery and a zoom range of 8:1 from a narrow spot of 7 degrees to a wide flood of 56 degrees. The 8-inch-diameter front lens produces a large, full beam that easily can be varied by the luminaire's zoom lens and beam size iris. The zoom consists of four groups of lenses—each independently controlled for accuracy while maintaining focus during zoom changes—plus edge control for gobo focus and gobo morphing. Bad Boy's two rotating gobo wheels are indexable, and each of the seven gobos per wheel is individually calibrated so the unit automatically indexes the orientation of each gobo regardless of placement.

Bad Boy has the swift, organic movement comparable to a unit just one quarter of its size and weight. Three-phase, high-speed servomotors with a clutch and dual optical encoders for pan and tilt provide repeatable, precision responses with a maximum speed of 4.1 seconds for 540 degrees of pan and 3.2 seconds for 270 degrees of tilt with an accuracy of 0.2 degrees.

Vibrant colors are a result of the Quantum Color system. This innovative color method features four color wheels with seven discrete colors on each wheel: one designer wheel with user-changeable color filters and three fixed color wheels organized into cyan, magenta, and yellow. Bad Boy's individual color filters allow for variation in both saturation and hue of the CMY colors, resulting in a much broader and vibrant range of saturated colors that are pure and homogeneous across the beam. Because no diffusion material is required, the output brightness in white and the brightness of the colors are maximized. Each wheel uses a high-speed servomotor for rotation, which provides for instantaneous color bumps and flashes as well as smooth control for timed color changes.

ELECTRONIC THEATRE CONTROLS (ETC)

Although others have tried to develop and market an ellipsoidal that is designed around a moving luminaire, ETC has a proven product in the Source Four Revolution (Figure 15.19). This is a zoom unit with full pan and tilt and a 15- to 35-mm zoom feature. A 24-frame color scroller and the exclusive QuietDrive motor control make it ideal for theatre or concert use. One of the most recognizable features of the standard ellipsoidal luminaire is the four shutters that can shape the light, and ETC has built them into a remote controlled module. Other modules are available,

FIGURE 15.19 ETC Source Four Revolution. (Photograph by Electronic Theatre Controls, Inc.)

such as a color scroller, rotating wheel module, and static wheel module, along with an iris module. This unique feature allows the user to choose which functions are needed. All modules are autosensing and require no patching or internal changes to the unit. The only limitation is that the unit cannot accept all modules at once, so there is some need to be prepared to make choices. The lamp is a 750-watt, 77-volt QXL with a modulated dimmer.

ADB LIGHTING TECHNOLOGIES

ADB Lighting Technologies started life as a Belgium company in 1920. It was bought in 1987 by the SIEMENS Group, and a management buy-out followed in 2002; the company is now headquartered in France. They have a very long history in theatre design of lighting luminaires, dimmers, and consoles. The WARP Motorized 800 W (Figure 15.20) had to be included in this chapter for its uniqueness alone. Yes, it is a moving luminaire, but it doesn't have to

FIGURE 15.20 ADB Lighting Technologies WARP. (Photograph by ADB Lighting Technologies.)

be. It can be purchased with or with out a moving yoke. More importantly, it is a profile spot, and, yes, is still unique. The company is not trying to market this as a touring concert luminaire, but this book is about all new technology in lighting. A unique feature is that the ring control, conventional shutters have been replaced by four integrated blades, each with 360 degrees of endless rotation capability. Their innovative (patented) ring control provides focus, zoom, iris, gobo, and shutters. The rings allow for accurate fingertip (manual adjustment) from any position around the luminaire or via remote control. It uses an 800-watt halogen axial lamp and operates

at 230 volts, 3200 Kelvin, which produces light output higher than with a 1200-watt conventional profile luminaire. All features beside the pan and tilt can be fitted with motors for all functions. If cost is not the main issue, this radical luminaire truly warrants further investigation.

THE FUTURE OF MOVING LUMINAIRES

The future is, to say the least, no pun intended, *bright*. With some of the luminaire discussed there can be no question that the versatility of moving luminaires has application not only for concerts but also in all forms of entertainment and architectural applications. The costs are dropping, and there is no comparing the design flexibility that is possible over conventional luminaires. The state-of-the-art quality and cutting-edge design of these developments are almost moving too rapidly to keep up with. When you can focus, color, and even dim a source of such brightness and high color temperature remotely, the time has come to broaden our thinking to not just concerts but all entertainment applications. I can see no area that these innovations, and along with others shown in the book are not at least something to consider in making your next production unique.

Our wish for a trouble-free streamlined system that uses as little cable as possible is becoming more of a likelihood as more computer protocols and transferable, slick computer designs are introduced to modernize our way of doing things. There is plenty to look forward to. More and more theatres, even high school and college theatres, now have moving luminaires installed. This field is simply moving too fast to stop it from taking over virtually all entertainment lighting in the not too distant future.

16

LED UNITS

T he newest entry into the concert lighting field is the light-emitting diode (LED). Some argue that it is the only real advance in lamp technology in the past 100 years. Its history can be traced back to 1907 to a researcher named H. J. Round, who worked for Marconi Labs and noticed that electroluminescence was produced from a crystal of silicon carbide when electricity was applied to it. A Russian independently created the first LED in the mid-1920s but the discovery was ignored in the United States and England.

The first reported visual-spectrum red LED was created in 1962 by Nick Holonyak, Jr., who was a researcher at General Electric. Holonyak is considered the father of the light-emitting diode. He later moved to the University of Illinois at Urbana-Champaign, where a former graduate student came up with the first yellow LED and a 10-times brighter red and red-orange LED in 1972.

To fill out the color spectrum, Shuji Nakamura, working in Japan for the Nichia Corporation, produced the first high-brightness blue LED. He later moved to the University of California at Santa Barbara in 1999 and was awarded the 2006 Millennium Technology Prize for his invention (Figure 16.1), but it fell to Isamu Akasaki and Hiroshi Amano in the late 1980s and 1990s to make white LEDs widely available. The so-called white LED is actually a mixture of the yellow and blue LED.

The first practical application of this technology was using red LEDs as indicators on televisions, radios, telephones, test equipment, and watches. As the technology advanced, the light output was

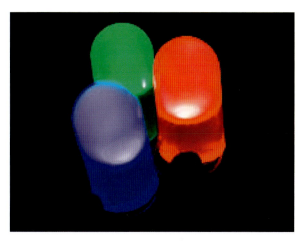

FIGURE 16.1 RGB LEDs. (Photograph by Piccolo Namek.)

increased to the point where they could be used for illumination. The modern LED is most commonly a 5-mm T1-3/4 and 3-mm T1 package. With higher power, it has become necessary to control excess heat for reliability purposes. Modern state-of-the-art, high-power LEDs do not resemble the early LEDs at all.

LED TECHNOLOGY

There is little doubt that this new and fascinating area of science has invaded the concert lighting field, and there is no turning back. The swiftness with which manufacturers of LED walls and luminaires have entered the entertainment markets is even greater than for the introduction of moving luminaires. How can we account for this? I would ask you to go back to the first edition of this book and to the second edition, where I observed that the next big advancements would be in light sources. Well, LEDs completely satisfy that prophesy. Now the problem is that we are not just dealing with an advance in luminaire housings, more sophisticated lighting consoles, or better truss structures. We have all had to start from scratch to learn a whole new technical language; beyond that, we must have the vision to use the products that are spilling out of warehouses faster than any lighting designer could hope to keep up with. All that the reader, as well as the author, can do is begin to digest this new terminology and hope to keep up. The following chapter is daunting, I know, due to the new terminology, power consumption, light measurement methods ... the list goes on. So, I will begin with as simple a description as I can.

Let's back up a little. What is a *diode*? A diode is an electrical device that allows current to move through it in one direction with far greater ease than in the other. The most common kind of diode in modern circuit design is the *semiconductor* diode. The term "diode" is customarily reserved to refer to small signal devices. A *rectifier* (as in *silicon-controlled rectifier*, or SCR) is also a diode used for power devices (Figure 16.2).[1] We should already be familiar with two types of diode: the Triac and the SCR, both of which are used to control the signal to modern electronic dimmers. So, the use of another diode to bring electroluminescence to a lamp seems rather logical.

Without boring you with a treatise on the electrical makeup of a diode, the facts are that, with certain characteristics of current flowing through the diode, electroluminescence can occur. The visible wavelength of that light (its color) depends on the band-gap energy of the materials forming the *p–n junction*. In the normal diode, this band-gap energy is not radiative and produces no visual emission. The materials used to create an LED have a direct band gap with energies corresponding to near-infrared, visible, and near-ultraviolet light.

An early problem was that the light produced is refracted back into the semiconductor, where it can be absorbed and turned into additional heat. This is a primary cause of LED inefficiency, as more than half of the light can be lost. The most common way

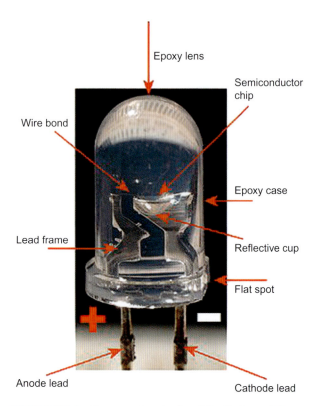

FIGURE 16.2 Component parts of an LED. (Photograph by Pschemp.)

to reduce this problem is by using a dome-shaped clear cover so the outgoing light rays hit the surface perpendicularly, thus minimizing the reflection. Some manufacturers use an antireflective coating. Some of this covering may be colored, as well; however, this is for cosmetic purposes and doesn't affect the color of the light emitted.

EFFICIENCY

One of the key advantages of LED-based lighting is that it can be highly efficient. But, a word of caution—the light output of LEDs cannot be measured on normal light meters used for incandescent and Mercury (Hg) medium-arc iodide (HMI) lamps. From a purely technical standpoint, white LEDs now match or exceed the efficiency of standard incandescent bulbs on a per-watt basis, but many people still question the actual color quality of the light. In 2002, Philips Lumileds manufactured a 5-watt LED with a luminous efficiency of 18 to 22 lumens per

---

[1]For more information on diodes and rectifiers, visit http://www.allaboutcircuits.com.

watt (lm/W) comparable to conventional 60- to 100-watt incandescent light bulbs, which produce around 15 lm/W.[2] Other companies have made announcements claiming higher and higher lumens per watt. Nichia Corporation has developed a white LED with a luminous efficiency of 150 lm/W. And, the author has no doubt that, by the time this book is printed, these figures will be several generations old in technological terms.

In December of 2007, scientists at Glasgow University announced that they had found a way to make LEDs more efficient by imprinting billions of holes into the LEDs using a process known as *nano-imprint* lithography.[3]

Unlike incandescent light bulbs, which do not require polarity, LEDs will only produce luminescence with the correct electrical polarity. When the voltage across the p–n junction is in the proper direction, a significant current flows and the device is said to be *forward biased*. LEDs can be operated on direct current (DC) but will only light with a positive voltage, causing the LED to turn on and off at the frequency of the AC supply.

Although LEDs are also known for their long relative life, some at 30,000 to 50,000 hours, there are at least a dozen ways to shorten this life or cause complete failure, including thermal stress. Manufacturers are working on ways to dissipate the heat from the diode in more efficient ways.

COLORS AND RGB

Conventional LEDs are made from a variety of inorganic semiconductor materials. Figure 16.3 shows the colors and wavelength of the most current 11 colors. Note that they also include infrared and ultraviolet outside of the visible spectrum. White light can be produced by mixing different colors of light. The most common method is to combine red, green, and blue; however, this method is very sophisticated and involves electro-optical design to control the blending

Infrared	$\lambda > 760$
Red	$610 < \lambda < 760$
Orange	$590 < \lambda < 610$
Yellow	$570 < \lambda < 590$
Green	$500 < \lambda < 570$
Blue	$450 < \lambda < 500$
Purple	Multiple types
Violet	$400 < \lambda < 450$
Ultraviolet	$\lambda < 400$
White	Broad Spectrum

FIGURE 16.3 Visual spectrum of LEDs.

and diffusion of different colors. In principle, this mechanism has higher efficiency in producing white light.

There are several types of multicolored white LEDs: di-, tri-, and tetrachromatic white LEDs. The key factors to consider are color stability, color rendering capability, and luminous efficiency. Often, higher efficiency will mean lower quality color rendering. As an example, the dichromatic white LEDs have the best luminous efficiency (120 lm/W) but the lowest color rendering capability. Tetrachromatic white LEDs have excellent color rendering but often have poor luminous efficiency. Trichromatic white LEDs provide a balance between both good luminous efficiency and fair color rendering.

We must define "white" light. In the common usage of light in entertainment lighting, color temperature, as defined by the Kelvin scale, is generally described as household lights, 3000 K; theatre and film lights, 3200 K; and outdoor film light, 5600 K. The problem is that what LED manufacturers call "white light" is actually around 3000 K or 4700 K. Depending on your application, this can be an obstacle for using specific white LEDs.

Using phosphors of various colors to create white light LEDs produces what is called *phosphor-based white LEDs*. Depending on the color of the original LED, phosphors of different colors can be employed and the resultant spectrum is broadened. This effectively increases the color rendering index (CRI) value of the particular LED; however, you pay for this gain with lowered efficiency. This is still the current most popular method for manufacturing high intensity white LEDs.

[2] "Seoul Semiconductor squeezes 240 lumens into 'brightest LED'," December 21, 2006, http://www.engadget.com.
[3] "A new type of super-efficient household light bulb is being developed which could spell the end of regular bulbs," *BBC News*, December 28, 2007 (http://www.bbcnews.com).

Avago Technologies	www.avagotechnologies.com
Bright LED Electronics Corp.	www.brightled.com
China Semiconductor Corporation	www.chml.com
Cree, Inc.	www.cree.com
Dialight	www.dialight.com
Edision Opto Corporation	www.edison-opto.com
Everlight Electronics Co. Ltd.	www.everlight.com
Highlight Optoelectronics Inc.	www.highlighted.com
IDEA, Inc.	www.ledidea.com
Kingbright	www.kingbright-led.com
LEDtronics, Inc.	www.ledtronics.com
Lumex	www.lumex.com
Nichia	www.nichia.com
Optek Technologies	www.optekic.com
Optoloco	www.optoloco.com
Panasonic	www.semicon-panasonic.co.jp
Photron LED	www.protron.com
Toshiba	www.semicon-toshiba.co.jp
Wilbrecht (LEDCO)	www.wilbrecht.com

FIGURE 16.5 Basic LED manufacturers.

Entertainment Applications of LED Technology

Once the Nichia high brightness blue LED emerged as a viable source, the entertainment industry jumped on a red, green, and blue (RGB) combination for new applications. Commercial products began appearing in the mid-1990s. One advantage that LEDs have in this business is that they do very well as primary colors, but they do not do so well for pastel or white light. Mike Wood said, "There is an oft-cited law concerning LED brightness called *Haitz's law*. The law forecast that every 10 years the amount of light generated by an LED increases by a factor of 20, while the cost per lumen (unit of useful light emitted) falls by a factor of 10. These are pretty steep curves, but, so far, not only have these forecasts been fulfilled

FIGURE 16.6 Styles of LEDs. (Photograph by Gophi.)

but … with the injection of large amounts of R&D funds from government seeking energy efficiency it's possible that they have actually been exceeded" (see Figure 16.6).

Wood went on to say that entertainment is a very, very small market in this race, but as we often have, we will benefit greatly from the R&D done by others. "In entertainment lighting we always want the newest toys to play with and are prepared to put up with more aggravation to use them than other sectors," he added.

FIGURE 16.7 LED PAR-64. (Photograph by Chauvet.)

The initial interest in LEDs was that we could package them with red, green, and blue (sometimes amber), and by controlling each LED we had infinite control over an almost unlimited color range. Yes, at first there was some "picking" as the lamps transitioned on and off, and in the cheaper products this still occurs, but for rock & roll purposes absolute smoothness was not number 1 on the list. Four generic types of products seem to comprise the bulk of the market: PAR, moving luminaires, LED walls, and wash luminaires.

THE LED PAR

The LED PAR would seem to be the simplest of the whole family—simply take the stock PAR housing, drop in the appropriate-sized bundle of LEDs in three or four colors, give it DMX and AC power, and you could replace all of your PARS … almost. The problem is that the LEDs are not very good at focused, calumniated beams. Yes, with haze you can see beams, but there is still debate about how much compromise can be given between the ability to have all the potential colors and reduced intensity. Therefore,

FIGURE 16.8 Color Blast 12. (Photograph by Color Kinetics.)

some decision is needed as to where the design can lose the very narrow spot (VNSP) or AC tight beam look that is so at the center of air light graphics (Figure 16.7). Some units have a strobe feature as well as an on-board package of preprogrammed effect such as chases and color sequences. Of course, we should not leave out the power advantage. A normal PAR-64 draws 1000 watts at 120 volts. Most of the LED PAR-64 size units draw 95 to 100 watts. In this day of "green" touring, that could be a factor.

WASH LUMINAIRES

In this category are small wash luminaires, such as the LED wash light by Color Kinetics called the Color Blast series, which was one of the first LED units to be used in television lighting (Figure 16.8). It is made up of 72 red, 72 blue, and 72 green LEDs and has seen both direct lighting and scenic applications. As a *wall-wash* luminaire it has been a favorite for architectural applications, as well as television scenic elements and concerts. There are many choices in this general design, including several ultraviolet units.

When looking for luminaires to wash a backdrop, there are two ways to proceed in the LED category. At least one company, Altman Stage Lighting, has used the more traditional one-cell cyclight using RGB and amber emitters. Others are producing 3- to 4-foot-long units that are more of a match to the old striplights. Color Kinetics offers several versions; one uses 12 circuit boards with 18 high-intensity LEDs per board (Figure 16.9), has forced air cooling, and yet consumes only 360 watts of AC power.

Of course, any packaging could be provided by most manufacturers on a custom basis, since, as

FIGURE 16.9   ColorBlaze 72. (Photograph by Color Kinetics.)

I have said, once the emitter package is designed the housing is a secondary issue. LEDs are without question our most creative entertainment illumination source today, but who knows about tomorrow. Andi Watson's account of a Radiohead tour where all the lighting was LED based because the group believes in a sustainable environment and wanting to cut their carbon footprint demonstrated many creative ways to use this new technology (see Chapter 23). Other bands are doing the same to one degree or another; however, I must restate one thing. This is new, high-tech stuff … and therefore not cheap! There are budget considerations that even the biggest artists may have a hard time swallowing for now. But, perhaps developments in the future will soften the financial blow and more LEDs can be developed in an economical package for entertainment.

See Figure 16.10 for a list of static LED luminaire manufacturers.

MOVING LUMINAIRES

Actually, the bundling for a moving luminaire and a PAR is not all that different. You replace the short-arc lamp with a bundle of LEDs, possibly HPLEDs, with fans to cool the clusters. Add even more complex programming possibilities, as the SHOWPIX

Altman Stage Lighting	www.altmanltg.com
Barco	www.barco.com
Coemar	www.coemar.com
D.T.S. Illuminazione	www.dts-lighting.com
Elation Professional Lighting	www.elationlighting.com
Irradiant	www.irradiantthg.com
Kupo	www.stage.com
Lighting & Electronics, Inc.	www.le-us.com
Martin Professional	www.martin.com
Phillips SSL Solutions (Color Kinetics)	www.colorkinetics.com
Pulsar	www.pulsarlight.com
Selador	www.selador.net
TwinkleWorks LLC	www.twinkleworks.com

FIGURE 16.10   Static LED luminaire manufacturers.

FIGURE 16.12   Legend 6500. (Photograph by Chauvet.)

FIGURE 16.11   SHOWPIX. (Photograph by High End Systems.)

American DJ	www.americandj.com
Coemar	www.coemar.com
D.T.S. Illuminazione	www.dts-lighting.it
High End Systems (The Barco Group)	www.highend.com
Irradiant	www.irradiantthg.com
ROBE Lighting	www.robe.cz
SGM Technology	www.sgmtechnologyforlighting.com
Tei Electronics	www.teilighting.com

FIGURE 16.13   Moving LED luminaire manufacturers.

has (Figure 16.11), and you have brought your game up several notches. Now you have pan and tilt capabilities, as well as the ability to make patterns in the LEDs themselves. This is when many manufacturers jumped on the band wagon by putting LEDs in their existing products and making whatever modifications were required; they already had the DMX technology, and it saved them from incorporating rotating pattern wheels, strobe shutters, and zoom mechanisms, which helps pay for the high cost of the LED bundle. There is also a group of products designed for wash use but which still have a pan and tilt feature (Figure 16.12) and from the outside look like any other moving luminaire.

Figure 16.13 provides a list of moving LED luminaire manufacturers.

LED Walls

One of the best applications of LEDs has been their use as projection surfaces. LED walls come in panels that are connected together, such as the open panel design by Element Labs shown in Figure 16.14, which can be built up to 18 × 24 feet, and you can still see through it. Element Labs also produces tile-type units (Figure 16.15) that cannot be seen through but are excellent for installations on walls or displays. Several companies make a system that is essentially a *roll-up* in 3-foot-square webs that can be connected

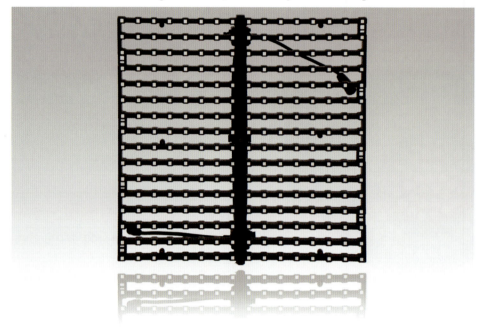

FIGURE 16.14   STEALTH display. (Photograph by Element Labs.)

FIGURE 16.15   Versa TILE. (Photograph by Element Labs.)

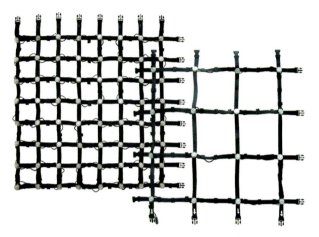

together. In the case of Chroma-Q's Color Web, the tri-color LEDs can be placed in 5- or 10-inch spacing on the web (Figure 16.16). The last general application is a tube version, shown in Figure 16.17, which is made by Element Labs; these are single tubes that come in three lengths (approximately 20, 40, and 79 inches) and two densities (16 or 36 pixels per meter).

There are many variations on these ideas, and investigation is absolutely necessary to determine what will match your needs in structure, pixel density, and size. Certainly one of the biggest advantages is that an expensive projector is eliminated, that was a problem to find a place to mount it so it doesn't interfere with the audience. Another advantage is that most are see-through so audiences on the sides and rear can still see the show.

Figure 16.18 provides a list of linear screen LED manufacturers.

FIGURE 16.16 Chroma-Q Color Web 125 and 250 LED visual effects surfaces. (Reproduced by permission of A.C. Entertainment Technologies, Ltd.)

FIGURE 16.17 Versa TUBE. (Photograph by Element Labs.)

Acclaim Lighting	www.acclaimlighting.com
Barco	www.barco.com
digiLED	www.digiLED.com
Element Labs	www.elementlabs.com
G-Lec	www.g-lec.com
Martin Professional	www.martin.com
Smart Vision	www.smartvision.com

FIGURE 16.18 Linear screen LED luminaire manufacturers.

17

DIGITAL LIGHTING AND MEDIA SERVERS

Concert lighting has certainly taken a giant step toward controlling more visual stage presentations with the latest addition of digital lights, media servers, and light-emitting diode (LED) walls. The lighting community has moved beyond practical illumination and effects-to-the-music lighting to incorporate projected graphic imagery, stills, and movie clips. Visual imagery and the mobile components that enable the lighting designer to create and manipulate these images have become staples of most touring concert lighting systems.

ADDING THE CONTROL OF VISUAL MEDIA TO LIGHTING

As if fulfilling the lighting needs of the artist weren't enough, lighting designers have inherited responsibility for the design and control of visual media with the introduction of digital lighting and media servers. A *digital light* isxa luminaire that uses digital projection techniques to modify the beam parameters, mostly shape, intensity, and color. A *media server* is a computer dedicated to the task of organizing and playing back video clips, still images, or three-dimensional images during a performance. The combination of these two components, with the help of LED wall displays (also driven with a media server), has presented the entertainment lighting world with powerful technology and control that can further enhance and assault the visual senses

and elicit emotion, more so than just lighting scene changes can.

Adding visuals was a natural progression and an unstoppable train that collided with concert lighting beginning in the mid-1990s. The use of visuals is still progressing with so much speed that this explanatory sentence about it will be out of date by the time this book is released. The visual media movement toward entertainment lighting has been fueled by clever innovation and adding computer protocols and telecommunications methods to the mix. This is all a result of efforts to improve moving luminaires and, using the DMX-512 protocol, to better manage them. Plus, there is a significant financial benefit; any time you can combine job duties and equipment function—in this case, by getting the lighting crew to pitch in and the lighting designer to control media imagery as well —a large part of the video crew can be eliminated. I'm sure this is a consideration, or at least an inadvertent consequence, that has pleased the accountants.

Projecting movie clips and still images has been used in live theatrical productions probably since the invention of the film projector at the turn of the 20th century. Bill Graham was famous for showing cartoons before his Fillmore concerts and other films during set changes between bands, but these cues were operated manually. Only with relatively recent tours has media control landed directly in the hands of the lighting director, along with so many options. Adding control of visual media automatically promotes the lighting designer and programmer positions to those of *visual artists*, which make them even more valuable to the product, but at a price. New tools, hardware, software, parlance, and technical information have been added to the lighting designer's inventory beyond luminaries, dimming, and control.

Medusa Icon M

First, as usual, we should add some history. The evolution of controlling a digital light from the lighting control console came by way of Light & Sound Design, a company originating in Birmingham, England. After establishing itself as an innovative company with such clients as AC/DC, Prince, and Madonna, they moved to offices in Los Angeles around 1981. A couple of years later Vari*Lite released the VL.1, and Light & Sound Design began working on its proprietary moving luminaire, called the Icon, which was eventually released for their rental customers in 2003, even though some of their Icon road tests began as early as 1993. Fast forward a few years to the 1999 Lighting Dimensions International (LDI), when the Medusa Icon M was introduced surrounded by a shroud of mystery and demonstrated only for small groups of people at a time (Figure 17.1).

FIGURE 17.1 Medusa Icon M. (Photograph by Production Group.)

Mark Hunt, imaging system and software engineer for the Icon M, explained that, "The Medusa was developed in strict secrecy. It was rumored that it would never work and, if it did, it would be too expensive. Medusa Icon M *was* the first digital light even though the name 'digital light' had not yet been coined as the official term at that time. It came about as a result of Bill Hewlett's* persistence and his recognition of the possibilities of Texas Instruments' DLP technology." Hewlett was the technical director of Light & Sound Design. The idea of a digital light was entirely his concept, and he joined the company to see the project through.

The Icon M used Texas Instruments' Digital Light Processing (DLP) micromirror technology. It contained the entire DHA Lighting gobo catalog onboard, with custom vector images or grayscale bitmaps, each with instant random access. The DLP technology enabled digital dimming, iris, strobe, and zoom functions. The Medusa never reached production, and when we asked Hunt why, he said, "The main issue was light output. We could never get enough light to compete with other luminaires, and it was felt that the fixture would have to be useful as a luminaire as well as an effect. The DL.1 rather proved our point." (More on the DL.1 later in this chapter.)

High End Systems' Digital Light (DL) Series

High End Systems started the move to digital-based lighting with the Catalyst moving mirror product and showed it a year after Medusa Icon M was shown at LDI. The original Catalyst combined a media server and moving mirror light with a stock video projector. Richard Bellview—High End Systems' chief technology officer and innovator of the first Intellibeam moving mirror luminaire (circa 1990) and more recently the SHOWPIX and SHOWGUN range of fixtures—added some first-hand account details: "I think that we give credit to Richard Bleasedale for providing the video server concepts for Catalyst. Mr. Bleasdale was the software writer and inventor for the media server portion of Catalyst. The Catalyst moving mirror head was conceived by Peter Wynne Wilson of WWG (acronym was derived from the two owners, Tony Gottelier and Peter Wynne Wilson). In putting the two concepts together, we basically showed people that video projection could be used as another entertainment instrument different than conventional video projection." The term *digital light* was simply invented at that time to get the lighting community's attention. Bellview explained, "We have used the term digital lighting to define DMX controlled products that produce images without a stencil, by some type of light valve control. The defining factor is if the instrument is controlled by a lighting console; otherwise, it's a projector." In 2003, High End Systems launched the Catalyst media server with the ability to apply more than 80 visual and color effects to digital media.

The DL.1 (second generation) debuted at the Professional Lighting and Sound Association (PLASA) tradeshow in London in 2003, which was designed to complement the Catalyst media server. The transition from the moving mirror system to the DL.1 was simply a projection system package with a moving yoke. It used the communications standards of both VGA (video graphics array), RGBHV (red, green, blue, high voltage), and S-Video inputs for movie, film, or graphics content but with DMX programming capability could be remotely pointed or focused. The LD could now switch the video input selection and choose content from multiple input devices simultaneously with color, brightness, contrast, and other video-related parameters, such as the keystone correction feature (keystone compensates the image on a surface for off-angle projection). High End Systems marketed the DL.1 as the lighting industry's first commercially available digital moving light. The DL.1 is out of production today.

The DL.1 was connected with data and RGB cables to a Catalyst media server that was generally mounted in racks (the number depending on the number of lights that it was driving) and positioned off the stage or sometimes suspended. The breakthrough innovation and technology used for the DL.2 was its onboard media server and a Sony camera mounted in the front of the light for live-action feeds. With regard to changes in the moving luminaire world compared to the fast-track video world, Bellview was asked to explain the differences relative to the DL series: "Let's talk about iterations, which happen very fast in the video world. We build an automated lighting instrument and five or six years later we are

making the same instrument with very little change. Video has annual or semi-annual changes, and everybody in video understands that there are advancements that displace their current products. Once we started tracking video in the lighting world, iterations took place based on video expectations. As the performance of the projection systems increased, DL.2 was created to put the server into the digital light and break loose of the video cables running down to FOH (Front Of House), to make it basically a plug-and-play operation." Substantive advances were made with the DL.2, but it is also now out of production.

In early 2008, the DL.3 was released with an on-board Axon media server, a higher light output of 6500 lumens (using a single 330-watt projection bulb and improved optics), new features, and additional libraries that offered more movie loops and high-resolution images. The integrated media server also supported the user's custom content with the ability to import three-dimensional objects, media files, and still images (Figure 17.2).

FIGURE 17.2   DL.3 digital light. (Photograph by High End Systems.)

FIGURE 17.4 DigitalSpot 7000 DT. (Photograph by ROBE Lighting.)

FIGURE 17.3 DML-1200 digital light. (Photograph by Barco.)

In June of 2008, Barco announced the acquisition of High End Systems. Barco is the leader in four product areas of the rental and staging market: LED video displays, LED products, large-venue projectors, and image-processing products. High End Systems was integrated into Barco's Media and Entertainment division, instantly increasing their live entertainment market share and intellectual properties. This business acquisition strategy is certainly indicative of the convergence of video and lighting segments as an area of increasing growth and development.

Prior to the acquisition, the DML-1200 (Figure 17.3) luminaire was launched in 2007 as a super bright moving digital light source and high-quality video projector. It uses four 300-watt high-pressure mercury lamps, and in light mode the fixture has a light output equivalent to that of a 1200-watt hard-edged moving light at 12,000 lumens. In video mode,

the light output decreases to 10,000 lumens. The unit features a fully sealed DLP engine and full-color DLP video. As an option, there is a built-in media player, based on the Hippotizer V.3 media server.

DIGITAL LIGHT DEVELOPMENTS

The High End DL Series did not have any competition, and the brand itself was the name associated with the digital light. However, at Prolight + Sound Frankfurt 2008, two established manufacturers emerged: ROBE Lighting introduced two digital light models, DigitalSpot 3000 DT and 7000 DT (Figure 17.4), and SGM launched its Giotto Digital 1500 (Figure 17.5). The challenge was to produce a luminaire that would be an effects projector but also have the intensity to compete as a moving luminaire. While digital software and control were showing steady advancements, there were cooling issues. The SGM Giotto 1500 is an attempt to tackle that problem. It includes a feature for lamp power reduction in the event of overheating, but the increased intensity is competitive with a moving luminaire and is achieved by using a Philips MSR Gold 1200 FastFit short arc, which is rated at 48,000 lumens.

SGM has taken a different approach from High End with its on-board media servers, Axon and Catalyst. The Giotto Digital 1500 uses DLP technology and has all-digital system graphic effects. The

FIGURE 17.5 Giotto Digital 1500. (Photograph by SGM.)

still images and AVI-MPEG4 format video clip libraries can be updated wirelessly.

ROBE's DigitalSpot 3000 DT is the first moving digital light that combines an LED-based wash light in the front of the fixture. The fixture is rated at 2700 lumens from a single 200-watt projection lamp and is used for smaller projects. The DigitalSpot 7000 DT is based on LCD and LED technology. It has one 330-watt projection lamp for 6500 lumens. Combining a digital projector and two 48-RGBW LED modules generates digital gobo effects and saturated colors with the one digital light.

With only three manufacturers releasing digital products to date, it will be interesting to watch the competition in this market unfold. As businesses go head to head in owning the digital market, LED options are plentiful and growing. Low- and high-resolution LED projection displays, both straight and malleable, and soft curtain and other shapes, such as tubes, strips, and tiles, offer more wide-ranging light-emitting options than do digital luminaires, which are limited to surface front or rear projection.

See Figure 17.6 for a list of digital lights.

MEDIA SERVERS

Media servers were designed to control media in a way that is intuitive for the lighting designer, director, or programmer through the DMX-512 protocol. Because media server parameters often exceed 512 channels in a show, and considering other conventional and moving luminaires, this pushes control to limits that require organizing show control through multiple universes (a universe equals one DMX grouping of 512 channels). Generally, a letter is used to separate each universe; for example, A512 would be conventional lighting, LED, and effects; B512 would represent moving lights; and C512 and D512 would represent digital luminaires and media servers.

High End Systems (The Barco Group)	www.highend.com
ROBE	www.robedigital.com
SGM Technology	www.sgm.it

FIGURE 17.6 Digital lights.

Whereas the digital luminaire is essentially a unique projection light source with beam shape, intensity, and color, the media server is a sophisticated computer that stores video clips and still images libraries. It can be used to manipulate any of those images with color, effects, and image correction to surfaces (curved or flat) through digital light sources, or similarly manipulate images that appear through various LED screens.

The lighting control manufacturers have adapted quickly to bring visualization, media, and lighting together in user-friendly, all encompassing consoles and software, enabling switching, programming, cue recording, and playback to come together in a logical way. It is a lot to be responsible for, and control is the key to managing the growing list of effects and functions that have been introduced to the lighting designer and director and push the media server frontier to produce live visual graphic presentations that audiences will enjoy.

MEDIA SERVER CAPABILITIES

Currently, about a dozen media server manufacturers offer products and accessories usable in this market. Only a few manufacturers are mentioned here as examples of the many features that media servers are capable of providing. Basically, the differences are variations in file storage capacity, graphics and effects capability, and the royalty-free content selection that is included with the media server package. However, a set of standardized hardware components is needed to fulfill the basic functionality of a media server and generate effects that all the makers have. The media server will have a computer, containing:

- A large amount of hard-drive storage for the content
- A high-performance graphics card
- A control interface, for either DMX or Ethernet

Ethernet is a standard communications protocol embedded in software and hardware devices, intended for building a local area network, or LAN. Media servers send a screen signal from the video output of the computer to projectors and light-emitting media, such as high- or low-resolution LED screens or fixtures. If it requires DMX controls, the media server generates the DMX data from the video pixels internally and sends the data to the lights. Each media server manufacturer will have distinct features that will ultimately be sent to the light source or projection, and many of those are standard fare among all the makers, such as keystone correction, rotate, color, and scale of images. Manufacturers sometimes change the labels for standard functions, using a descriptive proprietary name such as "Collage Generator," which the High End Systems Axon server uses (Figure 17.7), which does the same thing (blending several digital lights together to form one seamless image) as "Soft Edge Blending" by Coolux International's Pandora's Box media server.

FIGURE 17.7 Axon Media Server. (Photograph by High End Systems.)

With standard functions fulfilled, the media servers vary in features and price (normally in direct proportion to how advanced they are), including the library content offered, pixel mapping (a video scaling technique used in display devices), or audio data synchronization features.

MBox

An example of advancement, Production Resource Group (PRG) used the technology developed from the early roots of the Medusa Icon M for their design of the MBox Extreme. The MBox Extreme is a single server rack unit equipped with two video outputs that can be used to drive a projector, LED wall, or other light-emitting displays. This unit stores up to 65,535 videos and still images! Large production clients will use multiple MBox servers (up to 14 so far) to get the flexibility they want. The Madonna tour used a single server playing high-definition (HD) content then split the video out to drive several screens, lending perfect synchronization between screens.

Green Hippo

Green Hippo developed different products to satisfy small to large users. *Hippotizer Express* uses one standard definition output and larger applications by linking servers for multiple-machine interaction via *HippoNet*, which uses standard LAN technology to allow multiple servers and control computers to communicate and share elements to create one control grid. The *Hippotizer V3* system is made up of multiple building blocks called "components." Each component is specifically dedicated to running a control screen, outputting video to displays, and listening to DMX.

V4 Media Server

Catalyst refined their original version to produce V4 Media Server and software, that supports Mackie control hardware and other MIDI devices including Behringer BCF200 and MIDI keyboards. This is the first version of Catalyst that offers alternative control, but DMX still reigns king in concert lighting control. The Pro version of Catalyst has 6 on-screen output mixes and 12 independent assignable layers. Multiple

on-screen mixes mean that show operators can break down their set designs and run different effects over different sets of LED fixtures, LED wall controllers or projectors at the same time. The Catalyst has been discontinued since High End and Barco merged companies, but at the time of this writing, Catalyst V.4 and earlier models are still widely used in concert productions.

Pandora's Box

Pandora's Box MediaServer has given rise to the products *Media Server Pro*, *Media Server STD 8* video, and *Media Server LT 4*. Each product adds more features than the last, increasing the amount of video, graphic layers, and redundant arrays of independent disks (RAID) for up to 500 GB of hard disk space.

As you can see, the digital luminaires and media servers are as complicated and as large as your budget and imagination will allow. To create and produce sophisticated visuals on a smaller scale, the media server is already a capable tool for concert lighting visual expansion that can be assured of exponential product development.

CONTENT

Building a show using a media server requires content. One the most powerful features of a media server is the ability to combine and modify content in many different ways. This allows users who use only stock content to do so in an original way, such as taking a still image and manipulating it with an automatic motion effect, texture, or color.

The various media server manufacturers use different methods to provide content, and an important feature is the stock video clip library that is unique to each server. Generally, a content provider, such as ArtBeats (artbeats.com) or Digital Juice (digitaljuice. com), license QuickTime short film loops for everything from aerial views of New York City by helicopter, to underwater tropical fish swimming through clear oceans, to violent weather patterns or gentle nature. The licensing is for extended periods of time; for example, for the lifetime of all DL.2's will be permitted use on only certain clips licensed in bulk for that unit. Green Hippo allows users to choose DVDs

Arkaos	www.arkaos.net
Coolux International	www.coolux-us.com
Diagonal Research	www.diagonal.tv
grandMA	www.malighting.com
Green Hippo Ltd.	www.green-hippo.com
High End Catalyst	www.highend.com
Martin Professional	www.martin.com
Production Resource Group (PRG)	www.prg.com
Radical Lighting	www.radlitecom
SGM Technology	www.sgm.it

FIGURE 17.8 Media server and software manufacturers.

of content suitable for their show requirements. Once the content is purchased or licensed for a rental media server from a company such as PRG, the content is royalty free to use in any live form worldwide or broadcast that it may appear in.

Using the content responsibly is somewhat self-policed by the users and vendors, although there are ways to preclude custom content from ending up in someone else's show as a media server moves from one tour to another. Mark Hunt, Electronics and Software Engineer at PRG in Birmingham England, explained two ways that the company is addressing potential license breach problems. "All our content is encrypted and can only be played on a genuine, licensed MBox. We make our encryption available to our clients, so they can protect their content in the same way. This protects against casual theft from an unattended unit, and it keeps our stock content providers happy, too. All of our media servers are 'cleaned' between rentals, so there is no risk of us giving a client's custom content away accidentally." Every lighting designer will have access to the same media server or digital light tools. Content selection and timing will be the key differences between one show and another. See Figure 17.8 for a list of media server and software manufacturers.

18

CONVENTIONAL LIGHTING AND ACCESSORIES

From the earliest days of rock & roll touring, conventional theatrical lighting luminaires were a problem. It was not that they could not, from a design standpoint, do the job of giving a theatrical look to the stage, but they were not designed to take the kind of punishment of being loading in and out on a daily basis and trucked to another location all within 24 hours. Yes, theatre bus and truck companies predate the concert era, but they were traveling with the luminaires boxed up in crates and reconnected to a house pipe and circuited at each new theatre, often with weeks between moves. For this new type of touring, though, you could count on broken lenses, handles falling off, and lamps loose or off their bases when you opened the crate. Constant rewiring of individual cable connectors was also necessary. Also, the Fresnel and older plano-convex spot luminaires just did not have the lumen output needed to project heavily saturated color; however, they, along with the Leko (ERS), did offer creative alternatives to the simple blast of color from the PAR luminaire, so what was a designer to do? [*Author's note:* The term *Leko* is appropriate here because this was before the modern ellipsoidal reflector spotlight (ERS) was invented.]

THE PAR FAMILY

Chip Monck, it is generally agreed, brought the PAR-64 lamp to the concert field. He said he "stole" two from Four Star Lighting in Los Angeles, which was a big film rental house in those days. He took them to Altman Lighting's factory in Yonkers, New York,

where he and Ronnie Altman could see that they would not hold color so close to the "bottle." The first run of 1000 units of their new design added 6 inches to keep the color farther away from the lens. Of these, 500 were purchased by Bill McManus in Philadelphia, the remainder by Bob See at See Factor Industry in New York. Still, Bill said, "Halfway through the show, the color was burnt out by the great heat generated so close to the lamp."

Sometime before the first real rock & roll benefit show in 1971 (the Concert for Bangladesh), Chip was walking through the Altman Lighting factory with the founder, Charlie Altman, when he spied a pile of 8-inch snoots (extension holders that can be added to the housing) that had been made for the company's 8-inch ellipsoidals. He thought that adding this additional 6 inches of distance to the already incorporated 6 inches would be much better for the color. He remembers Charlie walking away mumbling something about "that fucking hippie." He continued to re-gel every night on tour but admits that the intensity of the saturated color could have held for three shows. Eventually, color media were reformulated and soon after were able to handle the heat for longer and longer times, with some exceptions.

The original use for the PAR-64 family of luminaires was for film lighting in a unit they called the Cine Queen. Chip's innovation was to put vibrant primary colored gels in front of the Cine Queen. The housing was simple, no moving parts, and it had a gel frame holder farther away from the lamp than did the Cine Queen which helped with some of the color burnout issues, but not completely. Bill McManus said, "There were no porcelain caps back then, and the wiring left a lot to be desired." But it worked. He

FIGURE 18.1 PAR luminaire family. (Photograph by James Thomas Engineering.)

used the 500 units that Altman built for him on Jethro Tull's Passion Play tour in late 1971. What he didn't know at the time was that Altman had also run 500 more of the design and sold most of them to Bob See at See Factor, another early concert equipment supplier . "I was so impressed that Chip Monck had picked a bottle that had so much punch over anything on the market, but it just wouldn't hold color," Bill told me. On that Jethro Tull tour, people went nuts. "We had 250 of the new PAR cans in a box beam truss supported with inverted CM chain hoists. It was the first tour to fly the whole system. The [Rolling] Stones had used Gallaway rams from the floor as support," Bill went on to verify Chip Monck's earlier remarks.

To be sure, the answer to many of the problems of shear lighting punch and road-ability was found in the parabolic aluminized reflector (PAR) lamp and housing. The details of how the PAR came into use for touring have already been well documented in Chapter 5. The PAR lamp had the highest lumen-per-watt rating of any lamp then developed, in the 3200 Kelvin range. Heaven had arrived for the concert designer. In a media where bravado was a key element in design, the PAR-64 promised a low initial purchase price and long road life.

This sealed-beam lamp family offers a range of beam spreads from very narrow to narrow to medium to wide flood. The aircraft landing light (ACL) can also be fitted into the standard PAR-64 housing. It produces a very, very intense narrow beam of light. The drawback is that it is a 24-volt lamp. Four can be ganged in series so U.S. 120-volt dimmers can handle them; however, if one goes out you lose the entire group. You also give up some individual control. Yes, ACLs really are for aircraft. I remember an aircraft supply company calling us and asking how big a fleet

of aircraft we had because we ordered so many lamps we had completely depleted their stock. And, yes, I know that 4 lamps at 24 volts only adds up to 96 volts. Remember that the voltage at the house disconnect can be anywhere from 110 to 120 volts but generally is well below 120 volts. Add the dimmers, which will reduce the power by at least 2 to 3 volts, and a 100-foot multi-cable to the lamp itself, and the over-voltage factor is expectable. In fact, it has the advantage of making the lamp burn brighter, admittedly also reducing the lamp life, but that is a small price to pay.

The PAR luminaire family (PAR-56, -46, -38, -36, -30, -20, and -16) also has housings for smaller lamps and reduced wattages in both black baked enamel and silver (Figure 18.1). These luminaires may not have any effect if they are housed in a truss 24 feet in the air, but they are useful for lighting objects up close, hiding in stage sets and small coves, or even exposed as a practical part of a set piece. Very common is the six-lamp PAR bar, which has six PAR lamps attached to a pipe with hangers and is prewired to a Socapex connector (Figure 18.2).

The Raylight (Figure 18.3) combines a reflector and lamp—the DYS lamp, which is a small, 600-watt mini-pin base lamp—that fit in the standard PAR-64 housing. It throws a very concentrated beam, very similar to the ACL, but operates on 120 volts. A 30-volt version of the lamp operates at 250 watts. Reflectors are also made for the PAR-56 and PAR-36, as well. The PAR is now a staple in theatre and television.

ETC's PAR and PARnel

Comparable to the size of a short-nose PAR-64, the Electronic Theatre Controls (ETC) Source Four PAR

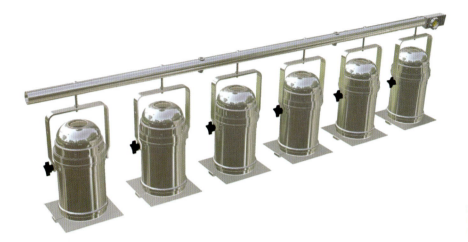

FIGURE 18.2 Six-lamp PAR bar. (Photograph by James Thomas Engineering.)

FIGURE 18.3 Raylight reflector fits in the standard PAR-64 housing with a DYS lamp. (Photograph by Creative Stage Lighting Co., Inc.)

FIGURE 18.4 ETC Source Four PAR. (Photograph by Electronic Theatre Controls, Inc.)

(Figure 18.4) started the trend of improved optics and a smaller luminaire that would complete with the PAR-64 in brightness and user convenience. Made of die-cast aluminum, these luminaires use the 575-watt High Performance Lamp (HPL), a real energy saver that, compared to the 1000-watt PAR-64, has a quality of light that is not distinguishably different. The unit is also rated for a 750-watt HPL lamp. Another unique feature is that the one housing has an interchangeable set of four snap-in lenses that alter the beam characteristics from very narrow to wide angle. Also, the football-shaped beam is rotated via a front ring that adjusts the beam just like the regular PAR-64.

The PARnel (Figure 18.5) is another style of Source Four luminaire. It is a mix between a PAR and a Fresnel that allows adjustment of the spot to the flood setting simply by rotating a manual knob on the luminaire to achieve the desired spread. Accessories include barndoors for extra light control. Altman's Star Par is a similar product, as is American DJ's version. These types of PARs are widely used as a conventional concert lighting rig and to augment moving luminaires, often replacing the PAR-64 altogether. They have gained wide use in theatre and television as well as for other vestal applications.

FRESNEL

The Fresnel is one of the workhorses of the theatre. Named for the inventor of the lens, Augustin-Jean

FIGURE 18.5 ETC Source Four PARnel. (Photograph by Electronic Theatre Controls, Inc.)

FIGURE 18.6 Fresnelite®. (Photograph by Strand Lighting.)

Fresnel, it is designed for soft-edge applications. It can be used for lighting acting areas as well as set and backlight applications. The size range is great, but for most theatre use the 6-inch version is most common. (The inches are an indication of the unit's physical width of the lens.) The 6-inch Fresnel can be lamped for 250 to 500 watts, and more robust units can handle 750- to 1000-watt lamps (Figure 18.6). The 1500- to 2000-watt 8-inch Fresnel is not in favor due to its size and weight. For video and film use, the sizes go up to 10- and 24-inch units. Their application may be limited in touring, but they do find special application from time to time.

ELLIPSOIDAL REFLECTOR SPOTLIGHT

The ellipsoidal reflector spotlight (ERS) is one luminaire that has more names than you can count on one hand: Leko, ERS, profile spot, ellipsoidal (okay, I only have four fingers). Leko is a name that has been seen on light plots and in contract riders for many years, often misspelled "liko" or "leco." Leko is a contraction of the names of the two men credited

with introducing the ellipsoidal reflector spotlight: Ed Levy and Edward Kook. It is the registered trademark for the luminaire developed (circa 1932) by the old Century Lighting Company, now Strand Lighting, a company within the Philips Group. This old Leko design included two plano-convex lenses placed back to back to create parallel beams of light. The convex side of the lens was made to focus at a specific distance. To find the focal length of the lens, you could take one out of the housing and do the old magnifying glass trick for finding the focal length by moving it closer and farther away from a surface. This lens combination provided a focus of 6, 9, or 12 inches, or narrower. Those measurements would then be translated into degrees of light spread (Figure 18.7).

A unique feature of this luminaire is that it has four framing shutters that allow you to shape the light as a circle or a square or many other geometrical shapes. The light output can be hard edged or softened. It is easiest to think of this light as the single-lens reflex (SLR) camera of theatrical lighting. The shape of the reflector and the compactness of the lamp filament allow the light beams to converge

FIGURE 18.8 Source Four ERS. (Photograph by Electronic Theatre Controls, Inc.)

FIGURE 18.7 360Q 6 × 12. (Photograph by Altman Stage Lighting.)

at a point called the *focal point* or *gate*, which is where, like a SLR camera, the image is turned upside down and backwards. So, a gobo pattern is placed backwards and upside down if the designer wants the image to appear as viewed directly. Other accessories can be placed at the gate such as an iris.

The Leko went through a radical redesign by ETC in 1992, when it was named the Source Four (Figure 18.8), and this version was followed by others with their own improvements. The luminaire is now commonly referred to as an ERS. It is used on tours to project patterns, and its many accessories include the TwinSpin (two patterns rotate counter to each other), color changers, and other toys. The ETC models started the trend to change how luminaires are designated, from inches to degrees, which provide a more accurate way for a designer to visualize how wide a beam will be simply by using a protractor to draw lines equal to the beam spread of the unit. ETC also divided the lens into more categories and has since expanded the choices even more: 5, 10, 19, 26, 36, 50, 80, and 90-degree lenses. Most other

FIGURE 18.9 Profile Spot. (Photograph by Selecon.)

manufacturers also offer most or all of these lenses. Several zoom versions of the ERS are available.

Originating in England, but now also available in the United States, the Profile Spot (Figure 18.9), with its distinct silhouette, offers essentially the same functions but with a radical design difference in how the lamp is accessed. These units are designed to

FIGURE 18.10   Cyclight. (Photograph by Altman Stage Lighting.)

FIGURE 18.11   MR16 striplight. (Photograph by Lighting & Electronics, Inc.)

make relamping easier via the bottom, not the back. Any ERS or Profile Spot can also find good use during filming of a tour as a "key" light on the band members and singers to cuelluminate cutaway shots the director wants (see Chapter 27).

CYCLIGHT AND FARCYC

Concerts have generally adopted one television luminaire, the Cyclight, which is widely used for lighting backdrops. The design of the reflector, a sheeting configuration, allows the unit to be placed close to a backdrop and light vertically up the surface for 10 to 14 feet (depending on model and distance). Their light output, if not their size, has made them a favorite; they take up a lot of room and do take time to set up, but the advantages outweigh many of their shortcomings. Several versions of Cyclights are made. For floor mounting, a series of 4 to 12 lamps in 3 to 4 circuits can be used (Figure 18.10). Overhead mounting favors the Farcyc, which is a square of four lamps and was first used at the BBC Studios in London. Currently, the trend is to use an MR16 lamp version or LED strips (Figure 18.11), which are sometimes smaller but can also be heavier. MR16 strips can also be seen at the front of many stages as uplight to fill in

chin shadows, particularly those of female performers or cowboys with hats.

THE NEW BEAM PROJECTORS

Beam projectors are a reinvention of an old standby. The beam projector is an open-face, lens-less unit whose purpose is to project parallel beams of intense light. The old units were not very efficient, but the latest generation has vastly improved on the design and light output (Figure 18.12). Another company, PANI Projection and Lighting, refers to their units as parabolic spotlights, but they are essentially the same; however, instead of an 800-watt, 120-volt lamp, they use low-voltage 250- and 500-watt lamps.

AUDIENCE BLINDERS

Not quite the slang name or use the manufacturers intended, but this is what audience blinders are generally used for—to illuminate or literally blind the audience. They are various configurations of the PAR-64, PAR-36, or MR16 grouped in 9 to 24 lamps. They found their way to touring via film (Figure 18.13), where they were called *fay lights*, because the PAR-36 had an ANSI three-letter code of FAY. An extremely unique take on the audience light is the Litepod (Figure 18.14), manufactured by Wybron, Inc., which surrounds a strobe in the middle with groups of lamps. The original idea for the Litepod came from Doug Brant and Justin Collie of Artfag LLC. The light alternates between a DWE 9-light luminaire, a 6-light luminaire with a single

FIGURE 18.13 Audience blinder. (Photograph by James Thomas Engineering.)

FIGURE 18.12 Beam luminaire. (Photograph by Wybron, Inc.)

Martin Atomic 3000 strobe, or a double Atomic 3000 strobe luminaire and can be fitted with a large-format CXI for unlimited color options.

See Figure 18.15 for a list of conventional luminaire manufacturers.

FOLLOW SPOTS

Truss-mounted follow spots are a common device for providing concentrated light on a moving subject. Coupled with a mounting bracket and chair with a safety belt (the operators also wear fall protection equipment), the follow spot can be placed anywhere on the truss rig but usually in the rear. The smaller

FIGURE 18.14 Litepod. (Photograph by Wybron, Inc.)

Altman Stage Lighting	www.altmanltg.com
Coemar	www.coemar.com
Creative Stage Lighting, Inc.	www.creativestagelighting.com
Elation Professional	www.elationlighting.com
Electronic Theatre Control (ETC)	www.etconline.com
Irradiant	www.irriadanthg.com
James Thomas Engineering	www.jthomaseng.com
Kupo	www.stage.com
Leviton	www.leviton.com
Lighting & Electronics	www.le-us.com
Luci della Ribalta	www.ldr.it
Optima Lighting, Inc.	www.optimalighting.com
PANI	www.pani.com
Selecon Lighting	www.seleconlight.com
SGM Technology	www.sgmtechnologyforlighting.com
Strand Lighting (The Philips Group)	www.strandlighting.com
Tei Electronics, Inc.	www.teilighting.com
Times Square Lighting	www.tslight.com
Tomcat	www.tomcatglobal.com
Wybron	www.wybron.com

FIGURE 18.15 Conventional luminaire manufacturers.

HMI units are most popular. Most use a 400-watt HTI lamp (metal halide short-arc lamp) in such models as the i-marc series (Figure 18.16), which uses 200- and 850-watt sources. Lycian's SuperArc 400 (Figure 18.17) uses the HTI-400-24 lamp. These units are only about 30 inches long and put out better than 300 fc at the normal throw from the truss—a very hot source that is very effective in concert lighting design. The HMI and HTI sources are 5600 Kelvin, so adjust your color media to account for the bluer light.

Front-of-house (FOH) follow spots have always been the weak link in a touring design because most productions are at the mercy of the units already positioned in the venue: their positions, type, wattage, age, and how well they have been maintained. Many of the high-end productions do carry their own FOH follow spots so that they are assured of balanced intensity and color. These productions may also place the follow spots on specially built and flown platforms so they also control distance to the stage.

There are several remote-controlled follow spot units on the market. These were designed to remove people from the trusses, and certainly this is a safety advantage. Wybron, Inc., has had their Autopilot series on the market for several years. These lights locate the subject in three-dimensional space, translate the position to pan and tilt for any number of moving luminaires, and follow the performer.

DMX Iris

You can buy mechanical irises and place them in an ERS luminaire to get the precise circle desired, but they tend to slip. A remote-controlled iris can solve the

FIGURE 18.16 The i-marc 850. (Photograph by Phoebus Manufacturing.)

FIGURE 18.17 The SuperArc 400. (Photograph by Lycian Stage Lighting.)

Altman Stage Lighting	www.altmanltg.com
Coemar	www.coemar.com
Irradiant	www.irradiant.com
Kupo	www.stage.com
Lycian	www.lycian.com
OmniSistem	www.omnisistem.com
Pani	www.pani.com
Phoebus Manufacturing	www.phoebus.com
Robert Juliat	www.robertjuliat.com
Strong Entertainment Lighting	www.strong-lighting.com
Tei Electronics, Inc.	www.teilighting.com
Times Square Lighting	www.tslight.com

FIGURE 18.18 Follow spot manufacturers.

problem and also give you another way to change the size of the image at different times in the show without using a moving luminaire.

SHUTTER

There is that occasion when you absolutely have to use an HMI source but you also need to turn it on and off on cue. Consider a remote-controlled shutter. Phoebus Manufacturing has been making all sizes to match just about any luminaire you can name for years. Phoebus shutters (Figure 18.19) are DMX controlled for a smooth fade or quick blackout and can be powered with 120 or 240 volts.

FAT LIGHTS

Here is one of those new terms. *Fat light* has come to mean an extremely powerful light source that produces a beam that cuts through anything, even the largest follow spots. In years past, some of these would be called *searchlights*, as homage to the old movie premier lights stabbing bright beams into the night sky. Of course, these are now more likely to be an attraction at a discount furniture store. Fat lights can certainly do that, but at a much reduced footprint and with a non-carbon-arc light source.

One of this new group of high-power moving luminaires is the Morpheus Lights' BriteBurst 2000E (Figure 18.20). This is a 1200-watt HMI source with a 15-inch-diameter aperture that uses the company's

ColorFader3 color mixing system. I know of one designer who calls this the best "fat beam" available. There are many single heads (Figure 18.21), such as the Synchrolite MX-3000D, a 2000-watt xenon source, that all rotate or move in preprogrammed patterns. Skytracker has been a leader in this field for years; they are now part of the Strong International family. Figure 18.22 shows one of the most recognizable multiple-head units used for festivals and business openings along with concert use. The Skytracker series products are not designed for attachment to a flown truss. Strong is best known for their long throw follow spots. Most of these units use 2000- to 10,000-watt xenon lamps. Synchrolite has some units that can be attached to the truss and provide more features than simple rotation, as does Novalight from OmniSistem. New contenders are two units called the BigLite and the LittleBig (Figure 18.23) by ZAP Technology.

FIGURE 18.19 DMX-controlled shutter. (Photograph by Phoebus Manufacturing.)

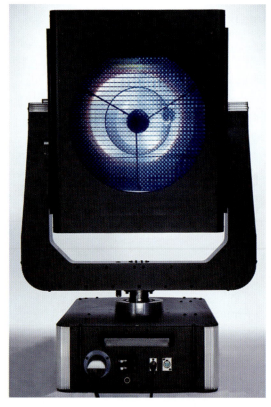

FIGURE 18.20 BriteBurst 2000E. (Photograph by Morpheus Lights.)

FIGURE 18.21 Synchrolite MX-4000. (Photograph by Synchrolite, Inc.)

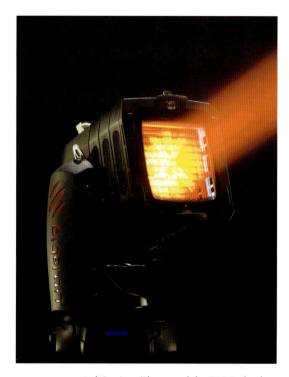

FIGURE 18.23 LittleBig Lite. (Photograph by ZAP Technologies.)

FIGURE 18.22 Skytracker STX4. (Photograph by Strong International Lighting.)

They are an unusual shape employing a single-arm design, and they provide CYvM color. Also placed in this category could be PRG's Bad Boy, High End's SHOWGUN 2.5 and SHOWPIX, Vari*Lite's VL3500, and a new entry, the Falcon Search Light.

See Figure 18.24 for a list of fat light and search light manufacturers.

COLOR SCROLLERS AND CHANGERS

Being able to create multiple colors from one luminaire is a very important part of moving luminaires. Showco, beginning with its first Vari*Lite luminaires, led the field in this area. Their approach was radically different from the old color wheel. Instead of the boomerang-type color changer, they used two dichroic prisms arranged to rotate inside to produce a palette of some 60 usatble colors. This capability has been expanded by Vari*Lite to, in their words, produce any color in the visible spectrum. Other internal methods for changing color have appeared. Some use two wheels with one having at least the primary colors (red, green, and blue) and a clear open space and another with at least the secondary colors and clear. This method is known as *additive color mixing* or RGB. CYM stands for the colors of cyan, yellow, and magenta and is the *subtractive color mixing* system many moving luminaires use to create color from "white" light.

High End Systems, Inc.	www.highend.com
Irradiant	www.irradianthg.com
Morpheus Lights	www.morpheus.com
OmniSistem	www.omnisistem.com
Phoebus Manufacturing	www.phoebus.com
Production Resource Group (PRG)	www.prg.com
Strong Entertainment Lighting	www.Strong-lighting.com
Syncrolite Entertainment Technology	www.syncrolite.com
ZAP Technology	www.biglites.com
Tei Electronics, Inc.	www.teilighting.com
Vari*Lite, Inc.	www.vari-lite.com

FIGURE 18.24 Fat light and search light manufacturers.

FIGURE 18.25 ColorRam IT. (Photograph by Wybron, Inc.)

Most *color string* color changers come in one of two forms. First, there is the scroller that allows the designer to request a string of colors from any media manufacturer to fill from 12 to 32 positions (leave one clear, of course) (Figure 18.25). The second is where dual premade color strings are designed to give the hue and saturation of the full spectrum by working in opposition to each other; however, this requires two control channels. At least two companies provide

this type of unit. Morpheus Light uses three strings in their ColorFader product to achieve CMY color mixing; in this case, three channels are needed to have full control (Figure 18.26).

One issue is fan noise. All of the units require cooling to keep the color media from losing its desired saturation. There is also noise when the units scroll from one color to another, so try to make such moves during applause or blackouts (people don't hear well in the dark!). Sizes vary to accommodate almost any luminaire, even large cyc and audience blinder units. Be careful, as specific plates that match particular luminaire accessory slots need to be ordered with the units.

A twist on these takes a somewhat different structural approach. The SeaChanger (Figure 18.27) is an internal color changing unit that uses three color dichroic wheels; it is placed between any lens train and the lamp and reflector on the ETC Source Four luminaires. One version already has a Fresnel lens assembly attached to the changer for a soft-edge effect. A four-channel DMX-512 with Remote Device Management (RDM) capability uses dichroic filters— CYMG (G for an added green filter)—to create an infinite color range. Another contender in this market is NexeraLX Luminaires by Wybron, a complete wash and two-zoom profile spot luminaire (Figure 18.28) that uses CMY color mixing via DMX-512 and RDM communication.

FIGURE 18.26   ColorFader3. (Photograph by Morpheus Lights.)

FIGURE 18.27   Dichroic profile spot. (Photograph by SeaChanger.)

See Figure 18.29 for a list of scroller and changer manufacturers.

STROBES

There are many styles of strobes ranging from low end to high end in cost, quality, and features. The StarStrobe by GAM Products (Figure 18.30) randomly flashes every few seconds; when several are linked together, it gives an effect similar to paparazzi camera flashes on the red carpet. Most tours that carry strobes want blinding power options. The

FIGURE 18.28   NexeraLX Profile. (Photograph by Wybron, Inc.)

Martin Atomic 3000 (Figure 18.31) is the most popular concert touring strobe because it has proven to be both powerful and rugged. It uses a xenon strobe lamp and offers such features as

Apollo Design Technology, Inc.	www.apollodesign.net
Luci della Ribalta	www.ldr.it
Morpheus Lights	www.morpheuslights.com
Pclights	www.pclights.co.jp
Sea Changer	www.seachangeronline.com
Strong Entertainment Lightings	www.strong-lighting.com
Wybron	www.wybron.com

FIGURE 18.29 Color scroller and changer manufacturers.

FIGURE 18.30 StarStrobe. (Photograph by GAM Products.)

FIGURE 18.31 Martin Atomic Colors. (Photograph by Martin Professional.)

variable adjustment of flash rate and intensity (dimmable from 0 to 100%). It is DMX controllable with preprogrammed effects, such as the continuous blinder effect. The Atomic 3000 has an optional color changer attachment with ten gels. The unit has a voltage auto detect from 100 to 260 volts. Diversitronics, Inc., specializes in professional strobe light manufacturing for the entertainment industry; their products ranges from their Finger Strobes for small spaces to the Mark series with three products and lamp options (2000, 3600, and 5000 watts). The company offers nine hyper modes with a single input

trigger and variable rates and intensities. Recently available is an ellipsoidal strobe module that converts a standard ETC Source Four by switching the lamp cap with a xenon lamp cap and using an external lightweight strobe module with DMX control for single flash, burst, and variable intensity and rate.

See Figure 18.32 for a list of strobe manufacturers.

[*Author's note:* There is a chance that the flash rate of a strobe can trigger epileptic seizures in someone viewing the effect.]

MOVING YOKE

This may be a small niche market, but when you need a mirror for a laser show that can be repositioned or you want a conventional luminaire to pan and tilt or maybe a mount for a television camera or

Diversitronics	www.diversitronics.com
GAMPRODUCTS	www.gamonline.com
Hungaroflash	www.hungaroflash.com
Irradiant	www.irradianthg.com
Kupo	www.stage.com
Martin Professional	www.martin.com
OmniSistem	www.omnisistem.com
SGM Technology	www.sgmtechnologyforlighting.com
Tei Electronics, Inc.	www.teilighting.com
Wybron	www.wybron.com

FIGURE 18.32 Strobe manufacturers.

projector, Apollo offers a unique side-arm unit (Figure 18.33). This can be a relatively, low-cost solution. In contrast, OmniSistem and City Theatrical use designs similar to standard moving luminaires.

PROJECTORS

Digital luminaires have started to replace fixed projectors for on-stage projection of live and recorded video images and effects, due to their smaller size, multiple features (particularly pan, tilt, and focus), and control server effects capacity, and road-ability. Spoiled by the digital luminaire's convenient, pan and tilt capability, and powerful options, fixed projectors are used mainly for image magnification (commonly referred to as I-MAG) screens off-stage and sometimes on-stage rear projection. Most of these units are very large. Video rental companies build double racks to accommodate two projectors to double the intensity of the images and so one can serve as a back-up. Most require a dedicated technician, and it is a very involved process to determine distance, adjust focus, and align the images of the two projectors to merge perfectly together. Even with variable lens sizes, finding space to place the units with the throw needed for the image size on-stage for rear projection has proven to be logistically difficult. From the front, this kind of projector has to be low enough with regard to screen height to stay within the parameters of keystone

FIGURE 18.33 Right-arm moving yoke. (Photograph by Apollo.)

compensation. The use of projectors for corporate shows, theatrical musicals, and ice shows is more of a possibility.

The leader in the ultra-big, nonvideo group for many years has been PANI Projection and Lighting (Figure 18.34). The company builds a full range of units that show large-format slides, endless loops, and special effects devices with excellent optics for crisp, clear images. However, they weigh in at 100 pounds plus. They use mostly 2500-watt quartz or 6-kW

GAM in 1983 which was also a deep-dyed color in a polyester base. His contribution was to add transmission curve and wavelength information with each color so designers could see how the colors balanced in transmission. Other countries were not asleep at the switch and color media can be found from Japan to Germany. What color media you like is determined purely by individual preferences, and many people mix and match between brands.

Three other ways to produce color in light have to be discussed here. First is the method that Vari*Lite first used to launch the modern moving luminaire industry; a 5600 K lamp was the source, and a series of dichroic prisms placed in the luminaire itself provided selective refraction. The beam-separating components offered a veritable rainbow of color choices. Other systems use two color wheels—one in or additive subtractive colors (RGB) and the other with secondary or subtractive colors (CYM) plus clear and possibly a diffusion slot—to offer a more limited rainbow of color choices than the dichroic prisms. Another use of dichroic filters is to rotate through a range of colors. At least one color scroller manufacturer goes beyond just placing selected color media on a roll. By combining two strings of color and using two control channels it is possible to overlay different combinations of the strings to get an amazingly wide range of colors. But, it comes at a price. The light must go through two media, and the reduction in intensity may be a problem depending on the application.

All of these methods are far more complicated than can be explained here. Many books are available to discuss the very important topic of the psychology of color. To the concert lighting designer, this is a major factor in the design.

There are about nine manufacturers of color media (filters and glass dichroic). The filters withstand the high heat that emits from the front of the luminaire; even so, while the more saturated colors may not burn a hole through like the earlier gels used to, they will burn the color out of the center of the media after extended use. The choice of color media is an important one that a designer must make to provide appropriate color options that will serve as the palette during the tour. More often, film and video applications require a full range of correction filters such as daylight, tungsten, arc conversion, and neutral density for light source control, all of which are also produced by the same companies. Even if there is no video carried on a tour, many promoters provide full camera crews and video screens. Normally, they will not come to you to change the way that you light your show. This presents an issue as to whether you and the artist's management want the 20-foot head of the star to look blue or have the correct skin tone. It will probably be up to you to adapt. Color correction filters should become part of your workbox so you are prepared for this situation.

GOBOS AND ACCESSORIES

Although acid-etched metal gobos have been around for a long time, the latest series of ERS units, beginning with the ETC Source Four, has a cool enough primary focal point to allow the use of glass gobos. This has opened a whole new world of color graphics and intricate designs not possible before. Custom-made gobos in either metal or glass are now common. These custom designs are also being made for inclusion in moving luminaires, some by the moving luminaire manufacturer. Rosco Laboratories' gobo wizard software allows the designer to create custom gobos using a computer and then print them on standard transparency film.

[*Author's note:* These transparency gobos only work as part of manufacturer-designated luminaires due to heat issues, so read the instructions before placing these in a standard luminaire.]

Another way to use gobos to introduce even greater fantasy and more patterns is to use a gobo rotator. Many companies make these in several versions—a single gobo rotating at a set speed or via DMX-512 control for variable speed. The TwinSpin by GAM Products allows two patterns to rotate in opposition to each other, at 2 to 15 rpm, controlled via a hand control or dimmer. Rosco makes an indexing feature that allows you to start and stop the gobo's rotation whenever you want. Other variations include the disc tray, loop tray, four-gobo tray, and six-gobo tray by GAM Products.

See Figure 18.39 for a list of gobo, color, and accessory manufacturers.

Apollo Design Technology, Inc.	www.apollodesign.net
GAMPRODUCTS, Inc.	www.gamonline.com
Goboland	www.goboland.com
Goboman	www.GoboMan.com
High End Systems	www.highend.com
Inlight	www.InLightGobos.com
Kupo	www.stage.com
Lee Filters USA	www.leefiltersusa.com
Omnisistem	www.omnisistem.com
Rosco Labs	www.rosco.com
SGM Technology	www.sgmtechnologyforlighting.com
Tei Electronica, Inc.	www.teilighting.com
YILONG	www.yi-optics.com

FIGURE 18.39 *Gobo, color, and accessory manufacturers.*

MAN LIFT

What does this have to do with lighting? A lot. If your crew isn't safe while focusing, then your design won't look the way you want it to for the show. Sure, a lot of the focus work today is accomplished by walking the trusses wearing properly certified fall production gear, but in many instances you must use a ground method. Large A-frame ladders are questionable because of the 6-foot rule … remember that you need to wear fall protection gear if you work over 6 feet off the ground. What avoids this issue is a certified man lift. One of the companies most often found in theatres and arenas is Genie Industries. When used properly, these are very safe for going as high as 36 feet. The Genie IWP Super Series (Figure 18.40) doesn't require outriggers up to 35 feet, 6 inches. Older versions require outriggers and can be difficult to place where needed when other equipment is on stage.

PERSONAL ACCESSORIES

The items below are things that the designer may find useful. Usually, they are one-of-a-kind, unique products that still must be recognized here.

FIGURE 18.40 Genie IWP Super Series. (Photograph by Genie Industries, a Terex Brand.)

Cases

At first look, why would a lighting designer care about road cases? Isn't it the job of the tour equipment supplier to protect their equipment for the tour? The simple answer is an absolute yes; however, you might want to consider having a case built for yourself and your master electrician. First, the master electrician needs to organize all kinds of spare parts, lamps, color media, patterns, plans, and paperwork such as circuit charts, dimmer schedules, etc. The designer normally wants a place to keep plans, paperwork, and … personal items, such as a photo of his dog, cat, girlfriend, boyfriend, baby … you get the idea—fun stuff, and maybe some candy for the crew. But, the real reason can be more utilitarian. What about having a place to set up your laptop and printer, to keep printer paper, to serve as a fold-out desk and chair to work on during load-in and focus? Designers need to be organized, and personal road boxes are the solution to this problem. There are lot of great road case manufacturers; Figure 18.41 is a photograph of a customized touring computer work station with table. This is just one example, but all of the companies will be quick to point out that theirs is a custom business—just tell them what you want and they'll create it for you. See Figure 18.42 for a list of road case manufacturers.

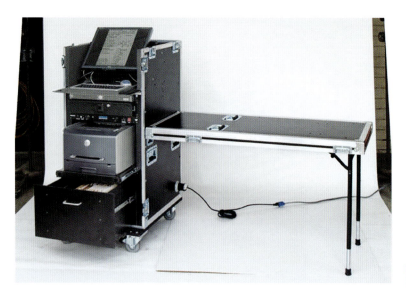

FIGURE 18.41 Custom road case workstation. (Photograph by Nelson Cases.)

Anvil Case Co.	www.anvilcase.com
Armande Cases	www.armandocases.com
A & S Case	www.ascase.com
Calzone	www.calzonecase.com
Encore Cases	www.encorecases.com
Federal Case Company	www.federalcasecompany.com
Janal Cases	www.janalcase.com
Nelson Case Corp.	www.nelsoncasecorp.com
New York Case Company	www.newyorkcasecompany.com
Stagegear, Inc.	www.stagegear.com
Viking Cases	www.vikingcases.com

FIGURE 18.42 Road case manufacturers.

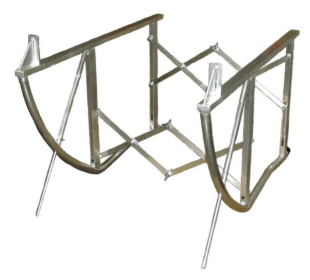

FIGURE 18.43 EZ-Tilt console stands. (Photograph by TMB Products.)

Console Cradle

No more setting up your expensive console on rickety banquet tables or stacking road boxes—the EZ-Tilt is a folding console stand that can hold up to 1790 pounds (Figure 18.43).

Color Checker

We use light meters in film and television and often a color meter to determine the Kelvin temperature so we can match different light sources, SeaChanger has a unique concept. The ColorBug is a wireless sensor that determines CIE color values or luminance and sends this information wirelessly to your iPhone or iPod. This device could be extremely helpful for checking that the color temperature is the same between groups of moving luminaires or follow spots.

DMX Lighting Control for iPhone

Synthe FX has developed a program that allows you to wirelessly control, view, and manage intelligent DMX lighting luminaires and consoles on your cell phone! Your iPhone or iPod Touch can use Artistic License's Art-Net protocol over the device's built-in Wi-Fi connection to communicate with other compatible hardware and software on your network. It can be used as a remote focus tool and provides quick channel setup for all of your DMX luminaires. A multi-track, touch-enabled mixer controls with accurate precision, whether you have one luminaire or a hundred. XY grid controls allow you to pan and tilt moving heads. The iPhone can also serve as a DMX data analyzer, as data can be displayed in real time as a touch-scrollable overview of all channels in the input universe. Selecting a specific channel shows the current value, as well as the past 24 seconds of data in graph form.

Launchers

There will always be special performances, such as on New Year's Eve or a national holiday, when the promoter or artist decides that a great ending to the show, short of indoor fireworks, would be to shoot off some confetti or streamers. Some are hand-held, one-time shots; others take CO_2 canisters and can be plugged into AC to be used over and over again. The SHOTMAX by CITC (Figure 18.44) is just such a reusable unit that can throw the contents 25 feet.

Cue and Desk Lights

There are many times when a headset intercom just won't cut it, and a visual signal would be better. TMB Products' ProCue comes as a single-unit LED luminaire that uses standard three-pin XLR cable to a 4-, 8-, or 12-way desktop station (Figure 18.45).

The lighting console probably came with built-in desk lights but somehow they never seem to reach over to your notes or script. The normal solution is to get a music stand light, or, worse, someone goes to the hardware store and buys a clip-on light and tapes some blue gel over the face. In both cases, the light probably can't bend down to where you really need it to shine and probably goes more into your eyes. So, do what the console manufacturer probably did—go to Littlite (Figure 18.46) and buy one of some 12 versions of their either incandescent or LED models, all of which come with nice goosenecks so you can place the light wherever you like.

FIGURE 18.44   Launcher, confetti, and streamers. (Photograph by CITC.)

FIGURE 18.46   LED gooseneck reading light. (Photograph by Littlite, Inc.)

FIGURE 18.45   Cue light system. (Photograph by TMB Products.)

19

SMOKE AND PYROTECHNIC EFFECTS

A very important ingredient in concerts (and television music award shows), especially with rock & roll and heavy metal bands, is some type of medium that allows the beams of colored light to be seen by the audience. This is especially important for *air light* or *air graphics*, or when moving luminaires are doing beam animation. The patterns created over and around the musicians would not be visible if we didn't have some means to accent the beams (Figure 19.1). *Theatrical smoke* or, more correctly, what appears to be smoke is actually not smoke created from combustible materials. Even though Rosco Laboratories received an Academy Award in the Scientific and Engineering category in 1984 "for the development of an improved nontoxic fluid for creating fog and smoke for motion picture production," the Actor's Equity Association resisted its use on behalf of its member dancers and actors, who felt that the smoke was damaging to their lungs (see Chapter 9). The Entertainment Services and Technology Association (ESTA) also joined the standards discussion and produced a pamphlet in 1996 called "Introduction to Modern Atmospheric Effects." New ANSI standards based on work by ESTA are currently under public review.

FIGURE 19.1 Smoke-enhanced visuals from Alicia Keyes' tour. (Photograph by Lewis Lee.)

SMOKE

Technically classified as a pyrotechnic as related to fire and safety codes, smoke is used in many theatrical, theme park, and corporate shows, as well as rock & roll shows. The release of solid particles into the air is controlled by laws in many locations. Smoke is lighter than air and tends to be thick and billowy; therefore, air currents take the smoke wherever the wind or the hotter rising air wishes to go—not necessarily where you would like it to stay. Placement of smoke units can be a real science. Remember that buildings often do not turn the air conditioning or heating on until the audience is about to enter. In addition to doors opening and closing, even the heat from the lighting will influence where the smoke is carried. Most large hotels, convention rooms, and, in rare cases, arenas consider smoke to be a fire and safety issue, because fire alarms will be triggered when any smoke is sensed. The use of smoke has to be brought up during the advance survey site visit to make sure the alarms can be disengaged during programming sessions and performances. In many cases, a fire marshal will have to be present to inspect the machines and often is required to stay until the show ends, creating another show expense.

There is hardly another product area that has so much competition, so use the provider data provided later in Figure 19.7 and contact makers for more details. Most of the high-end units are DMX-512 controlled. A few units have fans built in. Some units are designed to be flown on the truss accompanied by a fan to disperse the smoke (Figure 19.2). There are small battery-powered units, as well as large models enclosed in road cases. Chapter 9 provides a more in-depth discussion of studies concerning smoke and haze.

HAZE

There is a definite difference between *smoke* and *haze*. Haze is designed to keep a fine mist, often based on water droplets or a nontoxic chemical, floating in the air, especially above the stage, to allow the light beams to be seen without obstructing the performer or video screens. The idea is to suspend these

FIGURE 19.2 Smoke machine and fan with truss mount. (Photograph by CITC.)

FIGURE 19.3 DF-50 diffusion hazer. (Photograph by Reel EFX, Inc.)

particulates in the air for as long as possible. These units can also be DMX-512 controlled from the lighting console but are normally left to run for the entire production, unlike smoke. Reel EFX's DF-50 diffusion hazer (Figure 19.3) has been a leader in the field for many years and is widely used by touring lighting rental vendors. The fluid is used at a rate of less than 2 ounces an hour and produces a totally odorless haze that can only be detected visually. The system utilizes a triple-filtered system that breaks down the diffusion fluid (food-grade mineral oil blend) to

FIGURE 19.4 Arena Hazer in case with built-in fan. (Photograph by Chauvet.)

FIGURE 19.5 Low fog generator. (Photograph by Antari Lighting & Effects, Ltd.)

uniform 1-micron droplets or smaller. The unit is self-contained, requiring only the proper power rating and a sufficient amount of fluid. This unit, however, does use a chemical and has become the subject of discussion regarding the safety of haze (see Chapter 9).

Chauvet has a large touring unit called the Arena Hazer, which has a built-in fan (Figure 19.4) and is already in a road case for travel. Most, but not all, hazers use *glycol*, which is an alcohol; a few hazers use only water. The idea is to get the droplets to form either as a cloud (fog) or as a fine mist (haze) of fine, widely dispersed solid or liquid particles, which gives the air an opalescent appearance. How long these droplets stay in the air is the *hang time*.

FOG

Fog is described by weathermen as a "cloud that touches the ground." Theatrical fog tries to imitate that effect by having a billowy white cloud stay close to the floor. For many years, theatres and tours relied on a modified 50-gallon drum that had a dryer hose sticking out of it. A basket was used to dip dry ice (CO_2) into a drum filled with hot water after the slab of dry ice was broken up with a hammer. A small squirrel-cage fan was then turned on. This method presented a few problems. First, dry ice is expensive; second, it cannot be handled except with heavy gloves or the crewmen risk third-degree burns; third, as the water in the drum cools, it does not melt the dry ice as fast; fourth, well, you get the idea. It's a pain in the a--.

With all that going against it, you would think that great minds would have solved the problem quickly, but, no. There were attempts to take some of the dreaded dry ice and stick it in front of the chemical fog nozzle, cooling the fog to keep it near the floor. These were pretty feeble. Not until very recently have large fog units been adapted with cooling units. Antari has introduced the DNG-200 Low Fog Generator (Figure 19.5), which uses no dry ice but instead a heat exchanger and an air compressor to produce a large volume of ground-hugging low fog.

[*Author's note:* Almost all smoke, haze, and fog machines require the user to purchase a specific fluid to use in the manufacturer's machine. *Do not* use a fluid that is not specified for your unit. Read the labels carefully.]

Another approach that has proven effective uses only water. Interesting Products has marketed their liquid nitrogen fog machine (Figure 19.6) for years and has a proven track record of large output, reliability, and safety. This system is ideal for singers, as

FIGURE 19.6 Liquid nitrogen fog machine. (Photograph by Interesting Products, Inc.)

CITC	www.citcfx.com
City Theatrical, Inc.	www.citytheatrical.com
Elation Lighting	www.elationLighting.com
Fog Factory	www.fog-factory.de
Haze Base	www.hazebase.com
High End Systems	www.highend.com
InnerCircle	www.icd-use.com
Interesting Products	www.interesting-products.com
Irradiant	www.irradianthg.com
Kupo	www.stage.com
Le Maitre Special Effects, Inc.	www.lemaitrefx.com
Look Solutions	www.looksolutions.com
Martin Professional	www.martin.com
MDG	www.mdgfog.com
OmniSistem	www.omnisistem.com
Reel EFX	www.reelefx.com
Rosco Labs	www.rosco.com
SGM Technology	www.sgmtechnologyforlighting.com
Tei Electronics, Inc.	www.teilighting.com

FIGURE 19.7 Smoke, haze, and fog manufacturers.

it creates nothing but water—no chemicals, no fog fluids, no toxic odors. With dry ice, the vapor dissipates as the water becomes warmer; however, the nitrogen stays at the same temperature and pressure. It is good for very complicated uses, such as multiple positions from the same unit. The unit has not only DMX but also ACN and Remote Device Management (RDM) feedback signals. One cautionary note: Do not place personnel in an enclosed area with nitrogen fog without access to air.

See Figure 19.7 for a list of smoke, haze, and fog manufacturers.

FANS

Most of the effects created by fog, hazers, and CO_2 or LN_2 require a little assist to direct the effect to the desired position on the stage. Some units have built-in fans, but most require assistance. Several fans designed specifically for the theatrical market have DMX control. Reel EFX's RE-Fan II Turbo (Figure 19.8) is one example of such a unit. Whatever you use, always be aware of smoke detectors and alarms, especially in hotel ballrooms. They can be set off by any of the smoke-, haze-, or fog-producing agents. Second, even though all are said to be nontoxic, they all carry warnings that irritation to eyes and lungs can occur in sensitive persons. Be sure to check with the artist before using these methods to generate smoke, haze, or fog. As stated before, in many states and municipalities, they are considered pyrotechnic devices, and you will be required to obtain a permit. See Figure 19.9 for a list of fan manufacturers.

PYROTECHNICS

There is not much to add to what has been stated in Chapters 2, 8, and 9. This is a highly specialized area of expertise that requires special training and in many cases a license. Do not think that fireworks that you can buy on the street are usable on the stage. Even smoke bombs or other devices that only produce smoke are not acceptable. The all require ignition and are therefore a pyrotechnic. Do not take the chance that the band Great White did. It simply is not worth it.

FIGURE 19.8   RE-Fan II Turbo. (Photograph by Reel FX, Inc.)

CITC	www.citcfx.com
Fog Screen, Inc.	www.fogscreen.com
Kupo	www.stage.com
Le Maitre Special Effects, Inc.	www.lemaitrefx.com
Look Solutions	www.looksolutions.com
Martin Professional	www.martin.com
OmniSistem	www.omnisistem.com
Reel EFX	www.reelefx.com
Wybron	www.wybron.com

FIGURE 19.9   Fan manufacturers.

20

PROTOCOL, CONTROL, AND ANCILLARY ENABLERS

PROTOCOL

DMX

The Digital Multiplexing (DMX) interface is a low-voltage signal that distributes data between the control console and dimming. Its use in lighting was initially a simple way to use any console with any dimmer, a standard that originated in 1986 by the USITT. However, DMX uses grew beyond anyone's initial intentions, and it became the primary method not only for linking a console to a dimmer but also to link multiple signals sent and received from an offstage or truss location to entire lighting rigs of moving luminaires, hazers, strobes, light-emitting diode (LED) fixtures, and digital lighting.

Each DMX universe is 512 channels and is wired using a 3-pin or 5-pin connector. DMX is the official universal protocol for networking concert lighting as developed by the United States Institute for Theatre Technology (USITT) and the Entertainment Services and Technology Association (ESTA) and codified by the American National Standards Institute (ANSI).

Because DMX can be linked from one fixture to another in a "daisy chain" configuration, it reduces the amount of cable running to the distribution location. Each light does not have to have a separate line, but the link cannot exceed 512 channels, either. Some moving luminaires can have 20, 30, or more attributes that must be controlled, and once the maximum of 512 channels is reached another universe is required to control anything further. There is no limit to the number of universes that can be used, assuming that the capability is available via the lighting console or other device.

With a large lighting rig, or even a small one, comprised of widespread fixtures that require DMX control, it is not always possible or practical to daisy chain the entire system. The daisy chain can give rise to several troublesome issues when linking multiple DMX runs together. A lot of time can be wasted troubleshooting problems that often occur, even for a simple point A to point B DMX cable run. One bad cable in a daisy chain can disable an entire row of fixtures. Ways to converge DMX from multiple locations are discussed below in the Ancillary Enablers section.

ACN

The Architecture for Control Networks (ACN) is the new kid on the block that may replace DMX as the control protocol for lighting systems. It was developed by ESTA and its development body, Control Protocols Working Group. ACN is a suite of network protocols for high-speed, bidirectional communication over standard Transmission Control Protocol/Internet Protocol (TCP/IP) on an Ethernet network infrastructure.

A primary goal of ACN is to provide a reliable transport mechanism, which it does through Session Data Transport (SDT), which transports data to ACN lighting components and devices. Once a session has been initiated in SDT, data are efficiently packaged in varying types and sizes and transferred back and forth. ACN also includes Device Description Language (DDL), which is a text language that describes features of a device—for example, a moving luminaire with several attributes. DDL allows devices to tell controllers how they would like to be described and controlled, which eliminates the need for luminaire libraries that currently are separate attributes, such as pan, tilt, and focus.

ACN is most usable for controlling more complex devices such as media servers and audio mixers. It has also been proposed as the primary transport for HD-MIDI, a protocol that enables electronic musical instruments, computers, and other equipment to communicate with, control, and synchronize with each other. The gist is that ACN is an open-ended suite of protocols that is cleverly packaged and effectively used by network devices to transfer data with greater and more adaptive control in theatre applications. ACN provides a fast and efficient mechanism to transport the well-understood DMX protocol over Ethernet in an open, industry-standard way to control a variety of entertainment systems.

RDM

Remote Device Management (RDM) is a protocol enhancement to DMX-512. RDM works along with DMX-512 to provide two-way data communication allowing for remote device management. It uses the same wire pair as DMX-512 but requires bidirectional devices in the data path for the RDM traffic to get through. In other words, the luminaires or other devices must also allow for such two-way data to be transmitted. Most of the newer lines of luminaires and consoles are adding this capability. Basically, this allows you to use remote-control devices in the DMX line with all of your other controllers without interrupting them. This protocol allows configuration, status monitoring, and management of these devices. It has a small impact on the DMX refresh rate, but should not disturb the normal operation of standard DMX devices that do not recognize the RDM protocol. RDM was approved by ANSI in 2006 and is rapidly gaining popularity.

ANCILLARY ENABLERS

The idea is that we want a trouble-free, streamlined system that uses as little cable as possible. There are around 14 manufacturers that develop and produce reliable and convenient electrical boxes designed to split, isolate, adapt, convert, interface, and distribute

FIGURE 20.1 RS-485 splitter. (Photograph by Martin Professional.)

FIGURE 20.2 Truss-mounted Optosplitter. (Photograph by Avolites, Ltd.)

DMX-512 transmission signals. We refer to these as *ancillary enablers*. More computer protocols integrating into lighting control mean more electrical boxes to contain, manage, and organize the amount of DMX cable running to distribution locations. These ancillary enablers are essential tools that improve control, data efficiency, and cable management in concert lighting.

Optosplitter

Because DMX-512 is an electrical (electronic data) signal, it must be isolated from other data lines and protected from any power surges than can damage expensive lighting and other DMX controlled devices. The Optosplitter is a small electrical box (Figures 20.1 and 20.2) with multiple outputs that provide data line separation for DMX receivers such as dimmers, color changers, and moving lights. This is a *star* configuration as opposed to a *daisy chain* configuration. In a star configuration, each control cable is run to a central point—in this case, the splitter. In a daisy chain configuration all the devices are connected on one control cable, the output of one feeding the input of the next.

Terminator

A terminator is a resistor that matches the impedance of the cabling used to join pins 2 and 3 of the connector. It is a small, pocket-size, 3-pin or 5-pin connector that plugs into the DMX-512 out connector at the end of a DMX chain. Without a terminator, signals can reflect back down a cable; the initial signal combined with the reflected signal causes it to see two numbers at once, resulting in a faulty signal.

Adapter

An adapter is an enabler. It is a short cable (usually about 6 inches) with a 3-pin connector on one end and a 5-pin connector on the other (or a 5-pin to 3-pin double connector part), and it is used for switching cables at the distribution end or the receiving end. Adapters can be a timesaver if you have accidentally run the DMX cable the wrong way, or they might be necessary when using 5-pin cable but the Optosplitter has a row of 3-pin outlets.

The DMX-512 standard provides for 2 data links in one cable, which only uses 3 of the 5 pins for data, making a 3-pin adapter possible for a 5-pin DMX cable. Line 1 is on pins 2 and 3, line 2 is on pins 4 and 5, and pin 1 is a shield for the cable. A cautionary reminder: DMX standards require the use of 5-pin connectors for lighting cables so they will not be confused with the 3-pin cables that sound systems use. Sound cable has different impedance and capacitance specifications for data bandwidths required by DMX-512. Among a few adapter options not listed here, every lighting tool box should have a selection of adapters or spare connectors that can easily be used onsite to accommodate any adapter situation.

INTERFACES

An interface is an enabling electrical box for those times when a device is not DMX-512 compatible.

FIGURE 20.3 CTI's SHoW DMX wireless DMX. (Photograph by City Theatrical, Inc.)

There still are a few old proprietary protocols being used in the lighting market. If the owners haven't upgraded those devices, it would be impossible to use a DMX console with a proprietary dimmer protocol, for example, without an interface that was specifically designed to make the necessary conversion.

Wireless DMX

Considering that most front-of-house lighting console positions are between 100 and 300 feet from the backstage dimmers and power distribution centers, the advent of wireless DMX is certainly a welcome development, saving time and carrying long runs of cables. It works by having the DMX-512 control data from any standard DMX-512 console output attached to the wireless distribution system (WDS) transmitter. The transmitter broadcasts the signal to the WDS receiver (or receivers). The WDS receiver then outputs the signal as standard DMX-512 data, and the unit can be connected with standard cables to WDS dimmers or other devices, such as moving luminaires or effects.

With so many wireless computers, cell phones, and wireless microphone frequencies competing for air space in a venue, indoors or outdoors, what if the frequency that controlled an entire lighting system dropped out or was interrupted in the middle of a show because the frequency was overcrowded? City Theatrical, Inc. (CTI), thought of that eventuality and in October 2008 was granted a U.S. patent for its wireless technology (Figure 20.3). CTI's patent covers the transmission of DMX, RDM, and ACN via frequency-hopping spread spectrum (FHSS) radios, as well as direct sequence spread spectrum (DHSS)

21

THE DESIGNER'S WORKBOX

SOFTWARE PROGRAMS

Anything that you could possibly dream up or plagiarize from casual observations for new lighting or stage sets and animation cueing can get a test run in virtual space by using computers and available computer-aided drawing (CAD) and three-dimensional (3D) rendering programs. About 18 software programs are available to draw concert lighting plots (Figure 21.1), and some of those allow (by using plug-in software) the drawing of 3D models to react in real time and recording cues using your lighting console. Like anything else, the higher the quality and the more options it has, the more money it will cost, although it is possible to begin with inexpensive, basic but very effective programs to plot a lighting rig. A few that are accepted as industry standards are featured in this section as examples of the power of these programs.

With any of the lighting or crossover 3D drawing programs, you can use the tool bars to manipulate the resource library and bring your ideas to life. Software has been modified to the point where the user does not need to know any complex programming, and designers can become fairly proficient by working through online tutorials and some devoted practice. It always helps to have someone show you shortcuts instead of laboring through complicated instruction manuals. It is fantastic and easy—just bring your imagination, experience, and some patience to work through understanding the program's logic. Once you do, new worlds will open up.

CAD Programs

One of the reasons for having industry-standard lighting software is simply to keep us all on the same page—literally. A designer can send a work-in-progress VectorWorks, AutoCAD, or WYSIWYG file electronically and exchange ideas across the miles to anyone involved in the build process, which allows them to work on the same drawings. File conversion options are built into the software, enabling you to send (export) or receive (import) drawings to suit your program. A CAD program and license for personal or company use will be upwards of $1800 for a VectorWorks, $3500 for basic AutoCAD, $1000 for basic WYSIWYG, and $5000 or more for advanced programs that move into the 3D visualization realm.

To receive drawings and to draw in AutoCAD or any other proprietary software, you need to buy the license to do so. Even if you don't use CAD programs, other simple alternatives exist for exchanging work-in-progress drawings. Just visit the website of the software manufacturer and download a read-only viewer program, or drawings can be converted by the sender to the more widely used PDF format. In any case, these programs have evolved with the latest advancements in software technology and continually update their resource libraries with new lighting equipment in each new annual release.

After you have collected all of your sticky notes and pieces of scratch paper, the CAD drawing is the

FIGURE 21.3 Screen visualizer example. (Rendering by Joe Cabrera.)

way an assistant keeps track of the focus, work notes, and other details that are necessary while the show is in technical rehearsal. A unique feature is that information, such as a color change or circuit number, can be typed into LightWright and then read back into VectorWorks and the plot will change. The new version also makes it easier for the assistant to send notes to other crewmen so they can be ready to act at the next call. (See Figure 21.6.)

Virtual Magic Sheet

Virtual Magic Sheet (VMS) is a new program that has no current competition. The concept by Eric Cornwell of West Side Systems is truly unique. Almost all light plot programs will produce paperwork, and some will even give you a *magic sheet*, which is what designers use during technical rehearsal to keep track of the luminaire, circuits, color, and function. Up until now, that was a piece of paper, maybe hand colored. VMS allows you to do the layout on your computer: group like luminaires, name the group, and indicate the channel. It also gives a representation of

the color. Next, it will show scrollers, and, if you tell it the color string, it will show that, too. If you are using moving luminaires, tell it the color system (say, CYM), and it will change color and show the rotation of the luminaire. After you do the layout (which you can change at will), you can attach it to the console and it will graphically show you the intensity of the conventional luminaires, the position of the scroller (color), and, for moving luminaires, not only color but also the direction it is facing. It is a completely interactive magic sheet. The program is marketed by Goddard Design Company. (See Figure 21.7.)

Other Programs

Several other lighting programs are available, such as FocusTrack. The inventor, Rob Halliday, has said that its main uses are for tours or long-running productions. It is a light management system that can also incorporate photographs to show how the light should look. Field Template's SoftSymbols, Version 2, represents an evolution of CAD lighting symbols and

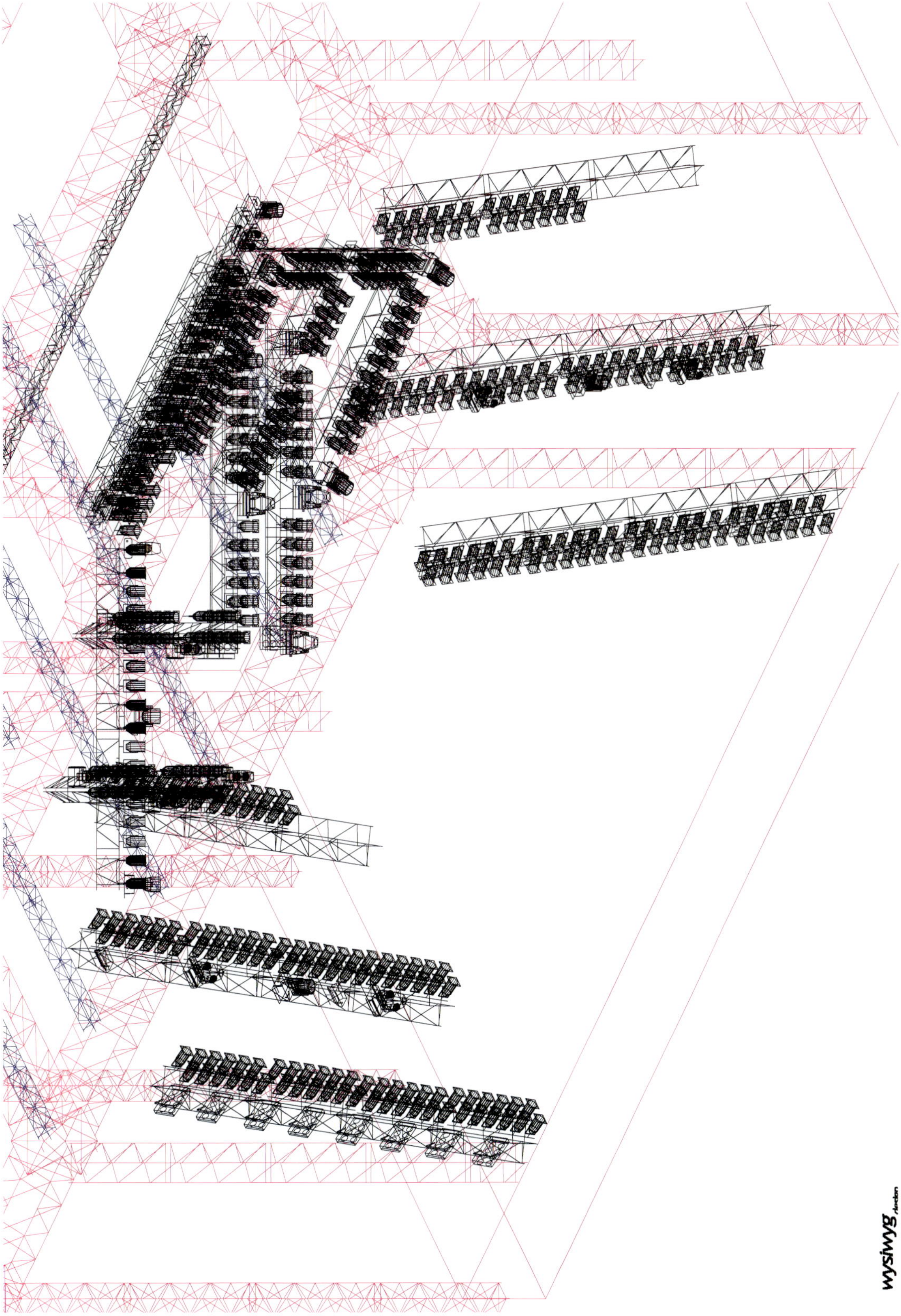

FIGURE 21.4 WYSIWYG wire frame perspective. (Drawn by Peter "Lumini" Johannsen.)

exactly as they would be in reality and relative to the stage dimension and height. You can assign all luminaires (moving and conventional) with DMX numbers and then see the effects of the lights on various surfaces (e.g., reflective or dull), view content from the media server, focus on 3D performer models and the shadows cast, and experiment with graphic light beam movement in the air by adjusting the thickness of the haze.

The advantages include savings in time and money by being able to critique a virtual rehearsal with full production luminaires, audio, and staging (the difference in cost between virtual and real is vast); the ability to label your console, select and record luminaires, focus groups, and palettes (color and gobos); and creating, practicing, and recording scenes without time pressures, interruptions, and the noise of a full production rehearsal. Visualization programs, however, should not be considered as complete substitutions for a physical hands-on lighting system build. Practice and interfacing with all of the other components involved in a live rehearsal situation will reveal differences from what was viewed on the computer that might require color, timing, and focus adjustments.

For larger productions, it is essential to request and conduct a live rehearsal. It would be safe to bet your last dollar that the lighting, video, scenic, and audio coming together for the first time before a concert tour will lead to unforeseen logistical problems that must be worked out. They could be complex, requiring new pieces to be made or some to be eliminated entirely. Nonetheless, visualization programs identify many problems prior to a major rehearsal and provide a constructive head start to finishing the live rehearsal race for time.

VISUALIZATION PROGRAMS

Among the handful of industry-standard visualization programs, Cast Software's WYSIWYG was the first to allow a DMX console or a compatible offline editor to be connected to the computer and WYSIWYG. Today, the company has secured the lion's share of name-brand lighting console manufacturers (about 35). WYG-it 2 is the hardware interface device used to initialize simulation software, enabling the user to connect a console to WYSIWYG and the physical lighting equipment. ESP Vision, by ZZYZX, Inc., provides a lighting simulation program that uses VectorWorks' Spotlight or RenderWorks programs as the basis for their plug-in software, Vision 2.3, along with their hardware interface, VBox. Martin ShowDesigner (MSD), Martin's Maxxyz console visualizer, is another proprietary onboard program. Avolites' Diamond 4 Simulator and Off Line Editor run on a good-specification MS Windows XP PC. Every function looks and operates exactly as the actual console would, and there is the additional benefit of the built-in Avolites Visualiser. The Avolites Stage Visualiser simulation program combines software and hardware, with drop-and drag-luminaire positioning (no need to put in the truss). Sound complicated? Avolites claims that the program is extremely easy to use and quick to learn (typical learning time, 30 minutes).

Although learning times may be short, calling them easy to use is something of an understatement. The basics of the programs might be easy to learn, but only by investing your time in these programs will you be able to create layouts for senior lighting and scenic designers or become proficient enough to produce timely reports and 3D renderings that interface lighting simulators with lighting consoles. Like anything else, once you've done the work, then you can claim that it was easy.

VISUALIZATION STUDIOS

With about 10 locations in the United States and 15 in Europe, visualization studios have become a cottage industry. Lighting console manufacturers have steadily increased the number of onboard simulators and offline programs available as standard features. You would expect that visualization studios would become a thing of the past as more designers and programmers gain access to onboard simulator programs, but that is not the case. Currently, simulators are not being used fully, and then there is still the problem of having the available time and talent required to render 3D luminaires and stage sets. Programmers

FIGURE 21.8　Inside Pre-Lite Studios. (Photograph courtesy of Pre-Lite Studios.)

mostly use the onboard simulators as a focus aid and to correct timing issues that may be lost on stand-alone visualizers. Onboard lighting console versions are generally best for very basic representations.

Studios are wired to computers with advanced visualization program plug-ins and large screen displays in purpose-built, controlled environments. Creating 3D modeling, maximizing use of the programs, and staying ahead of the equipment changes has led to the biggest addition to virtual programming—media servers. Tom Thompson, cofounder of Pre-Lite Studios, located in San Francisco and New York, explains: "We use media servers as both programming tools; as normal—output shown within visualizer via a capture device; and as a visualizer tool— providing a composited output of several images that are then shown in visualizer." It turns out that concert designers do not represent the biggest portion of Pre-Lite Studio's business; Thompson breaks it down to 50% corporate shows, 15% architecture, and about 35% concerts for his company (Figure 21.8).

The second alternative to fixed studio location work is a transportable visualization studio. Pre-Lite Studios can take their packaged systems to the client, providing many of the studio amenities with their onsite

systems, including high-tech computers, projectors and screens, and technicians. On the other end of the spectrum, larger operations like the Production Resource Group (PRG) offers designers the use of visualization studios in Los Angeles and London as part of the overall package for using their lighting company. At these sites, they provide 3D production modeling in-house so a designer can walk in and start programming.

Between the capital outlay required to buy quality software and the time it takes to model in 3D, it is an upside-down equation for many designers, and visualization studios are actually seeing an increase in business. However, once lighting designers become more familiar with visualization programs or, new software releases feature even more time saving and earier steps, we may see this dynamic change.

Other Tools in the Toolbox

Another section of your designer toolbox should be devoted to helping you learn how to sell yourself and then how to effectively work with the rest of the design team, artist, management, and crews.

Arkaos	www.arkaos.net
Avolites America	www.avoam.com
Cast Software	www.wysiwygsuite.com
City Theatrical, Inc.	www.citytheatrical.com
CM Industries	www.cmrigging.com
Daslight	www.daslight.com
Design & Drafting	www.ldassistant.com
DMX Creator	www.dmx512.com
DmxSoft	www.dmxsoft.com
Field Template	www.fieldtemplate.com
Focus Track	www.focustrack.co.uk
Future Light by West Side Systems	www.future-light.com
Jands	www.1.jands.com
John McKernon Software	www.mckernon.com
Light Converse	www.lightconverse.net
LumiDesk	www.luxart
LuxArt Concepts, Inc.	www.luxart.com
LxDesigner	www.lxdesigner.co.uk
Madrix	www.madrix.us
Martin Professional	www.martin.com
Phillips SSL Solutions (Color Kinetics)	www.colorkinetics.com
Rosco Labs	www.rosco.com
Shock Solutions	www.shocksolutions.co.uk
Stage Research	www.stageresearch.com
Star Draw	www.stardraw.com
Strand Lighting	www.strandlighting.com
Sunlite	www.nicolaudie.com
Tei Electronic, Inc.	www.teilighting.com
Thematics, LLC	www.seelightbox.com
VectorWorks	www.nemetschek.net
Zero 88	www.zero88.com
ZZYZZ, Inc.	www.espvision.com

FIGURE 21.9 CAD, visualization, and paperwork programs.

There isn't enough room in this book to cover two often-skipped aspects of design: interpersonal communication and salesmanship, but these are exactly what you need to learn (see the Bibliography for more information). You may become the creator of the greatest design that never gets seen if you aren't an expert at these two important elements of design. I have witnessed it too often—a designer who does not have any technical grasp of the equipment but can sell himself better than someone who comes in with a great workable concept but doesn't know how to sell himself. Don't let your career languish and don't become discouraged because you have not taken advantage of all the tools in your toolbox.

Another important tool is being connected in your profession. You can learn a great deal by attending conferences or joining organizations whose publications will keep you up to date on the latest lighting designs, people who are doing the work, and the newest equipment. Most of these organizations also hold seminars throughout the year, regionally or at their annual conferences, that can introduce you not only to new products but also to new contacts that can be invaluable to your career. Take advantage of the wealth of knowledge offered. Yes, a number of top designers choose not join these organizations or attend their conferences. That's their choice; however, I believe that any knowledge I gain is an asset to my career.

My suggestion is if you are still in college to join these organizations at a reduced student rate to see how you like them. Then, after school, you can choose to continue or not. For designers who have not participated in college, the prices can be stiff, but attending at least one Live Design International (LDI) or United States Institute for Theatre Technology (USITT) conference is an investment well worth the expense. Figure 21.9 provides a list of these and other organizations.

Finally, there is an overabundance of commercial publications, as well as journals from the many organizations, that you should keep up with. Fortunately, many of the commercial publications are free if you fill out a survey form. Subscriptions to publications by the various organizations are included with paid membership. These are valuable tools for the designer because many do feature stories on current tours that give you a lot of insight into the design and equipment used, how the designer went about creating it, who the tour suppliers are, and much more. This is an armchair way of seeing a lot of tours without spending the time or money to attend. True, there is nothing like being there—feeling the excitement, seeing how the cues work together to create a total connection between the audience and the artist, what the lighting designer adds as the layer over all that the music is saying—but not many of us have the time to go to that many places. I would say to young designers, though, that it is valuable for you to go to as many shows as possible. I know it is hard on the pocketbook now but what you will learn will prove to be invaluable down the road.

22

FESTIVALS, FAIRS, RACETRACKS, AMPHITHEATRES, CASINOS, AND LOCAL LIGHTING EQUIPMENT

FESTIVALS

Festivals have contributed an immeasurable range of venues to the worldwide list of possible performance spaces. They can include one-time community events that hire artists, local or famous, to raise money for a charitable cause, such as victims of Hurricane Katrina or the Ohio flood. Or, festivals can be annual worldwide events that draw thousands of fans over the course of several days, such as Live Aid or Farm Aid. Heavy metal, jazz, blues, rap, pop, Christian rock, country music—festivals that are a growing but seasonal venue, primarily active during the summer months. The Sweden Rock Festival (Figure 22.1) is held at a dedicated site about 3 hours north of Copenhagen in Sölvesborg. It started in 1992 with 9 bands; in 2008, 75 bands performed. This well-known festival is but one in 50 for Sweden alone (see www.festivalinfo.net). Milwaukee Summerfest is touted as the World's Largest Music Festival, spanning 11 days with performances on 13 stages. Not new to the genre, the Country Music Association (CMA) celebrated its 37th annual festival in Nashville in 2008. The 4-day festival attracts worldwide broadcast media and newspapers and 52,000 fans per day to see 106 acts perform for more than 34 hours.

These festivals are by and large extremely well organized, and the bigger annual festivals are often staffed with year-round employees. A touring group and entourage can expect that all of their production needs will be met and exceeded; however, only the headliners will be able to enjoy a festival-style lighting design to accommodate their production. The main reason is that darkness generally falls at around 9:00 p.m., even later in the northern hemisphere. If a festival is designed around a touring headliner (of the caliber of Kiss or Aerosmith), the bands performing before sunset will only be allowed to use a portion of their touring system. The poster shown in Figure 22.2 is from a German rock festival where both groups performed in 2008.

A festival-style lighting design is a generic design of five primary, full-stage, color PAR washes; usually between 6 and 12 ellipsoidal reflector spotlights (ERSs) for key light specials; a few bars of 4 aircraft landing lights (ACLs); several 8-light audience blinders; and 12 or more moving luminaires, normally a mixture of profiles and washes. There normally will be a straight 12-inch box truss upstage on two motors for backdrops and sometimes corporate sponsor logo signs that can be easily moved in and out to change with backdrops that the various groups carry with them. The festival system is intended to provide a generic lighting arsenal for all visiting lighting designers to use. With so many visiting bands and lighting designers, one generic system for everyone is the most cost-effective method for maintaining a lighting rig during the course of a festival season. There will be very few color filter changes, no truss configuration changes, and, consequently, no rigging changes. In addition, if the touring artist does not have a lighting designer, usually the lighting vendor's staff designer, who more than likely is the person who designed the system, will be appointed to create and operate the festival's system for that artist.

The festival lighting systems can be adapted or customized by limited conventional lighting color will be, power for audio situated near their distribution center, and troughs that make it easy to lay control cables to front-of-house lighting and sound positions. Most have catering rooms with fully equipped kitchens and ample room for chairs and tables. The dressing rooms are close to the stage, and many have showers and high-quality bathrooms.

Amphitheatres generally do not have in-house systems. Very few compromises need to be made in this type of venue, and generally the entire production can be used with room to spare. Controlling the atmosphere is sometimes tough due to the inevitable wind factor inherent to outdoor events. This can generally be countered by placing extra fans and haze machines on stage. Almost all amphitheatres have covered stages and roofs over part of the audience (usually the first 5000 seats before a grassy incline and festival-type lawn seating), which work as barriers to keep wind out of the performance space.

Aside from containing the haze atmosphere during a show, daylight lighting focus is probably the biggest challenge at amphitheatres, particularly for moving luminaires. You can stare into a conventional light even in the brightest sunlight and still accomplish a fast and accurate focus, but if there are set pieces and backdrops moving luminaires are more difficult to accurately focus. Roof shade often helps, but some of the amphitheatres face west, and direct sun streams into the stage area late in the afternoon. Overall, though, you can easily imagine that a tour of amphitheatres will make road crews generally very happy.

Casinos

Not all of the casino operations have new entertainment complexes, but casinos generally pay handsomely for the bands, more than a concert promoter would guarantee, and a band's management, based on that premise, is likely to add casino dates to a tour. But, some of the venues can barely be considered controlled performance spaces, and production riders will be minimally satisfied. The managers often just look the other way because of the big payday for the band. Sometimes the performance space will be larger conference room conversions with stages. Many of these can present real challenges for a touring lighting designer

trying to adapt a show to a 12-foot ceiling and barely enough space to fit band equipment or staging. If the drummer stands up on a riser, the sticks can touch a luminaire; if that doesn't happen first, the performer will surely feel the heat from the luminaires.

Contributing to the growing pains of casinos trying to be performance venues, often the room conversions and even new builds are designed by architects that do not understand that stage lighting dimmers should be separated from the architectural lighting. The house lighting control will take up several dimmers and DMX channels. The stage lighting system will take up the rest, leaving just a few spare channels, if any. If touring lighting designers have any lighting or effects of their own to augment the house stage lighting system, one can expect that a few wall sconces or house lights will overlap DMX channels, and they will flash on and off with the stage luminaires. It will take time to figure out a way around these problems; however, it can be done (while you are muttering under your breath), and expect that each situation will bring new and unique challenges.

Harrah's Casino, one of the largest chains that have arrangements with Indian casinos, along with a few other independents, are also promoting outdoor summer festival seasons and using their large parking lots to erect outdoor roof and stage systems. At outdoor shows, inclement weather is always a possibility, which makes this performance condition less than ideal because there is very little additional coverage beyond a four-post roof design. The parking lot could be on a slope and it is common for street light poles to be in front of or near the stage. The follow spot towers are hardly towers at all, and the front-of-house position has been known to barely be high enough to clear the tops of the audience's heads. As an aside, this should be addressed during the advance of the show, but the person doing the advance has to know to ask the question. There have been cases when the promoter has asked that the band forego the contracted amount of performance time and cut their song set short so the casino personnel can direct the audience back into the casino sooner. (All time away from the casino is a loss to the owners.) This is the attitude you have to work with, and you have to do the very best that you can under the circumstances. Playing these types of venues is like a world-renowned band warming up for the casino, which is the headliner.

LOCAL PRODUCTION EQUIPMENT

Local lighting and production companies come and go. They have to compete against large companies in major cities, or they may be the only one in a small regional market or town. Usually, concerts are only a small part of their business because they make their major income from supplying a wide range of everything from exhibits and conventions to corporate meetings and parties that stage entertainment events. The happy consequence is that there is no shortage of ways for the lighting industry to expand; *everyone* needs light, and entertainment lighting is in high demand. Local lighting vendors are simply providing the equipment to accommodate the demand, and diversifying is just smart business.

The main reason for touring groups to use local production equipment and services is pure economics. It is by far less expensive for a skeleton crew to travel than to pay trucking and tour bus costs to move a full production of luminaires, stage sets, sound equipment, and the production team from city to city.

Instead of transporting equipment and crew, as the larger touring shows do, the majority of touring artists rely on renting from local lighting vendors or on the use of in-house equipment. In fact, it is not the band that hires the local lighting vendor; it is the show promoter who contracts the vendor. This is because the concert artist's contract must be adhered to by the show promoter, and within the contract will be an addendum with lighting specifications.

The band's production or road manager will distribute the local contact information internally to the road crew to confirm that the local vendors understand and can deliver the equipment and the required staff. Essentially, the promoter pays the bills and the road crews implement the designs, including:

1. Providing a wish list of equipment for the show to be included in the contract.
2. Following up with advance communications with the local lighting vendor.

3. Supervision by the road crew department heads at the venue.

Every good plan will have drawbacks, and here the caveat is twofold:

1. In most cases, the band's agent will send the contract to the promoter; however, the agent's focus is far removed from production as his 10% commission doesn't depend on the production requirements in the contract. Lighting plots and stage layouts are likely to be old news, regardless of the production manager's attempts to update them.
2. Some, but not all, promoters are greedy! They may hire substandard equipment to cut corners.

Between the promoter's representative for production and the local vendor, they may arbitrarily decide to substitute some of the lighting requests with what may be deemed as comparable equipment but isn't; for example, they may try to substitute color filters without proper research or, worse, use tired old gels. It is also common for the detail of the requested bulb type to be overlooked or ignored—just to name a few known hurdles, with the unknowns presenting themselves on show day.

On the other hand, changes and substitutions may be totally understandable. Maybe it is because the venue is away from any major city and the local equipment is all that is available. In any case, by the time the lighting designer gets the contact number it is usually too late to rectify any significant equipment change unless the local lighting vendor is sympathetic and will make changes as part of the production package. The artist's management has already cut the deal with the promoter. Tickets are on sale, and the venue has received its deposit along with the other contracts it has probably negotiated, ranging from union stage hands to tour merchandising. A touring crew may have very little say at that point with regard to upgrades. You will be stuck with what you've got, and the challenge will be to use that bare minimum requirement equipment with lamps in disrepair, filters that don't match, and several channels that will have a wide PAR-64 bulb on one side and a very narrow one at the other. All you can do is face the challenge and adapt, improvise, and overcome. The show will happen, just not with the perfection that we aspire to. But, another day offers the hope of better equipment. Just be comforted by the fact that your check was good.

THE ADVANTAGE OF LOCAL LIGHTING VENDORS

It's not all bad news, though. For the most part, local lighting vendors provide services as good as or better than larger companies. As you have already read in Chapter 1, the larger companies were established by buying small companies to give them a national presence, or small companies merged for operating and market share advantages. The local lighting vendor could be an outlet of a larger corporate operation, or it could be an independently owned company that offers several advantages:

1. Reduced or no equipment transport costs
2. No hotel or *per diem* costs because crews are local.
3. Competitive rates for the equipment (they want to beat out the big company's price).
4. Generally well-maintained equipment (with exceptions, of course).

5. Relationships with and knowledge of other local venues.
6. Familiarity with new technology or willingness to refer to qualified outside media services.
7. Maintenance of an active list of qualified freelance professionals.

Through their diversified work, many local companies have obtained a substantive inventory of new technology and are able to purchase the newest technology as it becomes available. The competitive free market in entertainment lighting rental is alive and well. Finally, and most importantly, is to understand the touring mindset. The local lighting vendor is there to help designers achieve their goals by providing service with care and passion. There is not much time for error. A show comes in for the day, and the equipment has to be prepared precisely to the touring designer's requirements before the touring crew arrives. While the other events that the local production companies service can be lucrative, the first love of most local companies is staging the live concert. Many owners and their employees have come from lives on the road and have acquired an entrepreneurial sensibility from freelancing in the concert touring world. Touring is a hard life, and owning or working for a local lighting company allows former road crew members to settle in one place and continue doing what they love to do.

III

Designing with Touring Equipment

23
MASTER DESIGNERS

I n the second edition of this book, three designers showed projects, plans, and equipment lists and discussed how their projects came together. This is almost more important than actually drawing a light plot. The hardest part for many designers is dealing with the client, the performance spaces, and the artist's demands that affect the design. With this edition, my prophecy that touring equipment would be used for all theatrical applications of entertainment lighting (didn't think about architecture back then) has come true. Just since the second edition went into print, moving luminaires have played a key role in lighting Broadway musicals.

You could say that two of the designers featured in this chapter are presenting traditional concert designs while the other two are bringing something unique to the book. Paul Dexter has added the title of set designer to his credentials, a unique position that only a few lighting designers have assumed. Andi Watson is known for his breakthrough green tour featuring LED lighting, and Richard Pilbrow is a famed Broadway designer and founder of Theatre Projects, Inc. Andi's project, along with that of Jeff Ravitz, who was also featured in the last edition of this book, still represent classic tour models, albeit with differing equipment and structural designs. Richard, though, brings us all the way to Broadway with his latest project, "A Tale of Two Cities." While it may have a short life, for me it is one of the most exciting visual shows to appear on Broadway. He has utilized a massive "concert lighting" approach that he will explain shortly. Each designer has taken his

own route, to be sure, and that is what makes this business so exciting—the complete diversity of what we can do. What follows are the designers explaining their projects and answering questions about their work. In the next chapter, they discuss the business in more general terms.

PAUL DEXTER: THE HEAVEN AND HELL TOUR

The Heaven and Hell (H&H) stage set was born after a trip to Stanford University and seeing their chapel for the first time. First of all, it is magnificent. The stained glass windows were mounted in arched window frames. After scouring the planet for inspiration, who knew that it would come in this way? The centerpiece and the catalyst for the H&H set became three window frame arches, but without the stained glass for projection surfaces; the three arches were suspended and capable of rising completely out of view or being lowered to perfect viewing height. A DL2 (digital light) was dedicated to each arch and the images passed through a custom gobo to control projection light in the shape of the arch.

The band's image is dark, medieval, and known for Celtic crosses. The rest of the set developed from the central window frames and extended into castle ruins—22-foot-tall broken arch doorways using realistic stone façade ideal for breakup patterns and saturated colors. To add further to the overall and illusive dimension, each arch was hinged for an 18-foot-tall medieval castle door, complete with the detail of distressed wood and decorated iron bars.

■ **245**

FIGURE 23.3   Heaven and Hell stage production. (Photograph by Lewis Lee.)

FIGURE 23.4   Heaven and Hell stage production. (Photograph by Lewis Lee.)

years of designs for this client, but in different circumstances.

Q. Was there a budget?

Yes, but I was never given one. I asked for a figure, too. The management team replied with, "Let's see what you come up with." To me, that did not mean free financial rein with an open checkbook, but more of invitation to develop a self-governing plan. It also meant, too, that whatever I came up with, it would have to be a knock-out design. That

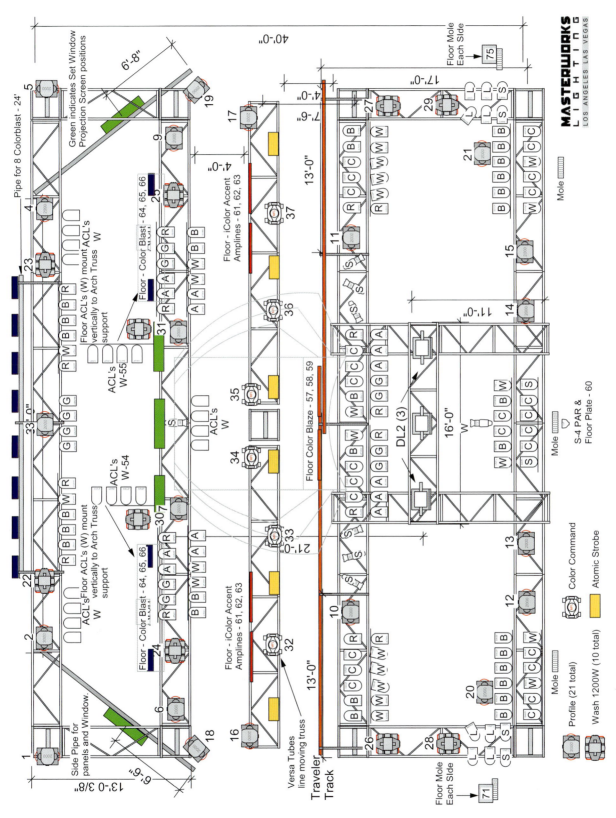

FIGURE 23.5 Heaven and Hell light plot. (Drawn by Paul Dexter).

FIGURE 23.6   Heaven and Hell rendering in SketchUp. (Designed by Paul Dexter.)

FIGURE 23.7   Heaven and Hell photorealistic rendering. (Rendering designed by Paul Dexter.)

Heaven And Hell Tour 2007 Gear

Video provided by Los Angeles-based CW Productions

Lighting and rigging provided by PRG

Moving Lights

21 Martin MAC 2000 Profile
10 Martin MAC 2000 Wash

Conventional Fixtures

150 PAR64 1kW Fixtures
8 Bars of 4 ACL Fixtures
5 Eight-Way Molefay
6 ETC Source Four Ellipsoidal 26°
1 ETC Source Four Ellipsoidal 19°
1 ETC Source Four PAR

Special Effects

14 Color Kinetic ColorBlast 12
2 Color Kinetics ColorBlaze 4'
6 Kino Flo 8' Fluorescent Tubes
6 Martin Atomic 3000 Strobe

Video Gear
3 High End Systems DL.2

Control

1 Flying Pig Systems Wholehog 3 Console
1 Flying Pig Systems Wholehog 3 DP 2000 DMX Processor
1 Flying Pig Systems Hog iPC

Dimming & Power Distro

1 ETC Sensor+ 96×2.4kW Dimmer Rack
1 Mains Feeder System
1 Power Distribution System

Trussing

2 4' Sections Total Structures Mini Beam Truss
3 8' Sections Total Structures Mini Beam Truss
4 10' Sections Total Structures Mini Beam Truss
2 4' Sections Total Structures D Type Truss
21 8' Sections Total Structures D Type Truss
8 Total Structures D Type Truss Corners
8 8' Sections 28"×28" Total Structures Intelli Truss

Rigging
6 ½-ton Columbus-McKinnon Chain Hoist
12 1-ton Columbus-McKinnon Chain Hoist

FIGURE 23.8 Heaven and Hell equipment list. (Provided by Jeff Ginyard.)

way, I could be confident that some leeway would be granted instead of trying to justify every penny.

Q. What was the type of venue for the production (theatres, arenas, stadiums)?

This tour and the live production were primarily designed for arenas, but as the show moved around the world it needed to accommodate every type of stage between European festivals and smaller venues, too, like the Pearl in Las Vegas. Needless to say, the whole production had to be cut considerably for that show.

Q. How many dates?

Nearly one year's worth, from January until mid-November.

Q. Who was involved in the initial creative meeting?

I showed my early concept to one of the managers to get some reaction artistically and to see if I was heading down the right financial track. I wasn't looking for a touchdown at that meeting, but I did get a first down.

Q. Was the artist directly involved in the design process?

Absolutely! One of the biggest thrills for me as a designer is to garner all the help and input that I can from the artist, no matter who I am involved with. They have to be comfortable on stage with their surroundings, and if I can facilitate that it is all part of a good design. Lead singer Ronnie James Dio has been much more involved in live theatrics than Tony Iommi or Geezer Butler and Vinny Appice, and I always work out his drum solos and riser setup well in advance. His drum solos are like a production within a production! While members have contributing ideas, Ronnie and I have been creating sets and lighting effects together for many years. He has always been there for me to bounce new ideas off as well as develop the cueing many shows into the tour. During the design process for the Heaven and Hell Tour, I made sure that Ronnie had an opportunity to share his ideas, and I integrated those changes in the evolving plans. I also sent updated PDF drawings to help Tony and Geezer to understand the stage set direction. It was a good thing because there was a concern for the position of Tony's guitar cabinets. They were elevated and he needed them to be on the carpet, so I had to change the layout to accommodate his unique sound. That change inadvertently compelled me to come up with a new façade idea for the cabinets, which has become one of the signature looks of the band.

Q. Was a set or staging already proposed before you became involved in the lighting design?

Fortunately for me, I was the stage set designer and the lighting designer on this tour, so I was the one proposing the stage design! I might as well tell you now that the reason that I took an interest in stage design and began to act on it early on in my career was twofold. The first reason was because I was hired as a lighting designer and not considered enough to be told that there was a set. It only took one time for me to build the lighting system for a tour rehearsal one day and show up the next day with a full set built on stage. What's this? I had no idea! I didn't have enough lights to cover it, and it was too late to get more. That was before the advent of moving lights, where you could cover multiple positions. After that, I started asking questions about staging before rehearsals started. Second, when I asked managers to meet the set designer prior to a tour, it was usually to see a scale replica model. By the time I was involved, though, the build process had started and it was too late to add any additional lighting in the set—a thought that many early rock & roll stage designers didn't consider. For me, adding lighting and electronic controlled effects in the set adds another layer to the overall stage picture. Set lighting in touring is fairly common now, but it wasn't always that way.

Q. Were other creative people brought in later?

Yes. I like to rely on all the help I can get. As production designer, you can easily spread yourself too thin—a likely prospect that I try to avoid. First the stage set: The builder was selected for the tour because they had proprietary material that I wanted. It looked like real castle stone. It was dimensional, lightweight but durable and roadworthy. The added surprise bonus was that their service went way beyond providing material and building a stage set to specification. They essentially became my partner during the build process, contributing creatively as well as engineering a cohesive set, all from the early beginnings of conceptual drawings. With regard to video content and lighting, there were hundreds of cues for both the LD and the projectionist/operator. It is important for me to allow the LD and associated operators to make the show their own, so their new ideas and suggestions for improvements were welcomed during the creative process and added to the show.

Q. Who was the lighting contractor, and did you have any say in the choice?

There has been a long-standing association with Nick Jackson, formally of Light and Sound Design and now Production Resource Group (PRG). He has been the lighting account rep since the early Black Sabbath tours, which has been a trusted professional and personal relationship. Needless to say, PRG was a shoo-in for the Heaven and Hell tour. Let me explain that, as a designer, I like to stay relatively impartial to the choice of lighting companies. I may have a favorite, but my job is to submit a plot to the companies bidding for the tour, which generally give three choices to the management. When it comes to choosing lighting companies, I let the managers make the calls. I will stay available to field technical questions that may help the potential vendors understand the requirements better and help the managers understand the quotation variances. The reason that I say that is because, in my years of doing this job, I have been accused of things that I am not devious enough to think of myself. It was all because one lighting company was selected over another one and I was put in the unsolicited position of being responsible for that decision. I just have to shrug my shoulders and move on—whatever! On smaller scale tours, you don't find competition so tightly controlled. It is like any other business, though— the more money and higher caliber of tour, the more political and competitive.

Q. What was the timeline from first meeting through to your first show?

I attended a lunch meeting with one of the managers in Studio City, California, in July and the first date of the tour was the following January. It was plenty of time to have conceptual design meetings, develop new ideas, select materials, and line up vendors.

Q. How does this timeline compare to your other clients?

For me, it is never a cut-and-dry timeline formula to develop live productions. I had a reasonable amount of time to design Heaven and Hell and the end results clearly benefited from that. But, most of the time I don't always get that time luxury. My work has been, by choice, very diverse and comparatively unequal to Heaven and Hell timelines. Oftentimes, the challenge is figuring it all out with less time and fewer resources.

Q. Was your first concept the one settled on?

Yes, it was. I don't think that anyone expected for me to come up with what I did, so they were probably shocked into reluctantly saying, yes, we'll have that. I found out later, though, that when the managers drafted the initial business plan for the tour, all that was included for the stage set budget was a drum riser and a backdrop. It ended up that a 53-foot truck was needed to haul the set.

Q. How much time did you have with the lighting rig in place before the artist arrived?

There was only one day to rig/hang the lights and one day to build the set. We started programming that second night, and the band showed up at noon the next day. Let me tell you that this band plays on volume setting 11! It was impossible to get anything done when they were in the room, which was between 12 noon and 8 p.m. Consequently, our only window to program was from 9 p.m. until 6 to 7 a.m. What a glamorous life we lead!

Q. How many days did you rehearse with the artist before opening?

There weren't any rehearsals to speak of, more like piecemeal, working out parts, like the opening, backdrop reveals, and coordinating projection cues. The total amount of days in the rehearsal studio was four, and at the end of the fourth day we were still programming while the stage set was being struck and packed. Cases were piling up on all sides of us, but we still kept going. Our first show was in Canada, and fortunately we loaded in one day prior to show day, so we had just enough time to at least feel comfortable that we had a cohesive show to present—a very close call.

Q. Do you use an assistant? How many people were involved in the drafting?

An assistant is a luxury reserved for other lighting designers. I am more of a blue-collar lighting designer, so I still do all my own drawings and I love it. It's because drawing prompts

me to address nuts and bolts problems and evolve a design with a first-hand understanding of the scale and relativity between stage set and lighting. For example, the Heaven and Hell set used the lighting truss as a subgrid for some of the set pieces, so I was able to accurately determine where set points needed to be. The truss structure was designed to accommodate that. If an assistant prepared my drawings, I probably wouldn't recognize all of the fine distinctions with the same depth that I do. Drawing exercises and pushes me to full completion of ideas and gives me a chance to work out all the practicalities, rigging, trim height, and positions. When I arrive onsite, there aren't too many questions that I can't answer.

Q. What program do you use to prepare the lighting drawings?

I use a VectorWorks program. I find the logic in Vector easy to grasp. Having started with stencils and vellum and drawing everything by hand, I have a real appreciation for drafting with a computer software program. It is simply click and drag.

Q. Do you do any preproduction visualization programming to save time in rehearsal with the artist?

This method is not something that I have spent enough time with to personally perfect. I know it works, and I have used it, but not under the best of circumstances. However, it does not preclude having to visit and edit cues when on site. Perception always changes with physical pieces versus looking at a computer graphic of those pieces. Certainly, with larger moving light systems, previsualization is a time saver because of the head start that can be made with organizing the lights in terms of setting up palettes, groups, and focus positions. Call me old fashioned, but I still like the process of building scenes and looking at cue changes in real time.

Q. In production mode, who calls the show, and how many board operators does it take?

In my experience, the lighting director operates the console and calls the follow spots. For this production, I wrote cues and passed out script books to the lighting director and the projectionist/board operator—so there were two operators on this particular tour. To get things started I traveled on the road for the first 10 days. I helped set up, worked through some of the inherent growing pains of a new production, and during the show stayed on headset. I was very involved at first, calling the spots and cueing everything else, but I slowly weaned off my involvement.

Q. What were some of the nightmares, problems, things that didn't go as planned?

Every new production has problems, but this one was void of any real nightmares. The learning curve was painful for some of the road crew, just getting used to new ideas and working out priority and order. There were two 16-foot-high gothic doors that turned out to be heavier than we thought, so of course having to deal with the doors on a daily basis caused some crew grumbling, but a new rigging method was soon worked out. For the most part, this set was large pieces, so assembly was not that complicated.

Q. Anything you want to add that would help other designers understand production at this level?

Yes. There are other crew members, too, besides lighting. It is easy to become buried in your work because there is so much to do on a large production causing you, not intentionally, to forget that others are under demands to complete their work, too. After all, lighting and rigging are first in, then there is focus, testing the haze and floor lighting. After the show, the lighting is the last to go in the truck. Let's face it; it is usually more concentrated hours than most of the road crew. While it seems that your life is harder than everyone else's, try to listen and understand the rest of the crew's situation and they will care (even though they don't show it that well sometimes) about yours. Managing your position with a larger production is a concerted effort, and it is easier when the rest of the crew are communicating and interacting with each other positively.

Q. Anything you want to add that you feel will be of interest to the readers?

We are in a business that does not require formal education to enter. There are no lighting MBAs or PhDs. There are certification programs available, but they are discretionary, not mandatory. All it takes to be in lighting is to have a wrench in your pocket and a good attitude and to

hang out with the right people—and you will soon get jobs. The touring industry is welcoming to anyone but has an unwritten way to qualify people and that is coping with the road lifestyle. You are either meant to be able to cope with the lifestyle or not. If you are not, it is okay and you will move on to a job inside or outside of the industry that will be more suitable for you—it isn't for everybody. If it *is* for you, what will help you rise above the trials that are dealt you on the road, tests that your road comrades will dish out, and even broken relationships at home it will likely cause will be the endless passion for what you do. It will be genuine. Those who are on tour around you will be of like minds with the same work ethos, and nothing else is quite like the fulfillment you will feel. This will keep you active in your work, both mentally and physically, and you'll like it! Subsequently, it will be a very organic way to grow in the business—you do what you love to do and the higher positions and money will follow.

The caveat is that, while there is opportunity to grow and climb quickly to higher positions, grow and gather experience *before* accepting a position that you may not be qualified for. For example, there are a lot of self-professed lighting designers out there working with unearned titles who have very little to offer creatively or are technically inept. It is not always their fault. Many have won the position by default because (maybe) the company needed a designer and nobody else was available. But, once a person is thrown into a position by default, after the gig is over you can't consider yourself, after working on a couple of shows, to be a *bona fide* lighting designer. Not only do you need to be technically adept in concert lighting, staging, and effects, but you also need to have some understanding of musical structure. You also should practice and develop your communication skills in order to work well with others. It takes years of practice and experiences.

For me, it is an integrity issue. Compare a lighting designer to a captain in the military. What if the captain had bought his stripes at the local uniform store and sewed them on his jacket, but he did not go through the proper protocol or suffer through a few wars to gain enough experience to really call himself a captain? That would be like a self-proclaimed lighting designer touting his name in business circles but offering very little in the way of substance. Could you put others at risk without proper experience? Yes! Our business does not have policing for this type of conduct so it is up to you to know when you are ready to accept more responsibility. All it will take is to have patience and do it better today than you did yesterday. That way, you will grow naturally to achieve your goals and not through false pretenses.

RICHARD PILBROW: "A TALE OF TWO CITIES," A BROADWAY MUSICAL

How does a veteran Broadway lighting designer get involved with a writer, director, and producers who have never done a Broadway show and not come out screaming? First, it takes the class of a man like Richard Pilbrow who quipped that he was surprised that anybody knew he was still alive. Well, master scenic designer Tony Walton did, and when he had the opportunity to suggest Richard he did. "Richard is always on the cusp of what's happening, much more so than any of his younger allies. The difficulty is luring him into it, so I always try to come up with something that's impossible to light. He moans quite a bit, but he really relishes the challenge," said Walton.[1] The two had worked most recently on a 2003 revival of "Our Town."

Tony had already designed six scaffold structures, 25 feet tall, that would move and reconfigure throughout the performance with one or two repeats. Added to Richard's orders were that no wires go to the units, so they were internally lit via battery-powered lights on 86 channels of DMX wireless receivers. The rest of the rig allowed for a lot of ETC Source Fours and 62 moving luminaires. In addition, 98 LED ColorBlaze units along with a number of other diverse lighting units make this production a true cross-blend of concert lighting and Broadway tradition.

Richard has said that one of the most exciting things the moving luminaires provided was the ability to change focus when the director decided to move any or all of the six modules. Without stopping the rehearsal, he could refocus one or more of the VL1000 spots positioned in the front of the house from the consoles. All in all, he only had 4-1/2 hours to write 350 cues using 1000 control channels. The combination of lighting was stunning and garnered rave reviews, even if the show did not have a long run. Richard, as you will read later, is no stranger to moving luminaires, and this production allowed the technology developed in concert touring to highlight his talent. Figure 23.9, Figure 23.10, and Figure 23.11 are production photos from the Broadway production. Figure 23.12 is the light plot, Figure 23.13 is the equipment list, and Figure 23.14 is the crew list.

About Richard Pilbrow

Richard Pilbrow (Figure 23.15) is one of the world's leading theatre consultants and a pioneer stage lighting designer. He is also an author and a theatre, film, and television producer. He has been responsible for many innovations in stage lighting, theatre architectural design, and technology. Richard was a pioneer of modern stage lighting in the United Kingdom. He was the first British lighting designer to design the lighting for a Broadway musical ("Zorba"). His Broadway credits include "A Tale of Two Cities," "Our Town" (starring Paul Newman), "The Life," Hal Prince's revival of "Show Boat," "Rosencrantz & Guildenstern Are Dead," and "The Rothchilds." In 2008, he was lighting designer for "The Sleeping Beauty" for the American Ballet Theatre at the Metropolitan Opera House.

In 1957, Richard founded Theatre Projects as a lighting design and rental company in London. Under his leadership, the company has grown to be an international theatre design–consulting firm with over 1000 projects in 60 countries. Fifty years later, as the founder and chairman emeritus of Theatre Projects Consultants, Richard is still consulting on some of the most significant arts projects in the world, including three major performing arts centers in the United States: the Dallas Center for the Performing Arts, the Dr. P. Phillips Orlando Performing Arts Center, and the Kauffman Center for the Performing Arts in Kansas City.

[1]David Barbour, "Do You Believe in Magic?," *Lighting & Sound America*, January 2008.

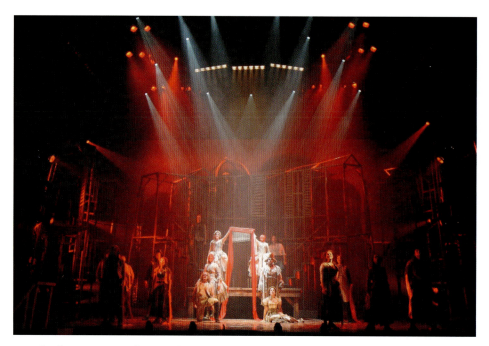

FIGURE 23.9 "A Tale of Two Cities" production. (Photograph by Michael Gottlieb.)

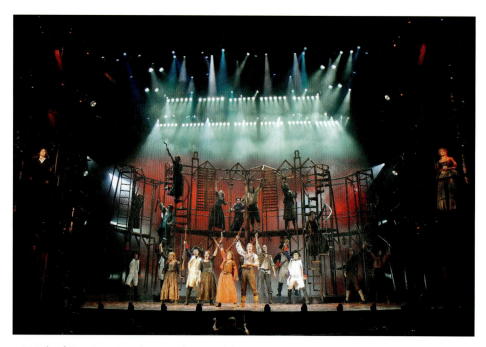

FIGURE 23.10 "A Tale of Two Cities" production. (Photograph by Michael Gottlieb.)

Published in 1970, his book *Stage Lighting*, with a foreword by Lord Olivier, became an international teaching text. A second book, *Stage Lighting Design: The Art, the Craft, the Life*, with a foreword by Hal Prince, was published in 1997 (reprinted in 2008). In 2003, Richard coauthored (with Patricia MacKay) *The Walt Disney Concert Hall—The Backstage Story*.

FIGURE 23.11 "A Tale of Two Cities" production. (Photograph by Michael Gottlieb.)

In 2005, Richard was honored as Lighting Designer of the Year by *Lighting Dimensions* magazine. In December 2005, in the premier edition of *Live Design*, he was named as one of the ten visionaries among designers and artists who were the most influential people in the world of visual design for live events.

His project is the bold and beautiful Broadway production of "A Tale of Two Cities" at the 1200-seat Hirschfeld Theatre in New York, which premiered in July of 2008.

QUESTIONS

Q. Who actually hired you?

Don Frantz, General Manager, Town Square Productions. Recommended by Tony Walton, Scenic Designer.

Q. How many designs had you done for this client in the past?

None.

Q. Was there a budget?

Yes.

Q. What was the type of venue for the production (theatres, arenas, stadiums)?

Broadway theatre: tryout at the Asolo Playhouse in Sarasota, Florida; performance at Hirschfeld Theatre in New York City.

Q. How many dates?

Florida tryout, November 2007; Broadway, July to October 2008.

Q. Who was involved in the initial creative meeting?

Scenic designer Tony Walton, director Warren Carlyle, and costume designer David Zinn.

Q. Was a set or staging already proposed before you became involved in the lighting design?

The concept of the setting was evolved prior to my coming on board. Previous attempts to get the show on Broadway had taken place over a 5-year period. The set consisted of a series of back- and front-lit cloths that depicted general locations (London, Paris, etc.) and six two-story-tall steel towers that formed up to 40 different configurations. Towers were dressed with properties and set dressing to represent different homes, bars, courtrooms, prisons, etc. They, and beautiful period-inspired costumes, were lit to form a series of strongly characterized locations and times of day and seasons of the year for the fast-moving story in a dramatic manner.

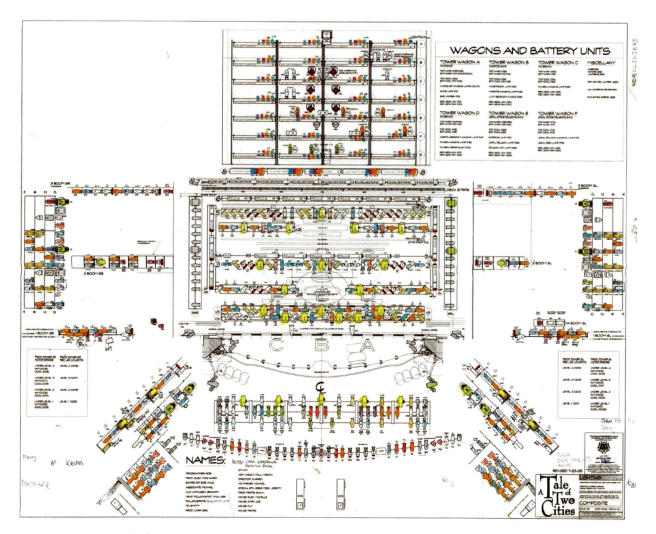

FIGURE 23.12 "A Tale of Two Cities" lighting plot. (Design by Richard Pilbrow.)

Q. Were other creative people brought in later?

Gregg Meeh for special effects.

Q. Who was the lighting contractor, and did you have any say in the choice?

PRG, New York. My recommendation in part because of their stock of DHA Pitching Light Curtains.

Q. What was the timeline from first meeting through to first show?

Tryout at the Asolo Theatre in Sarasota, FL, June 1987 to November 1987; Broadway run from February 1988 to July 1988 production and September opening.

Q. How does this timeline compare to your other clients?

Often similar.

Q. Was your first concept the one settled on?

In principle, yes. There was never a discussion with the producer or director as to what equipment I would use. My design came out of the needs of the fast-moving plot and scenic/lighting concept—my decision.

Q. If not, was the reason financial constraints or another consideration?

Budget was reasonable: $90,000 for preparation and $14,900 weekly.

Q. How much time did you have with the lighting rig in place before the artists arrived?

Broadway load-in, July 14; focus, August 1 and 2; first tech with actors, August 4 to 13; dress

A Tale of Two Cities - Hirschfeld Theatre NYC
Lighting by Richard Pilbrow
Equipment Summary for Press Distribution
October 9, 2008

Conventional Units

24	ETC Source Four-10° ERS
51	ETC Source Four-19° ERS
224	ETC Source Four-26° ERS
2	ETC Source Four-26° Enhanced Definition ERS
38	ETC Source Four-36° ERS
3	ETC Source Four-50° ERS
18	ETC Source Four-70° ERS
8	ETC Source Four Zoom 15-30° ERS
24	ETC Source Four PAR MFL MCM
11	ETC Source Four PAR VNSP MCM
6	Arri Junior 5K no lens/reflector
156	Birdie EXN (Painted)
16	GAM Stik-Up
2	PAR64 MFL
6	Mini-10

Striplights and LED Striplights

5	7'-0" 9-lamp 3-circuit T3 Strip 650w
28	8'-0" 40-lamp 4-circuit Ministrip EYC
41	Color Kinetics Colorblast 12 LED frosted
2	Color Kinetics Colorblaze 48 LED frosted
6	Color Kinetics Colorblaze 72 LED frosted

Followspots

2	Lycian 1290XLT Xenon Followspot
2	Lycian 1272 Starklite II HMI Followspot

Automated Lights

9	Vari-Lite VL1000TS
2	Vari-Lite VL2500 Spot
21	Vari-Lite VL3500Q Spot
4	Vari-Lite VL5B 8-Row
6	DHA 6-Lamp Digital Light Curtain
12	DHA 6-Lamp Pitching Digital Light Curtain

Color Scrollers

26	Wybron 7.5" Coloram II Scroller

Strobes and Effects

4	Martin Atomic 3000 Strobe
4	Diversitronics PAR64 SCM-64Q-DMX Strobe
3	Bowens 1000 Strobe
3	Wildfire LT-404F UV
7	TPR-F1-150-DMX-120-X Fiber Optic Illuminator
4	Rosco/DHA Double Gobo Rotator
2	GAM SX4 Film/FX
74	LEDtronics Red LED Array
36	LEDtronics UV LED Array
36	City Theatrical Flicker Candle

Atmospherics

3	LeMaitre Low Smoke Generator
4	LeMaitre Silent Storm Snow Machine
2	MDG Atmospheres Hazer
4	Look Viper II Smoke Machine
1	Look Tiny Fogger
1	Look Power Tiny Fogger
3	Cryojet
4	Jem Fan
3	Bowens Fan

Control

Strand Light Palette VL 3000
West Side Systems Virtual Magic Sheet
Cast Lighting WYSIWYG
Focus Track

Dimming

6	96 x 2.4kw ETC Sensor Dimmer Rack
1	12 x 6kw ETC Sensor Dimmer Rack
1	City Theatrical WDS Wireless Transmitter
7	City Theatrical WDS Wireless Receiver
50	City Theatrical WDS Wireless Dimmer
1	RC4 Wireless RC4Magic Transmitter
5	RC4 Wireless RC4Magic Dimmers
	LED Effects Stream/DMX512 LED Dimmers

FIGURE 23.13 "A Tale of Two Cities" equipment list.

rehearsal, August 14 to 18, with previews August 19 to September 17. The show opened September 19. The load-in electrics crew averaged 13 plus 2 effects plus 6 sound crew. Apart from hanging the rig they had to wire 90 channels of remote-controlled MR16s in the wagons that could only be done onstage after they'd been erected. Focusing conventional took a couple of days. All programming was done during tech except for three short morning sessions during late tech for programming "tidy-up." Total about 4 to 5 hours. 350 cues with about 1300 channels, including effects. Essentially, programming was the fast part thanks to WYSIWYG and Virtual Magic Sheet. We only very seldom had to ask actors to wait for us through rehearsals.

Q. How many days did you rehearse with the artists before opening?

First tech with actors, August 4 to first preview, August 19. Actors and crew moved the two-story units, so these changes had to be plotted and rehearsed. First preview, August 19 to opening night September 18.

Q. Do you use an assistant? How many people were involved in drafting?

Associate—Michael Gottlieb (drafting, technical supervision); assistants—Kathleen Dobbin (moving light program plotting) and Graham Kindred (follow spot plotting). Jay Scott was my associate on the Sarasota tryout but was not available for the Broadway run because of a

A Tale of Two Cities - Hirschfeld Theatre NYC
Personnel:

Lighting Designer: Richard Pilbrow
Associate Lighting Designer: Michael Gottlieb
Lighting Programmers: Robert Bell, Bobby Harrell
Assistant Lighting Designers: Kathleen Dobbins, Graham Kindred, Jay Scott
Production Electrician: Michael Ward
Assistant Electrician/Head Followspot: Paul Ker
Board Operator: Robert Hale
Followspot Operators: John Blixt, Thomas Burke, Bob Miller
House Electrician: Michele Gutierrez

Director: Warren Carlyle
Scenic Designer: Tony Walton
Costume Designer: David Zinn
Sound Designers: Carl Casella and Domonic Sack
Special Effects Design: Greg Meeh
Hair Design: Tom Watson
Production Stage Manager: Kim Vernace
Production Supervisor: Christopher Smith
Production Carpenter: Erik Hansen
Deck Automation: Russ Dobson
Fly Automation: James Sturek
Production Props: Dawn Makay

FIGURE 23.14 "A Tale of Two Cities" crew.

FIGURE 23.15 Richard Pilbrow, lighting designer. (Photograph by Donald Dietz.)

prior commitment on the Beijing Olympics. He led with technical planning and set the eclectics drawings and specifications in Sarasota. In addition, he was the liaison with Michael to get him up to speed to take over the show in New York.

Q. What program do you use to prepare the lighting drawings?

VectorWorks.

Q. Do you do any preproduction visualization programming to save time in rehearsal with the artist?

WYSIWYG and Virtual Magic Sheet.

Q. In production mode, who calls the show, and how many board operators does it take?

The stage manager calls show. Other crew include one master electrician (Michael Ward), one house chief electrician, one board operator, four follow spots (two front and two FOH high-side positions); during production and technical rehearsals, one lighting programmer (Robert Bell).

Q. What were some of the nightmares, problems, things that didn't go as planned?

No nightmares, challenges. Mainly pressures of a production with continually moving three-dimensional scenery moved by cast and crew into over 40 different configurations.

Q. Anything you want to add that would help other designers understand production at this level?

Key was very open communication between the director and design team and detailed preparatory work. Actual lighting preparation time without performers was about 4 hours for over 350 cues.

Q. Anything you want to add that you feel will be of interest to the readers?

A very complex and fast-moving production with numerous moving lights, principally employed as multiple specials to light into often moving, double-story scenic towers. Use of WYSIWYG and Virtual Magic Sheet enormously speeded up the process of handling over 1400 channels (including effects). Great design team, excellent electrics crew, and very cooperative stage crew organized by Chris Smith.

ANDI WATSON: RADIOHEAD IN RAINBOWS TOUR

At first, the experienced concert-goer might not detect the major change in lighting that is about to be presented in this current Radiohead tour: *green*. Not just the color, though; the tour is green in the sense of the lighting presenting a small carbon footprint. That was accomplished by using *all* LED lighting units to illuminate the stage, band, and audience. One of the reasons for this concept coming to life is that the group's lead singer is a member of the environmental group called Friends of the Earth. The production manager, Richard Young, is also involved to the point of contributing to a report on the tour, which showed that fans actually produce the vast majority of the carbon footprint.[2]

What could the band do to help? The logical starting point was the massive lighting rig. However, we can't forget about the miles and miles that the trucks carrying all the touring equipment pile up. That was figured at almost 40% of the band's portion of the carbon consumed. So, even though the lighting turned out not to be a major offender, it was an area whose footprint might be reduced. On the 2003 tour, the lighting required 900 amps per phase, or 2700 amps of power. When Andi Watson and the creative/technical team were able to lower that figure to 135 amps per phase, or 405 amps of power—less than 15% of the power consumed formerly. Andi described the show's look in a *PLSN* article as "a 3-D video display with very low resolution, with 12 pixels horizontally, 288 pixels vertically, and six panels deep."[3] The tubes are laid out in such a way if an audience member moves a few feet left or right the images will be totally different. "It is a very subjective user experience," Andi said.

The more traditional lighting on the band was provided by iPIX's BB7s. They are petal-shaped luminaires in a 7-cell, homogenized, 10-degree RGB light source. Andi believes that he was able to achieve matching colors because the LEDs matched better than multiples of most incandescent sources. All in all, 65,000 individual LED components illuminated the stage and projections, a singular achievement for a concert tour thus far. Figures 23.16 through 23.18 demonstrate the mass of LED luminaires used on this unique tour. Figure 23.19 is a 3D view of the lighting. Figure 23.20 shows one page of the light plot, and Figure 23.21 shows an equipment list, most of which was purchased by the band. Figure 23.22 lists the lighting crew.

About Andi Watson

Andi Watson (Figure 23.23) was born in a small village in the English countryside, and after leaving school he moved to Brighton to study Engineering at Sussex University. In his free time, he was involved in local concert promotion and worked at many shows, both as a stage hand and later as a lighting technician and board operator. After leaving Sussex University in 1985 with an honors degree in electrical and electronic engineering and computer science, he worked for a local lighting company doing pretty much everything from driving vans to designing computer memory lighting desks. In 1987, after answering an advertisement in the *Guardian* newspaper, he was offered a job at Samuelson's in London as a Vari*Lite technician working initially on VL1 systems. Within 8 weeks of his starting work, the Series 200 system was released and he found himself working directly for Vari*Lite on tour operating the first shows in Europe with VL2 luminaires. After working for 4 years at Vari*Lite and programming and operating world tours for bands such as INXS, The Cure, and Sinéad O'Connor, as well as numerous television shows, awards ceremonies, corporate launches, and raves, he left to pursue a career as an independent designer. Since leaving Vari*Lite in 1990, Andi has

---

[2]Claire Stentiford. *Ecological Footprint & Carbon Audit of Radiohead North American Tours, 2003 & 2006* (Oxford: Best Foot Forward, Ltd., 2007), pp. 1–36.
[3]Arden Ash. "Green, from Radiohead to Toe," *Production, Lights, and Staging News*, October 2008.

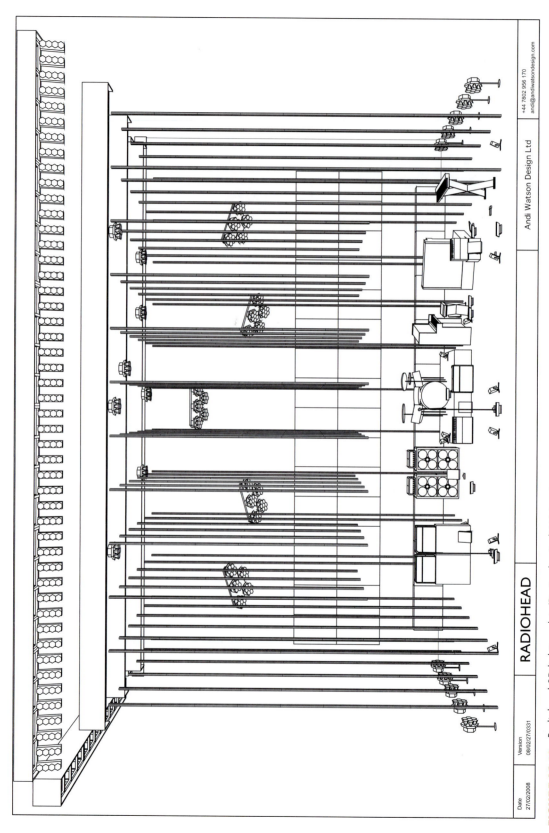

FIGURE 23.19 Radiohead 3D lighting plot. (Design by Andi Watson.)

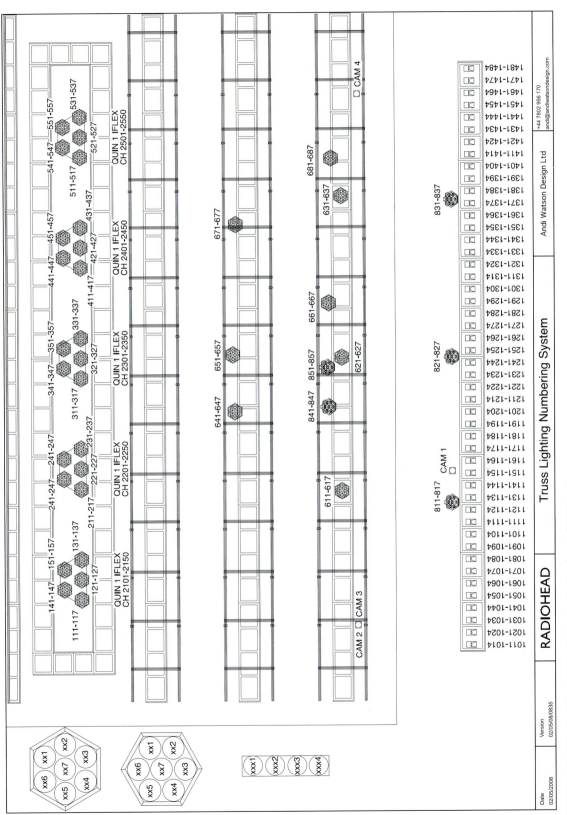

FIGURE 23.20 Radiohead light plot layer. (Design by Andi Watson.)

Radiohead In Rainbows Equipment List 050508

Front Truss

06 x 8' Upstaging PreRig Truss Sections
01 x 55' x 3'3" Black Velvet Truss Border
48 x iPix BB4 LED Fixtures
03 x Top Mounted Spot Seats
03 x Custom Spot Yokes for Color Kinetics Color Reach
03 x Color Kinetic Color Reach LED Fixtures
01 x Scenographic Remote Pan/Tilt/Zoom Camera
01 x Ultra High Power Infra Red 'Death Ray' Emitter

Versatube Truss 1

02 x 8' Master End Truss Sections c/wHeavy Duty Scenic Track System
05 x 8' Truss Sections 20.5"
01 x 52'w x 2'6" Black Velvet Truss Border
02 x 33' drop x 8' Black Tear Off Legs
16 x 6m Versatube Drops c/w Custom Rigging & Cable Management
08 x 9m Versatube Drops c/w Custom Rigging & Cable Management
24 x Element Labs Buffer Boxes
05 x iPix BB7 LED Fixtures
02 x Color Kinetics Color Reach LED Fixtures
03 x Scenographic Remote Pan/Tilt/Zoom Cameras

Versatube Truss 2

02 x 8' Master End Truss Sections c/w Heavy Duty Scenic Track System
05 x 8' Truss Sections 20.5"
01 x 52'w x 2'6" Black Velvet Truss Border
02 x 33' drop x 8' Black Tear Off Legs
16 x 6m Versatube Drops c/w Custom Rigging & Cable Management
08 x 9m Versatube Drops c/w Custom Rigging & Cable Management
24 x Element Labs Buffer Boxes
03 x iPix BB7 LED Fixtures

Radiohead In Rainbows Equipment List 050508

Versatube Truss 3

02 x 8' Master End Truss Sections c/w Heavy Duty Scenic Track System
05 x 8' Truss Sections (20.5")
01 x 52'w x 2'6" Black Velvet Truss Border
02 x 33' drop x 8' Black Tear Off Legs
21 x 6m Versatube Drops c/w Custom Rigging & Cable Management
03 x 9m Versatube Drops c/w Custom Rigging & Cable Management
24 x Element Labs Buffer Boxes

Quin & Video Box Truss

12 x 8' Truss Sections (20.5")
02 x 4' Truss Section (20.5")
04 x Corner Blocks (20.5")
01 x 56'w x 6' Black Truss Border
01 x 48' x 10' Black Trevira (Screen) Tear Off
01 x 48' wide x 8'10" high Nocturne V9 Video Screen
05 x Custom Quin Frames
25 x iPix BB7 LED Fixtures
05 x Colour Kinetic iFlex SLX (Diffuse Caps)
15 x Kinesys Hoists

Drapes Truss

07 x 8' Truss Section (12")
01 x 56'w x 33' Black Drape

Floor

10 x iPix BB7 LED Fixtures on Floor stands 2'–4'
07 x Pulsar ChromaFlood 200 Tricolour LED Fixtures
07 x Color Kinetics iW Blast LED Fixtures
15 x iPix Satellite LED Fixtures
02 x Scenographic Remote Pan/Tilt/Zoom Cameras
02 x Static Cameras (100mm Zoom Lenses)
01 x 'PianoCam' HQ Pro Camera
01 x 48'w x 4'h Nocturne V-Lite Video Screen
02 x Cirrus CS6 Water Cracker
04 x Le Maitre Stadium Hazer
08 x London Variable Speed Fans

Radiohead In Rainbows Equipment List 050508

Control

01 x Full Size GrandMA Control Console (Plus Tracking Spare)
02 x 17" Hi-Res GrandMA Monitors
03 x Neovo X-22W Catalyst Monitors (with Cat5 Balun Receivers)

02 x Fully Managed Twin Ethercon Data Snakes

06 x Ma Lighting NSP
03 x Element Labs D2 (Plus Live Spare)
01 x Element Labs Data Distro System

03 x Camera Control Stations with Video DA and Monitoring.

03 x Scenographic Catalyst Servers c/w Custom 3D Aware Software
    64 GB MTron SSD, Twin Active Silicon LFG4 Video Capture Cards
    NVidea 8800 Graphics Cards, Cat5 Video Baluns
    Luminex DMX8, Luminex GigaSwitch, Liebert GTX2-2000RT230 UPS

    Kinesys Motion Control System
    Nocturne Video Processing System for V9
    Nocturne Video Processing System for V-Lite

    Intercom (Min 10 Station Inc Spots),
    All Cabling, Mains, Rigging, Metalwork, Power Distro Etc Etc

09 x Color Kinetics iW Blast LED Fixtures for Production/Backstage

00 x Dimmers

01 x Andy Beller	Crew Chief
01 x Ed Jackson	System Tech
01 x Johannes Soelter	Carpenter/Lighting/Cameras
01 x Blaine Dracup	Lighting/Cameras
01 x Nick Barton	Lighting/Cameras
01 x John McLeish	Video Screens
01 x IanLomas	Lighting
01 x Travis Robinson	Lighting
01 x Tommy Green	Lighting
01 x Seth Cook	Rigging

FIGURE 23.21 Radiohead equipment list.

Radiohead Lighting Crew

Andy Beller-Crew Chief
Ed Jackson-System Tech
Johannes Soelter-Carpenter/Lighting/Cameras
Blaine Dracup-Lighting/Cameras
Nick Barton-Lighting/Cameras
John McLeish-Video Screens
Ian Lomas-Lighting
Travis Robinson-Lighting
Tommy Green-Lighting
Seth Cook-Rigging

FIGURE 23.22 Radiohead lighting crew.

Q. How many dates?

So far we have played about 50 shows, with another 10 scheduled for South America next year.

Q. Who was involved in the initial creative meeting?

Radiohead are an unusual client for me in the sense that prior to presenting my design ideas I did not have a traditionally creative meeting with

FIGURE 23.23 Andi Watson, lighting designer.

the band and management. Initially, I had spoken with the band's management regarding the band's wishes to make the tour have as little negative environmental impact as possible and had spoken with Richard Young, Radiohead's production manager, about the rather unusual scenario that we were involved in, with not air-freighting anything but the absolute minimum of band instruments. On the basis of this and knowing the expected venue range I then created the design concept. One of the huge implications of the band's desire to avoid air freight was that all intercontinental shipping was to be by sea container, and due to this it had been decided that we would actually use two complete identical systems for the tour, one initially based in North America and one in Europe. This meant that for any components not available locally we would have to create two sets, one for each system. Since, for various reasons, a decision was subsequently made for production to buy all the LED fixtures used on the tour as well as commission the complete associated custom hardware, this meant that we had to buy and build two complete sets of everything, obviously with both practical and financial implications.

Q. Was the artist directly involved in the design process?

I am very fortunate with Radiohead in that I have worked with them on a number of tours now and am allowed a huge amount of freedom in the design process. That is not to say that they would not tell me if there was anything they didn't like, but more that I think over a period of time we have developed and built up a visual language for their music which they trust me to apply to each album. As a result and also due to recording commitments, on this tour the artist's initial involvement was rather unusually limited to a series of e-mail exchanges. Having said that, of course nothing was signed off until all the band had seen the design proposal and were happy with it. I would never want to consider designing a production that didn't directly involve the artist in its conception, since I feel it is my job to create the environment in which they perform, and the best placed people out of anyone to help me understand how I can create that are the performers themselves.

Q. Was a set or staging already proposed before you became involved in the lighting design?

I am lucky enough to be the production designer for the tour so any "set" falls within my remit along with the lighting and video, etc.

Q. Were other creative people brought in later?

I had been fortunate to collaborate on several previous productions with Richard Bleasdale, the creator of Catalyst software, and at an early stage in my design process I had a series of meetings with him to try to figure out how to control and bring to life the LED array I wanted to build. It is always very interesting to work with Richard, and after I had explained the physical structure I wanted to create and what I wanted to do with it on an artistic level, he went away and very cleverly developed a way for me to do that. At around the same time, I was also having meetings with Chris Ewington from iPIX about the development of a high-power LED fixture that would be usable as a beam light. In addition, from very early on, Keith Owen and Dave Smith from Specialz were involved, along with production manager Richard Young, in the practical side of the design from the VersaTube deployment system, to transportation requirements, electrical, and signal distro, etc., etc.

Q. Who was the lighting contractor, and did you have any say in the choice?

The lighting supply contacts were John Bahnick at Upstaging in Chicago and Dave Ridgeway at Neg Earth in London. I think both Richard and I were very happy with this combination, and I feel that if either of us had been unhappy then a different choice would have been likely. Of course, the tour went out to bid, but this design was incredibly unusual in that due to a number of reasons the production bought a huge amount of the equipment involved. This meant that the lighting supply company's equipment list was minimal considering the size of the tour but obviously still as vital. On the recommendation of Richard Young, I went to see the V9 video screen and by doing so the choice of Nocturne as the video supply company was instantly decided.

Q. What was the timeline from first meeting through to first show?

I had my initial discussions with management, band, and production in late November 2007 and had preliminary meetings with Richard Bleasdale and later Richard Young in December 2007. In early January, there were a series of meetings involving myself, Richard Young, and Specialz as to how we could physically manufacture the system. At the same time, there were a series of comparative tests of various linear LED fixtures and LED screens and also meetings with iPIX regarding the development of what was to become the BB7 fixture. Concurrent to this were tests of the custom control software being written by Richard Bleasdale. This all culminated in a meeting at the end of January 2008 with the band and management at which I presented design drawings, a 3D CAD model, and also a demonstration of the array control software which resulted in the band approving the design. By the end of February, the choice of all the fixtures and the LED screen had been determined, the hardware had been developed, and the production of the iPIX fixtures had been given the go-ahead. In mid-March, the equipment from the U.K. that was to comprise the U.S. system was loaded onto sea containers for shipping to

the U.S., and at the same time a V9 screen system was shipped in a container from the states to the U.K. in preparation for rehearsals which began in Bray Studios outside London in the second week of April 2008. The 3-week rehearsal period was split between 1 week of build, 1 week of programming, and 1 week of band rehearsals. Toward the end of the rehearsal period, Blaine Dracup, one of the lighting crew, flew to Chicago to check the prep for the U.S. system that had to leave Upstaging in Chicago (before the end of our London rehearsal period) in time to be in Florida ready for our first show there. There were 2 production days to correct any issues with the U.S. system in Florida before the first show.

Q. How does this timeline compare to your other clients?

It varies greatly from one tour to the next. Radiohead have traditionally not given a huge amount of notice of going on tour while some of my other clients discuss their tours with me over a year in advance. Often I find that the eventual complexity of a tour has no bearing on how much time you are given to design and develop the system.

Q. Was your first concept the one settled on?

Conceptually, yes, although there were several changes to the actual hardware of the system throughout the design process. My original concept was to have the 3D video screen comprised of vertical LED strips and to only utilize LED light sources in the design. This remained intact throughout and by my choice was extended to also include the main horizontal video screen and in fact even backstage work lights and dressing room lighting. Thanks to Specialz, the original LED strip deployment method was simplified drastically which solved my one major design concern.

Q. How much time did you have with the lighting rig in place before the artist arrived?

We spent just under 2 weeks with the system before the band arrived. Of this, a week was spent completing the build, making improvements, finalizing the way we were going to use the camera system, determining camera placement, modifying the control software for the VersaTube array, adapting the manipulation methodology for the live camera feeds, and experimenting with the system. The remaining week was spent programming.

Q. How many days did you rehearse with the artist before opening?

The band rehearsed for 4 days, although in that time we only did one sample run through which was on the last day. Radiohead have always wanted the freedom to change the set list on a nightly basis and for this tour my "short list" of songs numbered about 65, which is a huge number to program! The band also has multiple stage positions and plays multiple instruments. To make things just that little bit more challenging, the songs can be played (especially at the beginning of a tour) in any position in the set list, meaning that the song that was played last in the set one night can become the opening song the next. This obviously means that it is very hard to create a visual journey throughout the show in the usual way and also means that every single song has to be visually unique. As a result, the programming schedule was incredibly intense!

Q. Do you use an assistant? How many people were involved in drafting?

I create all my own drawings and 3D renderings.

Q. What program do you use to prepare the lighting drawings?

I use VectorWorks Spotlight on a MacPro in my office and on a MacBook Pro away from it. Other critical software I use, apart, of course, from Richard's custom version of Catalyst, is Final Cut Studio (Final Cut Pro for video editing, Motion for motion graphics creation, and Compressor for outputting video content).

Q. Do you do any preproduction visualization programming to save time in rehearsal with the artist?

In this particular case, I did not use any since there were no moving lights at all, and I felt that for the purposes of demonstrating the concept my VectorWorks renders were explanatory enough when combined with demo'ing the custom version of catalyst on my MacBook Pro. For actual programming, no preproduction visualization software came close to being able to represent the VersaTube array, and the video manipulation was easier seen on a computer directly. Out of every tour I have designed, this one was perhaps the hardest one to imagine exactly in my head and it surpassed every hope I had for how it would look.

Q. In production mode, who calls the show and how many board operators does it take?

During the Radiohead show, I effectively call the show and run all the lighting and video via the grandMA. In addition, Andy Beller, my crew chief and Kinesys operator, triggers all the motor cues, and there are three camera operators at the side of the stage controlling the six remote pan/tilt/zoom cameras. Nick Barton is in charge of the master spreadsheet detailing all camera allocations for each song and cues within those. All the motor and camera moves were blocked and documented during rehearsals, although Radiohead are a true live band and no song is ever exactly the same as the night before so I call adjustments to camera shots, smoke levels, etc., throughout the show. My crew is absolutely amazing, incredibly dedicated, and professional and is able to cue themselves nearly all of the time.

Q. What were some of the nightmares, problems, things that didn't go as planned?

I think the only real major difficulties we had were at some of the European festivals where the stages were not really very friendly to the design, either because of their structure or because of the design of the rig already in them. Most of the festivals we played were very, very helpful and accommodating, but a couple were a little difficult. Outside of Europe, a lot of the festival appearances have actually been at events built around the production, so everyone has been really helpful. The only other slight disaster was in Milan, when one of our custom "set" carts containing two of the "Quins" (the upstage flying frames each containing five BB7 fixtures) somehow ended up disappearing off the edge of the stage. Fortunately, no one was injured, and amazingly apart from some seriously bent bits of metal the crew managed to get it all working again for the next show.

Q. Anything you want to add that would help other designers understand production at this level?

I think that one of the great successes of this tour from a production level was the return on

FIGURE 23.26 Bruce Springsteen and the E Street Band *Magic* tour (2007–2008). (Photograph by Todd Kaplan.)

Terrain, Inc., has been recognized as one of the most diverse and skilled design firms in the industry.

An active member of the Academy of Television Arts and Sciences, United Scenic Artists, International Cinematographers Guild (IATSE Local 600, as a director of photography), and the Illuminating Engineering Society, Jeff now specializes in all types of live entertainment being captured for television.

QUESTIONS

Q. Who actually hired you?

The tour director, George Travis, hired me over 20 years ago, after someone on the tour had recommended me.

Q. How many designs had you done for this client in the past?

Prior to this design, I have created eight designs for Bruce, including two solo acoustic tours and one with the Seeger Sessions Band.

Q. Was there a budget?

No, there's never been a stated budget for the Bruce tour, although I'm sure George has one in mind.

Q. What was the type of venue for the production (theatres, arenas, stadiums)?

This tour started in arenas and went into stadiums in Europe and the U.S. in the summertime.

Q. How many dates?

The tour did about 100 shows.

Q. Who was involved in the initial creative meeting?

It began with me and George initially and quickly involved Peter Daniel, the video vendor, since video had the potential to play a big part in this design.

Q. Was the artist directly involved in the design process?

No, Bruce was finishing the album and was not primarily involved. Our goal was to create some ideas for Bruce to comment on or, better yet, prepare a real-life mockup demo for Bruce to see—which is the best way for him to truly get an idea of what we're proposing. That is what happened, and in fact Bruce liked some aspects but mostly was cold on the concept, which guided our next steps with more clarity.

Q. Was a set or staging already proposed before you became involved in the lighting design?

No, although George did throw out some ideas he had in mind. Scenery is always an organic, committee process on a Bruce tour, between George, myself, and the production manager. I often take the reins initially to do a layout and start putting things on paper for our group to comment on. I do that because I can't really design much until I know more about the scenery—which for a Bruce

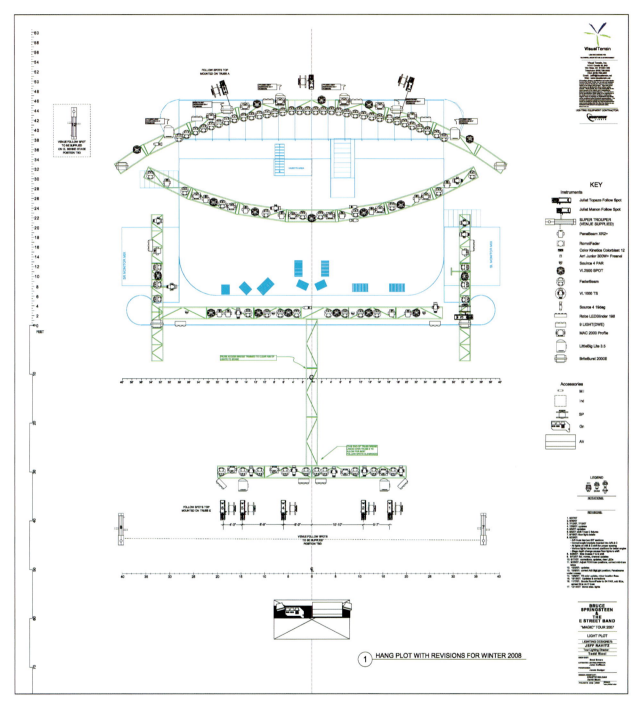

FIGURE 23.27 Bruce Springsteen lighting plot. (Design by Jeff Ravitz.)

tour is rarely more than a platform arrangement. So I'm usually in a fever to develop any semblance of a set so I can start the lighting design. I'll knock off several drawings in 3D and send them around to everyone. George Stipanovich, the production manager, then jumps in and starts to work on the drawing from a size and spacing point of view.

Q. Were other creative people brought in later?

George wanted Bruce to see what we were thinking about, but we know Bruce does not

Bruce Springsteen & The E Street Band
Magic Tour – Stadium Lighting Rig

Equipment

Automated:

83	Morpheus Lights **PanaBeam™XR2+** wash luminaire
80	Morpheus Lights **FaderBeam™** wash luminaire
20	VARI*LITE® **VL2500™** spot luminaire
16	Martin **MAC 2000 Profile** spot luminaire
16	Morpheus Lights **BriteBurst™2000E** wash luminaire
13	VARI*LITE® **VL1000™ ERS - TSD** spot luminaire
6	Zap Technologies **LittleBig™ 3.5 Xenon** wash luminaire

Conventional:

32	Philips/Color Kinetics **ColorBurst® 6** RGB LED fixtures
20	Morpheus Lights **9 Light**
20	Morpheus Lights **ColorFader3™ XLFader3™** CYM color changer
19	ETC **Source Four® PAR**
15	Morpheus Lights **ColorFader3™ MFader3™** CYM color changer
6	ETC **Source Four® 19° Elipsoidal**
2	ARRI **300W** *PLUS* Fresnel

Truss Spots:

6	Robert Juliat **Topaze** w/Morpheus Lights **RJE™** electronic ballast
3	Robert Juliat **Manon** w/Morpheus Lights **RJE™** electronic ballast

Control:

1	AVOLITES **Diamond II** console
1	MA Lighting **grandMA** console
1	MA Lighting **grandMA light** console

Rigging:

30	Morpheus Lights **30" FlipBox™** truss - 10' section
4	Morpheus Lights **30" FlipBox™** truss - 3' 4" section
3	Morpheus Lights **24" FlipBox™** truss - 10' section
14	Morpheus Lights **30" 8.4° FlipBlock™**- structural corner block
6	Total Structures **12" Medium Duty** truss - 8' section
4	Total Structures **12" Medium Duty** truss - 5' section
4	Total Structures **12" Medium Duty** truss - 5-way corner block
27	CM® **Lodestar®** chain hoist

FOH Spots:

4	Robert Juliat **Lancelot** (European Stadiums)
	Strong 4.5K **Gladiator IV** (U.S. Stadiums)

FIGURE 23.28 Bruce Springsteen equipment list.

relate to the computer-style renderings I produce. I'm not a set designer *per se*, and my rendering skills are limited. So, I brought in an illustrator to handpaint large-format renderings from my 3D versions. George also has a local friend who does pencil sketches, and he took the same 3D versions and redrew them in that style so Bruce could get his head into one version or another. Bruce had a few, not very helpful comments, so we really didn't know much more than before.

At this point, we were getting close to rehearsals and we still didn't have many answers or a set in construction. George Travis had to make

a decision, so we took the skeleton shape of our initial stage drawings, *sans* true riser fronts, which could give the show its own unique look. As has happened on every Bruce tour, the set really came together in rehearsals, which last about 3 weeks, thankfully. On my first tour, *Born in the USA*, we decided what we wanted scenically during production rehearsals, and then at the first show the production manager, Bob Thrasher, personally welded the entire thing together in the 2 days of setup we had before opening night. That's how it went, tour after tour. So, this one was true to form.

Bruce Springsteen and The E Street Band World Tour 2007–2008

Lighting Designer
Jeff Ravitz

Lighting Design Assistants
Kristie Roldan
David Mann

Lighting Vendor
Morpheus Lights — Redding, CA

LIGHTING CREW
Lighting Director: Todd Ricci
Automated Console Operator: John Hoffman
Crew Chief: Brad Brown
Fixture Technician: Travis Braudaway

Lighting Technicians
Carl Hughes
Bryan Humphries
Kevin Humphries
Troy Garcia

Programmer
Jason Badger

FIGURE 23.29 Bruce Springsteen crew.

FIGURE 23.30 Jeff Ravitz, lighting designer.

Once we got to rehearsals, temporarily using an old stage from a previous tour while the new stage was still under the knife at Tait Towers, we were at least all together. Our tour carpenter, Aaron Cass, is a whiz. He's got a solution for every creative challenge; he quickly understands what we're trying to accomplish and translates it into practical construction and touring terms, and he's creative in his own right. He also will not accept second best. As a result, a lot can happen in a short time during rehearsals, and the show begins to take shape rapidly. This is still frustrating for me, because I've got a lot on my mind at this point in rehearsals and I really, really wish the set were further along so I could … light it! But, alas, the mystery and wonder of the Springsteen alchemy have something to do with this last-minute race to the finish line.

We also brought in an old friend of mine, Bruce Rodgers, a talented and top-tier set designer. We wanted big ideas from Rogers, which is his strong suit. We also wanted him to help us with great scenic finishes for the riser façades. In the past, our fascias have been very homemade and looked it—fencing and other Home Depot materials. We made the best of it, but I really wanted this stage to have a quality appearance, while still retaining the funky Springsteen-esque look of the back streets. The album had a more finished sound than the previous few records—why couldn't the set follow suit, in order to give me a fun playground of surfaces and dimensional edges to light? Between me pushing for those light-friendly elements, Bruce Rodgers giving us some experienced input, Aaron communicating it to the set shop (and then ultimately going there to supervise), and George T. and George S. telling us if we could afford it and truck it, we ultimately had the best set ever for a Bruce/E Street Band tour, in my opinion.

Q. Who was the lighting contractor, and did you have any say in the choice?

This tour has had the same lighting equipment contractor since 1988, and I did participate in that original decision. Our previous contractor, Tait Towers, had sold off its lighting division, and at this time in history automated lighting was becoming somewhat *de rigueur*. I had just used Morpheus Lights on a John Mellencamp tour, with great success. Morpheus had a system that allowed for all the moving lights to store in the trusses for travel, so the setup was quite fast and we were able to have a system that was almost exclusively automated. Bruce's shows were legendary in length, and the automation allowed for incredible variety throughout the show—not necessarily with visible movement. The tour director liked the concept, so Morpheus got the tour and has been with us since.

Q. What was the timeline from first meeting through to first show?

The first talks began in March of 2007 and the tour opened in October of that year:

March 2007—George Travis calls to discuss an idea he has about a customized pan/tilt fixture utilizing LED technology. Conversations begin between me, Paul Weller of Morpheus, and Peter Daniel of Pete's Big TVs, the video supplier.

April and May 2007—Continue discussions and research into available technology and equipment.

June 2007—Meet Peter Daniel at InfoComm Convention in Anaheim to research screen and yoke technology. Begin first staging and scenic development. Research new follow spots. This tour has traditionally flown a follow spot bridge to maintain consistent follow spot angles and intensities from venue to venue. I wanted to upgrade the old Lycian Starklight spots we had been using in favor of a more adjustable follow spot. Juliat spotlight demo, Lycian spotlight demo, and Selecon spotlight demo. Handpainted scenic renderings created from 3D CAD drawings for presentation to Bruce. Tests and prep at Pete's Big TVs shop in Delaware for LED demo and mockup. Communication with audio vendor (Audio Analysts USA) regarding PA configuration to coordinate possible conflicts with lighting positions. First truss layout drawings begun; first-draft scenic and truss configuration drawings sent to tour director. Demo of Element Labs' Kelvin BRICK for possible use as uplight.

July 2007—Hippotizer media server demo at my office. We had initially wanted to integrate video into the set elements for this tour, and the Hippotizer was being considered as the engine to drive the video content. Scenic concepts becoming more defined. Band risers will be rounded in shape, inspiring and confirming ideas for curved trussing. Revised riser drawings are created to reflect rounded fronts. Other themed elements are introduced. Revised lighting truss configuration to reflect curved design are sent to Morpheus for feedback. Instead of fabricating a curved truss, Morpheus suggests an angled spacer between standard truss sections to create a curved shape, which offered more flexibility to vary truss lengths and greater truck packing efficiency. Engineering begins on angled spacer; these spacers will be under great stress in the rigging process and must be designed to withstand the maximum load. Begin industrywide search for lighting crew chief to replace previously departed one.

August 2007—At my suggestion, a full production team and vendor meeting/think tank takes place in Phoenix (most central location for all involved). All drawings are shared, equipment is demonstrated, more questions are raised, and some decisions are made. Continued evolution of light plot and scenic elements. Interface with tour lighting director and programmer. Riggers begin evaluating overall rigging plan. Complex overhead cable management design takes shape. Build of lighting system begins at Morpheus. Build of scenic elements, known at this time, begins at Tait Towers. Receive prerelease copy of new album. Listen to music and create detailed music and cue breakdowns as a guide for programming and for the tour LD and board operators running the show.

September 2007—Load-in to rehearsals in Asbury Park, NJ. Bruce begins to take an active interest in scenic elements. Onsite mock-ups of proposed riser fascias, colors, and other scenic choices are conducted for Bruce and us, and final decisions are made. New drawings are generated and final set construction commences. Daily band rehearsals. Notate onstage positions, movement, and nuances. Update musical notations to reflect actual live arrangements. Focus lights. Experiment with colors, gobos, focuses, movement. Run cues that have been programmed. Continue programming after band leaves. Ongoing system setup. Ongoing evaluation and resolution of production efficiencies and deficiencies. Load-out of rehearsal hall and move show to Continental Airlines Arena for final, arena-sized rehearsal with full lighting, PA, and video systems. Full setup of production. Determination by tour audio engineer that previously approved front-of-house lighting truss will, in fact, obstruct PA in a detrimental way. Entire curved-truss element of front-of-house truss is rebuilt onsite to be straight. Other curved trusses are retained, as designed. Dismantle entire front-of-house trussing and cabling. Redesign position of each light on truss in an attempt to retain carefully studied angles to light each band member. Rehang and recable entire truss. Lose 6 hours of valuable rehearsal and programming time. Continue focusing, programming, and rehearsing. Perform private rehearsal benefit to test show on live audience. Load-out of Continental Airlines Arena. Load-in to first stop on tour, Hartford Civic Arena, Hartford, CT. One day of setup and final tweaks. Opening night … *on the road.*

Q. How does this timeline compare to your other clients?

This far exceeds the pretour and rehearsal time I get for any other tour clients. For most others, we get a few weeks to a month or so of design and prep, and 1 to 7 days of production rehearsals.

Q. Was your first concept the one settled on?

Yes, the curved truss concept was one I had been toying with in my mind since the previous tour, and it was approved. At production rehearsals, it was altered a bit for logistical reasons, but the main components remained throughout the tour.

Q. If not, was the reason financial constraints or another consideration?

An idea that everyone supported, endorsed, and signed-off on was ultimately determined to be a problem for good audio coverage.

Q. How much time did you have with the lighting rig in place before the artist arrived?

We set up the lighting system at the rehearsal hall. It took two days to get things up and working, in the roughest sense, before the first band rehearsal. The band would come in around 9:00 a.m. daily. That's early for rock & rollers, but the drummer in the band works for a network talk show hosted by Conan O'Brien, and he needed

to be at the studio by late afternoon. That put us into a daytime schedule for band rehearsals. We could start programming by 5:00 p.m., and if we were making good headway we could be out of there around 1:00 a.m. to get some rest before the next day's rehearsal. Still, a long day, and if rehearsals had been only for a few days, we would have been forced to work a much more grueling schedule.

Q. How many days did you rehearse with the artist before opening?

We probably got around 14 days of rehearsal with the band. That's a generous amount, although I've heard of others that rehearse twice that, and this band has done more in the past. A great deal of this time is spent going through the oldest songs in the repertoire and trying them out for viability in the show. A large percentage of those songs never see the light of day, so that time, for me, doesn't result in as much progress as when we are rehearsing songs that we know are going onto the set list. On the *Born in the USA* tour, I counted 150 songs that were played over the course of the 18-month run. There are more, now. So, 14 days can get used up very fast.

Q. Do you use an assistant? How many people were involved in drafting?

Yes, I have an assistant who works with me throughout the design process. For this tour, Kristie helped me do a lot of the research we were conducting into new equipment and technologies. She also helped me draft. With CAD drafting, I will usually start messing around with ideas myself on my own computer. I may do a rough, unfinished plot on my own before I turn it over to her. At one point I want her to be the keeper of all the paperwork so updates or requests for documents can go through her for greater efficiency. I sent her to a class on advanced computer rendering, too, so she can really help when we need to generate those kinds of drawings.

Then, she accompanied me to rehearsals along with one other assistant, Dave Mann. Both of them have been with me on previous Bruce projects, and they know the drill. The cue book for this tour, as you can imagine from my description of rehearsals, is enormous. Old songs need to have their written breakdowns and cue sheets updated, and new songs need to be added to the book. We created a cue book for the tour LD, Todd Ricci; the moving light operator, John Hoffman; and for ourselves. The actual programmer, Jason Badger, who left shortly after the tour opened, didn't get a book. He was completely on his computer and the grandMA console. Nevertheless, it was hundreds of pages, times three. And there were a lot of corrections and updates over the course of the rehearsals and the first few shows, so Kristie and Dave were completely crazed and busy the entire time with paperwork and helping me keep track of everything during focus and programming.

Q. What program do you use to prepare the lighting drawings?

I use VectorWorks Spotlight with RenderWorks, which I just upgraded to version 2009.

Q. Do you do any preproduction visualization programming to save time in rehearsal with the artist?

On this tour, we didn't know enough about the show going into initial production rehearsals to benefit from pre-viz. However, when we made the transition to outdoor stadium shows, we knew we would only have a couple of days of setup, and most of that would be taken up with the job of simply getting the enlarged system in and working just in time for opening night in Dublin, Ireland. So, we availed ourselves of an ESP Vision studio in Los Angeles to make sure all the new lights were prefocused, colored, and cued into the show as much as possible. We sent the studio a 3D file of our stage, set, and lighting system. I do everything in 3D from the very beginning so I can really study the angles and also to have different views ready to send to the tour manager or others during the design process, so that part was already pretty close to what the studio needed.

The pre-viz for Dublin worked like a dream. We had almost no time onsite with the actual lighting system to really play around and program, so when we turned it on and tested our studio-generated cues, all we had to do was tweak the focus a bit for that particular stadium (since we created a generic stadium in 3D for the pre-viz sessions).

Q. In production mode, who calls the show and how many board operators does it take?

Our tour LD called the spot cues for 12 follow spots and operated a console that controlled all the conventional dimmers as well as the dousers for some of the searchlights. Our dimmed fixtures were still automated lights, but they had tungsten lamps, and we liked the response we got when they were controlled more manually, old-fashioned, rock & roll style.

A second operator worked the grandMA full-size automated console that controlled all moving lights positions, color, gobos, and intensities. The two operators had 4-inch-thick books of all the musical material we knew about that could possibly be played during a show, all arranged alphabetically. When the set list was delivered to the console platform minutes before the house lights went down, the two operators scrambled to pull those songs from the book and arrange them in the right order. Todd jotted down the conventional console memory page number for each song in big red numerals onto his set list so he could advance to that song quickly at the end of the previous song. Since the show order changed nightly, there was no good way to get this more preset and organized.

During the show, we had a dedicated intercom station backstage to someone who stayed close to Bruce's guitar technician. If an unplanned song got thrown in (all the time), the guitar tech was the first to know because Bruce would need the correct guitar for that song. Then, our backstage spy relayed the new song change to front-of-house and the two board ops ripped through the cue book, which was sitting close by on a side table, to find that song's cue sheet and music breakdown. Some songs may not have been played in a while and the music breakdown, showing details of verses, choruses, and solos, helped the two operators find their place quickly.

I would love to decrease the paper and put all this onto computer screens for them, but with the quick changes that occur, we haven't found a good way to access the right song as fast as flipping through the cue book, which is marked and divided better than a New York phone book!

Q. What were some of the nightmares, problems, things that didn't go as planned?

Three things I can think of: The first is the aforementioned crazy set list situation. Some nights Bruce didn't follow the set list at all so the boys were flying around all night trying to get their consoles to the right songs, cues. Sometimes Bruce fooled everyone, band and crew alike, when nobody could hear the "audible" he called out in the darkness between songs, so the first note of an impromptu song change could have been an enormous train wreck with everyone playing—or lighting—a different number!

The second major thing that happened on this tour was the initial approval and acceptance by every department of the curved front-of-house trusses, which also were our front follow spot positions. The two trusses were separated by some open space that had been designed to accommodate a center sound cluster, which never actually happened. The two curves created a graceful arc shape and were very wide. They tied in the visual theme of the curved trussing onstage perfectly and the entire lighting system had mass and scale to fill the void of an arena space. Everyone had the opportunity to see the position of these trusses in all the 3D views we showed them on paper and via computer files. However, in the band rehearsal hall, there wasn't sufficient space to assemble them completely, so when they were first put together, for real, in our second phase of rehearsals in an arena, the audio people felt they encroached on sound space to the point of altering the direct path of sound to certain parts of the arena. Sound *rules* on a Bruce tour, so our crew had to stop everything they were doing to dismantle the front trusses and reconfigure them to be a shorter, single, straight truss. A shorter curved truss was not going to work for many reasons. But, all the angles for each light had been carefully planned based on the shape, size, and position of the two curved trusses. Every light needed to be repositioned on the truss, which meant the beautifully arranged and measured cable harnesses had to be rebuilt. For speed, this was not done as carefully as it had been in the shop, so the tour traveled with a very messy setup for quite a while until a tour break allowed for a more thorough rebuild of the harnesses.

Finally, one issue that got very serious was the video I-MAG. When Bruce first used I-MAG, reluctantly, 24 years ago, it was simply to allow people very far from the stage to see him better. It was a live show and had been designed with no regard for television propriety, when the I-MAG came into our lives, and nobody really cared about perfect color temperature or angles. Now, all these years later, it had become very important to design the lighting with television in mind. It was stressed to me that skin tones needed to look natural, and the balance of foreground to background had to be right, all with attention to good dramatic lighting appropriate to each moment as we would normally do. I have spent the last third of my career specializing in television lighting, with a great deal of it for concert broadcasts, so this was not an alien task to me. However, this show was not predictable, neither in its show flow nor its stage blocking, so accidents and unplanned moments often occurred when two people got into the same light for a few seconds, causing temporary overexposures, or a colored show light got onto someone's face. This turned out to be a subject of much discussion, consternation, and redesign despite enormous care being taken to approach this show as seriously from a television standpoint as any television special, or live rock show might.

Q. Anything you want to add that would help other designers understand production at this level?

The evolution of touring production is continuing to grow towards the integration of lighting, video and scenic elements. The lighting designer already has become the lighting *producer*, conducting, if you will, the operations of several board operators and cue callers. Programmers have become strong designers and contributors to the overall cuing process, and there might be two or three programmers on one show, plus a video programmer. The lighting designer oversees all of these, makes macro decisions and although always involved in every decision, may allow the micro decisions to be handled largely by others on the design organizational flow-chart. This is the only way to get through this process, because it's simply gotten so huge.

Q. Anything you want to add that you feel will be of interest to the readers?

The importance of video art and technology cannot be minimized. Digital lighting fixtures that are actually projectors in a pan and tilt yoke will be the luminaires of the future, eliminating metal and glass gobos and the color mixing systems we use today. These fixtures will naturally prompt the desire to have more complex video content created.

Video scenery, using high- and low-resolution screens, projectors, and LED fixtures and components that are themselves scenic and lighting elements all in one, will become more and more affordable and will become integrated into the smallest of shows. The lighting designer often is charged with the responsibility to create the content and operate it simultaneously with the lighting cues. However, this will continue to become even a more separate area of expertise, with the lighting designer being responsible for the overall visual presentation and managing other designers, lighting and art directors, and content creators.

Lighting designers who have not been trained or gotten experience with proper video lighting will need to acquire these skills, as I-MAG becomes more and more like a broadcast TV experience. The expectation will be to design live shows so they look great onstage to the live audience but will also make visual sense onscreen. There is a fine line to be walked, here, and all designers must begin to approach show design with this in mind.

In designing the Bruce Springsteen and The E Street Band *Magic* tour, we kept all these considerations in mind. Even though a Bruce show emphasizes the band performance and strives to never overtake it with production elements, we attempted to keep pace with contemporary technology and aesthetics and bring a large-scale, exciting production to the fans, all the while remembering that our job is to make sure the focus is on Bruce and the band at all times.

24

DESIGNER'S PERSPECTIVE: ART VS. BUSINESS

T hose of you who have not read the first two editions of this book will be well served to go back and read this chapter, particularly because it will give you insight into five leadings designers' perspectives on their work and imagination—Leo Bonomy, Chip Largman, Jeff Ravitz, Peter Morse, and Willie Williams. Each one is an extremely individualistic, creative, self-made entrepreneur. I wish there was space in this edition to present all of the wonderfully insightful thoughts these people expressed, so I encourage you to borrow or buy a used copy of these earlier editions. I am sure you will agree with me that this may be one of the most important sections of the book. Ten questions were put to the four designers featured in the previous chapter. Because Richard Pilbrow is not exactly in the center of the concert field, his responses are from a slightly different perspective but are still valuable to concert lighting designers. I think you will find the designers split on several issues, such as education and being a technician first.

ANSWERS FROM FOUR DESIGNERS

The following are general questions not specific to the productions highlighted in the previous chapter.

Q. What is the most important consideration when you first approach a design situation?

Paul Dexter

To see that the artist and management are presented with a series of presentation ideas that will visually help them understand precise design direction. It is similar to a feasibility study in business. It creates a foundation to work from. Once I have confidence that my design has been clearly communicated and the artists and management sign off on it, the real work to develop that idea can be started. I know that I am not wasting my time or theirs, and together we accomplish that because we have established a clear direction first.

Richard Pilbrow

I see my job as assisting the author/director/designer to *tell the story*. I try to use light dramatically to evoke time, place, and atmosphere … to bathe the actors in the specific light of the character to underline the progression of the tale.

Andi Watson

Understanding a client's needs and desires and balancing those with productions. I always try to talk personally with a band before I even start doing sketches or drawings. If I am designing a concert production I immerse myself in the music completely, listening to it until following it is an almost unconscious action. I can then have a meeting with the band and talk in terms of their music and their world as well as in more general ways. For me it is important to create a performance environment as perfect as possible for that particular client as opposed to something that is generic. Of course, there are also the needs of the production to take into consideration which is where the budget, truck space, freighting needs, routing, venue types, crewing levels, etc., etc., all have to be brought into the equation.

Jeff Ravitz

It's a number of things, honestly, and it can vary by the situation. So, the answer is that "my list" is what I call to mind when I start a design. All of these must be considered as a whole in evaluating the first step:

1. Venues
   A. What size and types of places will this tour or show appear at? If it's a tour, there might be different kinds and sizes of buildings on the itinerary. A theatre tour could go into legit houses with diverse equipment and positions, an in-the-round place, a small arena that has no equipment, etc., all on the same tour. But, that's a worst-case scenario, but generally a show would be booked for theatres, arenas, sheds, fairs, or stadiums. This certainly guides my hand.
2. Style and content
   A. Type of music
   B. Type of presentation
   C. Band
   D. Featured performer with backup band
   E. Solo performer, etc.
3. Budget (this affects a lot of practical decisions)
4. Scenery
   A. Is there a set designer?
   B. Am I expected to be the set designer?
   C. Do they not want a set at all?
5. Is there video or I-MAG? Are there other production elements I must coordinate with, artistically and logistically?

Q. What one area of the design will you not compromise on?

Paul Dexter

There are always going to be design compromises, and I maintain flexible standards. I think that the challenge of design is figuring out alternative options when a problem with the first idea doesn't work. Any other way means that you are operating in controlled environments, that you can control people's whims, and that you can control the cost of those whims. In our business anything can happen! For that reason, a lighting designer needs to be flexible.

Not compromising, for me, is generally a reaction to a situation or a person who will try to convince me of something that I know better. I am very uncompromising when it comes time to the make decisions about the show. Someone might say to me, well, maybe we shouldn't put that scenic up today; we have a long bus ride tonight, and we want to get out of here quick. Absolutely not! I learned a long time ago that no one else really understands the importance of having all of the visuals operational during the show, besides the LD. If you think about it, everyone on the road has something to do with sound—technicians for guitars, drums, and amps and sound engineers. We are outnumbered, and the easy thing to do would be to acquiesce and say okay, let's not put that effect or scenic up—for whatever their reason may be. That reason is usually because someone else thinks it will be convenient for them not to, not for you or the integrity of the show. But, notice how you feel when the house lights go down and all of the sudden you are without all of your visual tools. Who is going to feel like the idiot? And, it is all because, at 9:00 a.m. that day you wanted to be accommodating to people who are thinking about unrelated reasons—like, a production manager who wants an easier load-out time or a sound person who doesn't understand your goals but wants to influence your show by suggesting that you eliminate something that may inconvenience them. That is one design compromise I will not make.

Richard Pilbrow

If the audience can't see the actors' faces when they should, nothing is right.

Andi Watson

I try to avoid compromising my artistic integrity and instinct. Generally, when I am designing a production, I have a very, very good idea in my head what the end result should look like and that invariably means that a certain video screen, a particular light, one type of fabric, or a specific motor control system will be the "correct" one to specify. Due to various factors, there is often pressure to substitute those items for others. I am always happy to investigate alternatives, but if they are not able to do what I need them to do I will fight very hard to resist. Often I will reduce the number of the "correct" components rather than accept something that will force me to change my design concept or my ability to create on stage what is in my head.

Jeff Ravitz

The element that I think makes or breaks a working design is the cuing: how the lights are used after everything is set up, focused, and programmed. Cues must be created for the proper visual composition of any given moment in the show, and the timing also must be perfect to advance to the next moment. An artfully executed cue can overcome the limitations of low budget or imperfect options. It doesn't work in the reverse.

Q. What is the most important factor in a successful design?

Paul Dexter

The most important factor is organization. Organization will be in the form of written documents and lighting plots that show the completed steps of your preparation and research. It is the final step before a designer will order the physical build of a lighting system. The build will be completed with the help of lighting vendors and freelance technicians. Mistakes and questions will be mitigated by good organization. The build will most likely find problems with your design or there may be substitutions. But, because they are organized with lucid written plans, it will help you find the time to field the inevitable troubles along the way. All the stars line up with preparation and research with organization. If you have to the best of your ability exhausted those areas and have properly organized with written documents and plans, you will have a successful design.

Richard Pilbrow

I believe in a seamless end result—scenery, costume, lighting, and sound integrated to support the story.

Andi Watson

Designing something that works on all the levels it needs to work on. Depending on the production, what constitutes a success can vary massively. It can be very hard to please everyone all of the time, if not impossible, but that is what we as designers are repeatedly asked to do. Remembering that everyone has their own perception of the production is important, and a lighting tech's view of a design is almost certainly massively different from the band's management or agent. To create something that everyone involved in is positive about, proud of, and wants to see is a real result.

Jeff Ravitz

That it works as a unique visual expression of the performance it is supporting—with *supporting* being the operative word.

Q. How heavily does the budget figure in your lighting design?

Paul Dexter

In my experience I have never been involved with a production that didn't consider budgets. Everybody is looking to take short cuts, and the concert touring industry is still one of the last great barter and bargain frontiers left in the business world. The closest comparison is probably used car salesmen. Hypothetically, if I were to present a design for $20,000, you could bet your last dollar that it will be countered with an offer of $15,000. The ironic twist is that some of the best designs follow the old adage that necessity is the mother of invention. In this case, it may be less money that was offered by the client, so what do you do? A good designer can think of an alternative that works as well as or better than the more expensive idea—there are still bargains out there. We are never working with a set menu of goods and services or have to wait for items to go on sale. Everything is negotiable! I look at budgets as necessary, but I allow for the initial negotiations to be countered with lower amounts. By the time all concerned parties agree, the budget is where it needs to be and can deliver the design I anticipated. It is all a big game.

Richard Pilbrow

It's simply a reality that must be coped with.

Andi Watson

I am very lucky in that budgets are less of an issue now to me than historically they have been. When I started designing, I worked on a lot of small tours where there was an almost nonexistent budget. Perhaps that was a good experience to have in that it taught me to appreciate that a light should only be on a plot if it has a reason to be there, but at the time it seemed very frustrating. Budgets always have a major impact on what you can or cannot do, and there can be a very fine line in between there being just enough or not enough to do what you want to do with the people you want to do it with.

Jeff Ravitz

It's all about choices, and sometimes a leaner budget forces me to think simply and pare the design down to its essence. The right design often comes to me when I know that all I can afford to do is exactly what is necessary and nothing more. That being said, it would be naïve to say my designs don't benefit from the advantages of having enough equipment and lighting positions, the best quality gear available, the most skilled crews—and enough of them, plus the time to properly design and rehearse. Those things take money, and although I've been lucky to create some magic on a shoestring, I can't stake my career and reputation on a shoestring. The important thing is to remember to keep to the essence even when the sky's the limit.

Q. What tells you a design has succeeded besides your own satisfaction?

Paul Dexter

When there are no further questions or additions. I let my designs evolve with the client. If I have included all of my input and experience and addressed all of the client's concerns and artistic contributions and there are no more further questions, the design has succeeded.

Richard Pilbrow

I always hope that the lighting will reveal and model the actor wherever he/she might be on the stage in an appropriate manner. Hopefully, every moment with every character onstage is modeled as it might be in a fine photograph or movie. That's the ambition for me. (And I hope the director and my design colleagues share my view.) Finally, of course, a positive audience response is what we're all working for.

Andi Watson

For me, a design is only a success if my client is happy, my crew are happy, the audience response is positive, and most importantly I am happy. If I am not happy with a design in any way it means to me that it isn't finished. It can sometimes be very hard to get to the point where you can take your hands off the controls, hand it to someone else, and consider it "finished."

Jeff Ravitz

If I happen to hear that a non-industry professional in the audience enjoyed the lighting, and that it suited the artist's performance, that gives me a pretty good indication of having hit the mark. It's to them, more than anyone, that I aim my design intentions. I want to be certain the performers are comfortable and happy with the design, but they don't see it like the audience does, and industry folks often overthink it. The real audiences just let it flow over them as part of the overall show, in the best of circumstances.

Q. Not withstanding the featured design discussed earlier, which would you prefer to deal with: artist directly, road manager, production manager, or business manager?

Paul Dexter

I prefer to deal with the person who is the nicest and most reasonable and has the power to make final decisions, but it doesn't always work out that way. Concert lighting design is an art that is multifaceted with technical and business parameters that involve the entire team and you have to communicate effectively with everyone. There are no superfluous positions on the road. But, making the artist happy is the ultimate goal, working in tandem so that lighting accentuates the performances, not overshadows them. If the artist is not good to work with or is disrespectful then I won't last long with that job.

Richard Pilbrow

Stage lighting for a Broadway musical is all about teamwork. You deal with management on business issues. In the case of TALE, we had two producers, Barbara Russell and Ron Sharpe, who were totally supportive. The director is in charge, and whatever you as a designer do must support and embellish his dream for the show. Obviously, in a musical, the music must be a vital driver to the handling of light.

Andi Watson

When you are working on a production design it is inevitable that you will deal with all of these people regarding the aspects of their work that affect you and *vice versa*. For example, from a purely artistic angle, I always try to collaborate with the artist directly, but I would never normally discuss my fees with them. Visas, travel, etc., but seldom found myself influenced artistically by their views on my choice of backdrop fabric. I think that once you are past the initial conceptual phase of the design it is vital to work very closely with the rest of the production team to ensure that their views and experience can be integrated in the realization of the project.

Jeff Ravitz

In theory, I prefer dealing with the artist. The concept for the show should come from what they've created—music, lyrics, or whatever they are presenting—and they should have the first say at what their show should look like. I want to establish a working relationship with them and have them be comfortable exchanging ideas with me. That sometimes doesn't work when the artist doesn't really have a visual sense. They either don't have any idea what they want or they go off on tangents wanting things for which technology doesn't exist—but they can't afford to fund the R&D—or they want things that exist but are beyond their budget. Sometimes they just want effects, things they've enjoyed at others' shows but may have no appropriate place in theirs. I'm fine dealing with the tour manager or personal manager when they can articulate what they, or the artist, wants. Some of them come from a production background and can be really astute and helpful in guiding me.

Q. Will you adapt (redline a drawing) a design to changing venues or do you feel that you need to start over?

Paul Dexter

Before a tour starts, an itinerary will be fairly revealing with regard to the venues. The design should consider any drastic alterations to the touring system that might be needed, with a reduce rig options such as an "A", "B" or even "C" alternatives so the need to start over would be rare—if at all. If that were the case, presumably you would have a console with a memory of the show. Simply assign the conventional lighting to the same patch numbers and position the moving lights as close as possible to the touring rig. Hopefully, there won't be too many of those types of venues, but when there are use as many effects as you can to perform the best that you can.

Richard Pilbrow

This show is pretty complex and if it should tour it would have to be adapted. I'd use the same principles, because they work, but I would start over in terms of the rig detail.

Andi Watson

I would only start over if there were a complete change of the tour. If only the venues are changing but the set list is staying the same, then the music, the emotion, the dynamic, and the resulting visual treatment will be very near the same. In that case, if I am happy with the existing design it would seem ludicrous to abandon it just for the sake of it. If it is deemed that an identifiable change is required for whatever reason then there are many ways of incorporating the same essence in a package that appears at first sight to be different.

Jeff Ravitz

I prefer to adapt a design that I believe has been developed and perfected for that particular show. If the act is positioned onstage roughly in the same relationship to one another, and if the music and performance are the same, then I will want to try to keep the core of the current design and make adjustments to suit the venues, as necessary—if possible.

understanding of rigging and how mechanics works is a valuable advantage when having discussions about hanging a system or fabricating custom set pieces. I think that for me personally, being a technician, then a programmer, then a designer was a wonderful way to observe how to do some things and how not to do others. When I left to become freelance, working as a designer on small tours with little equipment taught me about light and color and I learned from my mistakes as much as from my successes. Since that is the route I took it would seem to me to be an advantage not a disadvantage.

Jeff Ravitz

I do believe that it's a valuable experience to be a technician. It allows you to understand more of the flow from the power tie-in to the actual light doing its ultimate job. It is an opportunity to work with many working designers and to learn from them. And when that technician becomes a designer, he or she not only has a greater knowledge of how the whole process works but will also have a more genuine respect for the technicians working to bring the design to fruition.

Q. What do you, as a designer, feel is the most important personal quality needed to be a successful touring or Broadway designer?

Paul Dexter

Humility. If you are successful in this business at the level of Broadway or touring designer with major live productions, then clearly you have paid some dues. Your experience will shine through professionally. But, I have a very low tolerance for false pretenses or if someone is trying to pull the proverbial wool over my eyes. Conversely, I have a lot of patience for someone who is new or someone that shows enthusiasm and ambition. A designer at a higher level has some responsibility to treat others with respect and dignity. A lower professional position does not mean that they are not a good person, with a family or a story of their own. Not everyone that you meet will be able to excel and understand artistic levels of performance that a high-caliber lighting designer will or a professional crew will. Some people are just simple minded. That can often be mistaken as stupidity or a weakness, or that they are getting in the way. Try to distinguish those who don't belong because their character traits may be dangerous or even pretentious from those who are there for the right reason but may be of a simple mind. Always show humility.

Richard Pilbrow

For me, design is all about having a vision in your head of lighting's role in a theatrical venture and then possessing the ability to translate that vision onto the stage. Vision and the ability to employ light dramatically in the service of the play. Sufficient technical knowledge to get the right tools for the job (or have somebody invent them!). Political skills to survive in a very hectic and competitive marketplace. A cheerful disposition. A production period can be long, tension-filled, and arduous—you need to have mates around you.

Andi Watson

I think there are many important qualities to have. First, to care more than anyone else about the design that you have created. To have an independent creative vision and have the determination to fight for it when necessary. To be completely receptive to the world around you and take inspiration from everything you see, from the sublime gorgeousness of a sunset to the chaotic beauty of a foreign airport. To have the confidence to follow through on your instincts whilst at the same time listening and taking advice from others who can help you. Although ultimately as a designer it is your vision that you follow, it is never a solitary experience and can only work when you are part of a team. To have respect for those you work with. To always do things for a reason and to be able to explain your reason. The ability to ignore hunger and tiredness also seems to come in handy.

Jeff Ravitz

Maybe there is a difference between being a *good* concert designer and a *successful* one. To be a success in this field, which in this instance is defined as *getting the job*, requires a winning personality that has to be tempered to just the right degree. You will have to impress artists, managers, and other key decision makers that you are the best choice, and your portfolio may take second place to the impact of the vibe they get from the combination of ability, power, charm, and leadership qualities they sense. Exuding these attributes takes confidence that ultimately may be born of the self-assurance that comes from having ultimately pleased previous clients with great design work. However, poise can sometimes trump experience to get the contract.

After asking all the questions above, I felt that Richard Pilbrow's insight into the Broadway world raised further questions:

Q. Do you use the same assistant all the time?

Richard Pilbrow

Dawn Chang has been my assistant and/or associate since the early 1990s. I have used some others when she is not available, as was the case for "A Tale of Two Cities," when she was unavailable working in Macau for 18 months. Michael Gottlieb had been my associate on the "Magic Flute" in 1993 and some other shows.

Q. When was the first time you used WYSIWYG on a show?
"The Life," in 1997.

Q. When was the first time you used moving lights?
1993 for the "Magic Flute" at the Los Angeles Opera (VL5s), 1994 for Hal Prince's "Show Boat" (VL2Bs and DHA Digital Pitching Light Curtains) and "Busker Alley" (VL6s), and in 1997 for "The Life" (VL2 and VL5s plus WYSIWYG).

Q. How did you hear about Virtual Magic Sheet?
Eric Cornwell has been a friend and occasional assistant since the 1990s, I believe in "Show Boat." I started using VMS in 2004 on "Where's Charley?" at Goodspeed in Chicago. I'd certainly try to never light without it in the future. To me, VMS is a breakthrough. To be able to see on a single screen your entire plot, in color and changing in real time as the stage changes, is astonishing. No longer do you experience those dreaded moments when you wonder to yourself, "What the hell is that light over there?" It's immediately obvious and gives me unprecedented command over my resources—very rapidly.

Q. What program does your assistant use for paperwork?
My assistant/associates usually use Light-Wright file; I amend and alter as needed. In the theatre an assistant ensures it is up to date. LightWright is also used to patch up the Virtual Magic Sheet.

25

ARTIST'S PERSPECTIVE

Designing visual production is about selecting the right tools and special effects that will ultimately interpret the artist's music and image. The more that concert artists can express their images and convey their theatrical ideas, the more visual designers can understand which of the available tools and effects to apply, what colors are suitable, and how best to script the show cues. As you have read in Chapter 24, not all artists have the same level of involvement in their productions; when they do get interested, their input can vary widely.

WHAT TO EXPECT

An artist's comment can challenge a new design concept in a way that might not have been conceived by the designer. Busy with contemplating technical parameters, the designer is working with a point of view of making the production feasible by compromising between the wish list and keeping it real and within budget, satisfying practical illumination requirements, selecting appropriate types of lighting fixtures, keeping it consistent with the rest of the production's needs, and fitting it all into a truck. There's a lot to consider!

But, enter the artist, who now comes forward with an entirely new group of needs and most likely stuff that you would have never thought about. If the worlds of lighting and staging design and artist perception can happily collide, the upshot can elevate a good show to an even higher level of an extraordinarily great show! It is the visual or lighting designer's responsibility to make sure that this happy collision actually takes place.

Some artists may not be able to describe the specific equipment that will achieve the end result they are looking for, so it will be up to you to read between the lines with regard to what their explanation means technically. You are the expert, and they are going to rely on you to deliver. It will be your notes from the presentation meeting, ideas, prospective equipment list, and implementation of the artist's ideas that will finally convert concepts and research into working reality.

ENGAGING THE ARTIST

One important and often overlooked part of developing a production is how to engage the artist. There is a social divide between road crew and artist (I recommend reading *Roadie, A True Story*, by Karl Kuenning). If you don't believe that, notice who travels first class. Make sure that you say "hello" to the artist as you continue on back to your economy-class seat. That divide needs a bridge, and in this instance it comes in the form of help to arrange a meeting between the artist and the visual designer. Usually a member of the artist's management team will facilitate such a meeting.

Remember that a famous recording artist is living life with a totally different mindset. They will be coping with all the special circumstances that surround stardom—band member disputes, preparing for a show, being late for a rehearsal, a manager's last-minute coercion to attend a radio interview. Their lives are very different from those of the road crew or designers who are responsible for the technical and even artistic aspects of developing and controlling a visual live production. We all have different sets of problems and circumstances, but, make no mistake, you're entering the artist's world, not the other way around.

ESTABLISHING A RELATIONSHIP

There are some guidelines—mainly, knowing when to speak and when to listen. The best guide that you will have in almost any meeting with an artist is simply to employ good manners, poise, and common sense. Generally speaking, the study and diligence required to design concert lighting and organize equipment will not include a course in etiquette and intuition, but they are necessary tools to have in your toolbox.

Making a strong impression is not the purpose of this meeting of the minds. Play it straight. The days of rock & roll royalty have subsided, and today's stars are business savvy and ostensibly very open to new introductions. Don't get too familiar right away. Show respect, and keep the conversation focused. In time, things will loosen up, but only after you establish yourself and become part of that touring family. There is some work to do first, before you gain the trust of the artist and are viewed as a valuable contributor to the team.

Develop ideas with the artist by explaining some of the initial ideas that you have. It is always best to show ideas using visual aids; if the artist has not been paying much attention before the visual aids came out, chances are this is when the ideas will begin to flow. You could walk out a hero, and your next step will be ordering all the things that were on your list, with only a few minor and easily achievable suggestions to fulfill.

HANDLING REJECTION

On the other hand, the ideas that you show may be immediately rejected. Regardless of how much you thought about it and prepared your presentation before entering the meeting, artists know what they want. Be prepared for a no-uncertain-terms, candid response. If that's the case, get past that initial feeling of intimidation right away and ask questions about what they might like to see. Keep in mind that it was your presentation, even though it was rejected, that was the catalyst to stimulating this new direction! Be sure to seize the opportunity while you are with the artist to get as much information as possible, right then and there. Consistent with the social divide, artists are not usually accessible for subsequent casual meetings. Follow the ideas that the artists are trying to express, take lucid notes, and endure the rest of the meeting with good spirits. When you leave the meeting, don't despair. Give it some time. Consult with others if that helps. Chances are, though, that with concerted efforts and some time to work through their ideas and mesh them with your own, it can actually be fun to puzzle together another presentation. Something completely unexpected and exciting will likely emerge.

THE ARTIST'S PERSPECTIVE

Concert lighting, effects, staging, and high-definition video media play a significant role in the live show, not only visually for the audience but also psychologically. As much pride as we may take in our work as designers and technicians, it does not compare to what the artist is experiencing. The artist is the one on stage. The artist is the one exposed to potentially millions of loving concert goers and, oftentimes, caustic reviews. There is a lot at stake, and most of the time the artist wants to be assured that the show is the best it can be. What the artist might say when planning and performing their live presentation is anyone's guess. Each situation is different, and every artist is *very* different. Speaking directly to the artist about the process of creating a live production should bring a real insight to designers and crew as to how much or little an artist can become involved in their show. This chapter presents interviews with performers and recording artists who are known for their elaborate stage productions.

KEVIN CRONIN, LEAD SINGER, REO SPEEDWAGON

Kevin Cronin (Figure 25.1), lead vocals and guitar for REO Speedwagon, has written the number-one

FIGURE 25.1 Kevin Cronin. (Courtesy of REO Speedwagon.)

hits "Can't Fight This Feeling" and "Keep On Loving You," as well as the classics "Roll with the Changes," "Keep Pushin'," "Time for Me to Fly," "Riding the Storm Out," and "Take It on the Run." REO Speedwagon's self-titled first album debuted in 1971; later albums included *You Can Tune a Piano but You Can't Tuna Fish* and *Hi Infidelity*, which sold 10 million copies. The band had sold over 40 million records altogether. REO Speedwagon continues to record new CDs (e.g., *Find Your Own Way Home* in 2008) and to film live concert DVDs (e.g., *Live in the Heartland* for Soundstage in 2008), in addition to creating impressive live productions on tour for sold-out performances worldwide (Figure 25.2).

ESTABLISHING IDEAS

Q. When you are aware that a new tour is in the works, when do the stage production ideas begin to form for you?

Ideas come all the time … sometimes seeing another artist's show gives me an idea of how I might take that concept to another level, sometimes a new song brings on a production idea, sometimes I am bored silly on a transcontinental flight and begin imagining our stage set for next year.

FIGURE 25.2 REO Speedwagon live. (Photograph by Kenny Williamson, courtesy of REO Speedwagon.)

Q. Do you watch other shows or concerts to stay current on what is out there, or do you rely on your lighting or production designer for new technology updates? Or is it a little of both?

I would say a little of both, but when you form a relationship with an LD/production designer, and you trust his or her vision for your music, the whole process becomes much more enjoyable.

Q. Does the band do the research and discuss a concept first and then ask the designers and production team to develop those ideas? Or, would you say the designer presents ideas to you first and you develop them from there?

I never limit myself to one particular process for creativity. It matters not where an idea comes from; it is the quality, originality, and practicality of the idea that are most important to me.

Q. Do you have any preconceptions about new lighting and scenic designers when you meet them for the first time? What are some of the things that you hope this person is going to contribute to the live show?

I am very loyal to my team. When I form a relationship it takes a lot to make me look for new people. When I need to make a change I rely on a combination of good references, a common vision for our music and its presentation, and basic personal chemistry.

People That Create Ideas and Effects

Q. How important is it for you to see visual aids, ranging from rough sketches to 3D drawings, when a new concept is being presented?

I like to talk about ideas conceptually at first by trying to express something that I am seeing in my head or feeling musically. After I spend some time talking, I really appreciate seeing a computerized 3D mock-up of the idea. Back in the day, we just shot from the hip and hoped for the best. Now we can really see what our stage and lights will look like, complete with little stick figures of ourselves.

Q. Are you concerned with budgets first or ideas first?

I am always interested in ideas first, but the budget needs to be considered. I find it helpful to have a general understanding of how much we can responsibly spend on a tour production ... and then go on and exceed it, but not too much.

Q. Can you give a viewpoint on the advantages of a long term designer/artist relationship?

We have been fortunate to be working with the same production designer for a number of years now even though he is vastly overqualified for this position ... it helps to have incriminating video footage of your LD! REO has had the same manager, with no contract, since the band began in 1971, so obviously we understand the advantages of long-term artistic relationships. At the same time, there are instances where a band may want to shake things up simply for the sake of seeing what happens. But my experience is that, as long as I am evolving musically and the designer is operating at a high level, a long-term relationship with an LD allows us to grow together and improve upon what we did the previous tour.

Lighting to Convey Drama

Q. Let's say that the design is done, it is in rehearsal, and you like what you see. Do you leave it all to the designers at that point, or do you take an interest in further developing the theatrical presentation with them?

I enjoy the whole process of presenting a concert. I like to come up with new ideas myself, and I seek out the input of our LD ... after all he is actually seeing the show from the audience's standpoint. Again, for me it is all about teamwork.

Q. Do you watch the stage from the front at rehearsals, or do you use a video camera and watch later to see how the show looks?

I watch videotapes of our shows for ideas, I look at the stage from out front during rehearsals, and as the tour goes on I check out the look during sound checks.

Q. Do you like it when the LD comes to you with updates and new ideas to try?

I love when our LD comes up with new ideas as the tour progresses.

Q. Some will argue that too much going on visually is distracting to your live performance. Is there a line for you between enough and not enough?

For REO, we have always been pretty much a no-frills type of band, but I still like to give the fans the best visual show we can. There is a delicate balance between tastefully complementing the musical performance and overshadowing the band with effects. Each artist needs to understand where the line of good taste lies, and having a designer whom you can trust is a real asset in these decisions. A long-term relationship comes in handy in these situations.

In The End...

Q. Designers for many mid- to large-sized production tours generally don't tour with their designs but leave them in the hands of capable touring LDs after the show is up and running. Budget aside, would you prefer to have the designer stay on the road to continue to contribute his ideas to grow the production, or are you all right with having another LD stepping in?

It is always desirable to have the guy who designs the system also running the system, but sometimes it doesn't work out that way. We have been pretty fortunate over the years in this area ... like I said, incriminating videotape.

Q. Most effects and imagery happen around the stage with lighting air graphic patterns through haze and behind you with imagery and effects. How can you really tell, when you are on stage, if things are going right or terribly wrong?

It is impossible for an artist to ever really know what he or she actually sounds like or looks like to the audience. This is one of the great frustrations of being a touring band. We have to trust our sound men and LD implicitly. Again, the entire touring team has a job to do—to exceed the audience's expectations. When the band, stage techs, sound men, and LD are all working in sync, and having fun doing it, the ultimate winner is the audience—and that is how it should be.

Q. What would be your final most important bit of advice to anyone who wants to tour with a rock & roll band?

The touring lifestyle is very unique and can be extremely challenging. I don't think you can make it unless you are passionate about what you are doing and have a flexible attitude toward life. There are temptations at every turn, which

can be a lot of fun, as long as a person can keep the priority on the work. Getting hired to be on a touring crew requires equal amounts of ability and personality, so people skills are just as important as technical skills. Also, be prepared to throw your expectations out the window ... nothing can prepare you for life on the road. You listen and learn as you go.

RONNIE JAMES DIO, LEAD SINGER, HEAVEN AND HELL AND DIO

Ronnie James Dio (Figure 25.3), lead vocals and writer/producer for Elf, Blackmore's Rainbow, Black Sabbath, Dio, Heaven and Hell, pioneered classical metal work since Deep Purple's Roger Glover and Ian Paice spotted him in 1972. With his vocal pyrotechnics on two Black Sabbath albums, *Heaven and Hell* and *Mob Rules*, Ronnie drew rock royalty's attention during metal's heyday. With the 1983 release of his *Holy Diver* album followed by *Last in Line*, he was packing stadiums with groundbreaking

FIGURE 25.3 Ronnie James Dio. (Courtesy of Niji Management.)

stage productions. Ronnie reunited with Black Sabbath for the 1992 *Dehumanizer* album. The remainder of the 1990s saw Dio releasing *Strange Highways* (1994), *Angry Machines* (1996), and a live offering, *Inferno/Last in Live* (1998). 2002's *Killing the Dragon* was a classic Dio masterpiece, and the *Evil or Divine* DVD was filmed live in 2002 at New York's Roseland Theater. A tour with Deep Purple and the Scorpions followed. Today, Ronnie Dio writes, records, and tours worldwide with the classic reformation of Black Sabbath: Heaven and Hell featuring Ronnie James Dio (Figure 25.4).

ESTABLISHING IDEAS

Q. When you are aware that a new tour is in the works, when do the stage production ideas begin to form for you?

The ideas begin for me while performing during the current tour that I am on. The current show is always a great template for what's to come next.

Q. Do you watch other shows or concerts to stay current on what is out there, or do you rely on your lighting or production designer for new technology updates? Or is it a little of both?

I rely on the lighting/production designer for what new innovations are available. My technical expertise is limited.

Q. Does the band do the research and discuss a concept first and then ask the designers and production team to develop those ideas? Or, would you say the designer presents ideas to you first and you develop them from there?

Everything starts with the designer's research. After being informed as to new and interesting concepts I am able to incorporate them into my own weird perspectives. A great part of the stage presentation is involved with the album art. So, right away I have a path to follow. The designer and I are can now collaborate within the same world.

Q. Do you have any preconceptions about new lighting and scenic designers when you meet them for the first time? What are some of the things that you hope this person is going to contribute to the live show?

Yes. I've been spoiled by one designer for most of my productive years, so any affiliation with a new one is a real trauma for me. That designer is Paul Dexter and so I'll avoid any new traumas and "dance with who I brung."

People That Create Ideas and Effects

Q. When meeting a lighting or production designer for the first time to discuss tour plans, are you able to tell right away that they are going to

FIGURE 25.4 Heaven and Hell live. (Photograph by Lewis Lee.)

be a good match, and have you ever made a decision not to take them with you on tour, then and there?

Yes. I have been able to tell right away.

Q. Suppose you had to choose between taking a lighting designer on tour that very clearly demonstrates great technical knowledge but you are uncomfortable working with this person (for whatever the reason) or hiring a lighting designer who knows less than the other person but shows more ambition and enthusiasm. Whom would you choose?

For a start, I wouldn't work with someone that I wouldn't like, but if the person that I did like had the capability to become better and I could be a help I would use him right away.

Q. How important is it for you to see visual aids, ranging from rough sketches to 3D drawings, when a new concept is being presented?

I really like to see it—up close and personal.

Q. Are you concerned with budgets first or ideas first?

Always ideas first. A good designer can solve most of those monetary problems.

Q. Have you ever had an idea that met with resistance from designers who should know how to put it together?

I have never had resistance to an idea, because luckily I have always had control.

Q. Do you change designers often, not necessarily by choice, or have you had the same designer for years?

I try never to change the designer that I like.

Q. Can you give us your view on the advantages of a long-term designer/artist relationship?

To me it's the communication factor. Once again, I must have a relationship with the designer and that relationship means everything.

Lighting to Convey Drama

Q. Let's say that the design is done, it is in rehearsal, and you like what you see. Do you leave it all to the designers at that point, or do you take an interest in further developing the theatrical presentation with them?

I insist upon having theatrical input. Coming from a frontman's perspective, I must take an active interest in what an audience sees and what I think they should see from me.

Q. Do you watch the stage from the front at rehearsals, or do you use a video camera and watch later to see how the show looks?

I trust the eyes of the designer and others who have input. I don't need to watch what I'll never see as an audience member.

Q. What about after you are on the road? Do you take an active role in working with the lighting designer to try to stimulate more ideas to improve the live visual presentation?

I've always insisted upon a conversation with the LD following each show. Each show sparks my imagination, as I feel it should to the LD. So far, that's always been the case.

Q. Do you like it when the lighting designer comes to you with updates and new ideas to try?

I love it.

Q. What is your best experience with visual excitement on stage?

Pyrotechnics. Everything else from a stage perspective is just flashing lights and deafening music.

Q. What is your best dramatic, show-stopping performance moment?

Extreme contrast. Black and white rocks. The rush of being so completely in sync with lights and music is an incredible bonding and dramatic moment between audience and band.

Q. How about your worst?

Mistakes. When you have an elaborate production that is supposed to work every night, and usually does, it's the cock-ups that are always most impressive, and believe me there have been too many to document.

Q. Is there a right way and a wrong way to use video content, moving graphics, or live feed during your concert?

Only if they get in the way of musical presentation. I find them sometimes to be just silly diversions.

Q. Some will argue that too much going on visually is distracting to your live performance. Is there a line for you between enough and not enough?

I think that if you let the music dictate the performance then there won't be a distraction problem.

In the End …

Q. Designers for many mid- to large-sized production tours generally don't tour with their designs but leave them in the hands of capable touring LDs after the show is up and running. Budget aside, would you prefer to have the designer stay on the road to continue to contribute his ideas to grow the production, or are you all right with having another LD stepping in?

I think the designer who is worth his or her salt would be much more productive at the drawing board. That person has already gone through the hell that was necessary and deserves a break. Choosing an LD to execute the pre-planned ideas can always be difficult. But upon finding the right LD I think you should allow him to develop his own style.

Q. Most effects and imagery happen around the stage with lighting air graphic patterns through haze and behind you with imagery and effects. How can you really tell, when you are on stage, if things are going right or terribly wrong?

You can spot the obvious but are oblivious to the rest. Once again, the performers are mainly part of their own presentation and can't really see it from an audience perspective.

Q. Are there effects that you like to use on every tour that you update and use again? What are they, and what makes it worth repeating?

Pyrotechnics. They always work within the structure of extremely heavy music that I need to hear, and they always add the punctuation mark that is necessary. I think that it's the only effect that most of we stupid musicians can understand. Most of the time, we don't understand the incredible subtlety of the designer/artist.

Q. What would be your final most important bit of advice to anyone who wants to tour with a rock & roll band?

Do your job and keep quiet! The road is a great teacher, but school's never out.

GIL MOORE, VOCALIST/ DRUMMER FOR TRIUMPH

Gil Moore (Figure 25.5), vocalist/drummer and one third of the Canadian hard-rock power trio Triumph, has been writing positive-perspective lyrics and songs with guitarist Rik Emmett and bassist Mike Levine since the band formed in Toronto, Ontario, in 1975. They defined and epitomized arena rock with outstandingly produced live shows as Moore insisted on utilizing state-of-the-art lighting, laser, and pyrotechnic effects. Triumph's steady climb toward global notoriety was fueled by nonstop sold-out tours and breakthrough records: *Rock and Roll Machine*, *Just a Game* ("Hold On"), *Progressions of Power*, and *Allied Forces* ("Magic Power" and "Fight the Good Fight"). *Thunder Seven* produced the hit "Follow Your Heart." "Spellbound" and "Somebody's Out There," from 1986's *The Sport of Kings*, became major hit singles. Taking a hiatus from recording and touring, Gil created the Metalworks Production Group and the Metalworks Institute of Sound

FIGURE 25.5 Gil Moore. (Courtesy of Triumph Music, Inc.)

and Music Production, and his recording studio Metalworks was voted the number 1 studio for 12 years in a row at the Canadian Music Industry Awards. Triumph was inducted into the Canadian Music Hall of Fame during the JUNO 2008 Awards Ceremony by the Canadian Academy of Recording Arts and Sciences (CARAS). Triumph reformed to play at the Sweden Rock Festival in 2008, and more live touring is in the works (Figure 25.6).

ESTABLISHING IDEAS

Q. When you are aware that a new tour is in the works, when do the stage production ideas begin to form for you?

I think about production ideas all the time; it's just something that drifts in and out of your consciousness. I find it particularly useful to think about new ideas just before falling asleep.

Q. Do you watch other shows or concerts to stay current on what is out there, or do you rely on your lighting or production designer for new technology updates? Or is it a little of both?

The aspect of technical expertise can be overrated, and the people person aspect of the LD is incredibly important because the LD is the core of the look of the show and certainly a core member of the production team. I like to go with people that are good leaders and affable to work with in stressful situations and I think that is more important than technical knowledge.

People that Create and Implement Design Ideas and Effects

Q. How important is it for you to see visual aids, ranging from rough sketches to 3D drawings, when a new concept is being presented?

I don't think that visual aids are important at all if you are able to grasp concepts—you either have that ability to picture them in your mind's eye or you don't.

Q. Are you concerned with budgets first or ideas first?

I always like to start with the idea—to heck with the budget! There is more than one budget to an idea, depending on how the idea is scripted.

FIGURE 25.6 Triumph live. (Photograph by Peter Johannsen.)

Q. Have you ever placed your trust in a designer because of a great presentation and the band authorized him to build the set or lighting rig, only to end up with chaos?

We once built a monstrous truss (and this was before the era of moving lights). It was a moving light truss; each section had 27 lights in it, and every second light was a strobe light. They were able to pan and tilt and do all sorts of tricks, all running off an Apple computer. It was a great idea, but it cost a fortune, and it was really unwieldy. When we got it on the road, I think we used it for two or three shows and then that was the end of it. It was too big, too cumbersome, and too hard to utilize.

Q. Have you had an idea before that met resistance from designers that should know how to put it together? Did you pursue it anyway because you just had to have it? What happened?

We've had resistance from time to time about the size of a lot of the PAR can hangs that we've wanted in the upstage truss—we have always wanted that massive look in the rear. For a while, designers argued that it had been done to death, but we just looked at it like the bricks and mortar of our presentation—it always had to have that industrial-strength-looking upstage truss. We always got our way.

Q. Do you change designers often, not necessarily by choice, or have you had the same designer for years?

I don't think that it make sense to change designers all the time. Once you lock into someone that you work well with creatively, my advice to anybody would be to stick with your designer.

Q. Can you give us your view on the advantages of a long-term designer/artist relationship?

The advantage of a long-term relationship is that trust takes time to build up, and over time the artist trusts the designer and *vice versa*, and you start to be able to really collaborate in a special way.

Art and Using Lighting and Imagery to Convey Emotion and Drama

Q. Let's say that the design is done, it is in rehearsal, and you like what you see. Do you leave it all

to the designers at that point, or do you take an interest in further developing the theatrical presentation with them?

With Triumph, we never stayed out of the creative process, we never left everything (so to speak) to the designer. We always looked at it as a collaboration to keep improving and tweaking.

Q. Do you watch the stage from the front at rehearsals, or do you use a video camera and watch later to see how the show looks?

Many times I would watch the show from off stage. We would also use video cameras, and we brought our people out that were interested in us, like agents and so on, who would watch us and give us feedback about what they were seeing.

Q. What about after you are on the road? Do you take an active role in working with the lighting designer to try to stimulate more ideas to improve the live visual presentation?

Once the tour starts there is no reason to stop tweaking the show. We thought it was great to have a creative meeting at the end of shows or over breakfast on the road to try to come up with ideas on what was working well, what needed to be emphasized, and what needed to be cut.

Q. Do you like it when the lighting designer comes to you with updates and new ideas to try?

When the LD comes up with new ideas or updates it's always good. It is always good to try to make the show better, and there is no time like the present to just implement something.

Q. What is your best experience with visual excitement on stage?

Some of the best excitement when you are actually on stage performing comes from lasers—they tend to look phenomenal from on stage. Some of the other effects are best observed from the audience, but I always found that lasers look great right from on stage.

Q. What is your best dramatic, show-stopping performance moment?

I can't point to a specific spot in a performance that is the absolute highlight of a performance. All I can say is it's the gelling of the musicians that is unifying everybody's playing into one piston that is pumping over and over again in sync. Whatever that is, whatever that magic is—that pulse is amazing when it happens.

Q. How about your worst?

The worst thing about performing is when the band just cannot seem to sync up. It's like a bunch of moving parts that are disconnected. It happens to every band, I'm sure. It is no different with Triumph. It feels horrible.

Q. Is there a right way and a wrong way to use video content, moving graphics, or live feed during your concert?

I don't think that there is a right or a wrong way to use video, graphics, or I-MAG during a live concert. It is interpretive, and it depends on the artist and the music. It has to be tailored to the situation.

Q. Some will argue that too much going on visually is distracting to your live performance. Is there a line for you between enough and not enough?

I've never thought that it was possible to overshadow the music with the theatrics or the digitals; it is basically impossible to overdo it. With the exception of pyro, which sometimes, if it is overstated, can be too much.

In the End ...

Q. Designers for many mid- to large-sized production tours generally don't tour with their designs but leave them in the hands of capable touring LDs after the show is up and running. Budget aside, would you prefer to have the designer stay on the road to continue to contribute his ideas to grow the production, or are you all right with having another LD stepping in?

The trend for designers to assign the tour to a new LD once the tour is out and running is something that I've never been too enthused with. I like the LD working from beginning to end because he is such an important member of the team. I figure the LD is almost a band member.

Q. Most effects and imagery happen around the stage with lighting air graphic patterns through haze and behind you with imagery and effects. How can you really tell, when you are on stage, if things are going right or terribly wrong?

I don't really think the performers do know precisely whether things are going as planned. We really have to rely on the LD or someone else in the audience that we trust to tell us how the effects and imagery are coming off.

Q. Are there effects that you like to use on every tour that you update and use again? What are they, and what makes them worth repeating?

We've always liked to use lasers, and we use them differently on every tour. I think that they are worth repeating because they are so unique. Same goes for pyro; we've always been a pyro band, and it's always been identified with Triumph that we've had a fairly extensive pyro show, so I think the fans would be disappointed if we didn't repeat.

Q. What would be your final most important bit of advice to anyone who wants to tour with a rock and roll band?

The best piece of advice that I can give anybody who wants to go out on the road is you gotta have the attitude that you're going to stick with it. It's a little bit like going to war. It looks like fun for the first couple of days, but after you've been out there for 2 or 3 weeks it seems like 2 or 3 years, so you've got have a lot of guts to stay out there.

from a practical perspective. The ease of editing, the ability to capture the performance's ambience, and how much of the concert lighting can be readjusted are some of the considerations in choosing either film or video.

FILM

The logistical problems of filming live concerts often determine the choice of media. Usually, five to nine cameras are set up to cover the action of a live concert. Film cameras must be reloaded frequently, since 16 mm and many smaller 35 mm cameras can only take 400-foot loads, which translates to a maximum of about 10 minutes at 24 frames per second (fps). Even the most popular major motion picture camera, the Mitchell BNC, can only accommodate magazines up to 1200 feet, and that is while the camera is mounted on a tripod. For hand-held work, the magazine length drops to 400 feet. Some cameras can take 4000-foot magazines, but they are not generally used in this type of photography. A concert does not stop for film reloads; therefore, each camera is down for several minutes during critical performance time. Seldom are retakes possible in live performances, so any missed action is lost forever. Shots that the director counted on may turn out to be out of focus upon development of the film.

VIDEO

Video offers greater flexibility. A video director is in direct communication with each cameraperson and in real time sees the image each camera is getting. Duplication of shots can be avoided. The director sees the big picture, as he is able to view all the cameras at once. True, video assist is now common on motion picture cameras, as is intercom communication to the director, but the picture quality does not give an accurate representation of the final image or focus because the video assist camera does not have the resolution of the film camera lens. But, and possibly more importantly, video allows *online* editing, electronic cutting of individual cameras directly to a master tape or to large screens on stage during a live concert.

Depending on the budget and capability of the mobile video truck, the *program* (line cut) is double-recorded with time codes for *offline* editing use. The program is the real-time mixing of multiple video camera shots or taped feeds onto one master tape or onto a live line transmission. Offline refers to the time when the director can view the tapes without incurring the heavy cost of having an editor and expensive equipment in an editing suite, another cost savings. One or two *iso* (isolated) tapes are recorded for cutting in the postproduction editing session. Depending on the union situation, either the technical director, who also switches the line cut, or the assistant director will be responsible for switching cameras onto the iso feeds. Iso feeds are switched as straight cuts. No dissolves or fades are done on the iso tapes, but they are possible on the line cut, as are split screens and other electronic effects, which again saves time in editing.

Iso feeds are used for three reasons. First, they cover other action in case the online camera has technical problems or the cameraperson loses focus. Second, they allow the director to concentrate on the main action, although the director can ask for a particular shot to be isolated. Having that extra tape avoids forcing the director to make quick judgments on unplanned shots. Third, the iso feeds can also be used to lay in audience shots or other cutaways for the final edit. They can cover a composition error that is not seen until editing or can be used as the second image in a split screen or other effects on the final cut.

Each director has his or her own way of switching iso feeds. Some keep the wide camera on iso throughout the performance and switch only close-ups to the program feed. Others have the assistant director (AD) keep an eye out and switch cameras into iso that are not the same as the on-the-line (program) camera.

THE DEBATABLE "LOOK"

The "film look" is highly regarded as being the equivalent of our concept of fantasy. Video is often talked about as too slick and real life, too 5 o'clock news, for entertainment. You have to decide which image is right for your artistic goals. A cross-media trend is now developing in which performances are shot on film and then transferred to video tape for editing and viewing. One film format has an aspect ratio of 1.33:1 for both 16 mm and Academy 35 mm, the same that is used for television; however, wide-angle, Cinemascope-type lenses should not be used because too much of the image is lost when viewed on the home television receiver. It is hard for a cameraperson to keep in mind the television "safe area" when shooting, and that is why you see boom mics in pictures on television series shot on film more often than when they were shot on tape. Here, again, with video the director and technical director can *see* this happening, whereas with film it will not be apparent until screening the next day, when it is too late to go back and reshoot.

The National Television Academy technical specifications state that as little as 35% of the image on standard 35 mm film will actually be received in the home via television. The "safe area" for video represents about 70% of the total aperture area of the 1.33:1 aspect ratio of the film. You can see why it is so easy for critical action to be cut off in film viewed on television.

Sony markets an HD video camera outfitted like a film camera. The F33 is a multiple-frame-rate camera that has a price tag of $150,000. Because the more standard Sony 1500 HD used on the last Bruce Springsteen tour falls within the price range of $79,000, I think it is a good bet that the F33 comes closer to producing a film look than any other video camera.

LIGHTING CONSIDERATIONS

The inevitable conflict between the recorded media's lighting needs and the obligation to the live audience, who paid good money to see the concert, is always a tough fight. The concert lighting look has to be broadened for these other media. A compromise must be reached or the recorded product will suffer. Your job as lighting designer is to enhance the artist's image. If the tape or film is bad because you would not compromise your concert lighting, the artist is the loser.

When I first started doing rock video in 1972 as the lighting director for *Don Kirshner's Rock Concert* series, I had already logged 6 years of concert tour lighting. The concert lighting director is, in fact, the concert director. We direct the audience where to look; we produce the visual picture. In video or film, the concert lighting director becomes subservient to the director, who chooses the image, framing, and other shots to be recorded; therefore, the concert lighting must be broadened to facilitate these needs, something the live show usually avoids. The video lighting director has to use a broader brushstroke when lighting.

How can we best handle the live show so video or film and live audiences are equally happy? I feel there is no need for the concert to be completely relit. Rather, *balance* is the key. The best video cameras have a contrast ratio of 32:1, while film has greater latitude (from 64:1 to 128:1). Video gives you five *f*-stops, as opposed to film, which gives you eight or more. As a result, video lighting ratios should not exceed 3:1 in the overall picture balance. The 2:1 ratio is used to teach video lighting, but concert lighting can exceed this to help give the lighting that raw-edge quality.

WHAT THE CAMERA SEES

The best way to check how the camera will see the stage is to purchase a *contrast filter* for about $20. The Tiffen Company's model, which is widely used in film production, works well in video if you get it with a 2.0 neutral-density filter. Hold the glass to your eye and watch a live concert to see how much of the detail will be missed by the video camera's reproduction system. This same procedure can be used with film once the film stock and its properties are determined. A different filter, with the appropriate higher contrast, will accomplish the same thing for film.

Another factor of balance is color. The contrast filter will also show you how some color combinations are lost when recorded by the camera. Remember that the video camera has definite limitations. What the limitations are depends on the camera. The current king is the high-definition camera used for sports, most news broadcasts, game shows, and variety shows. Reality shows do not spend the extra money, mostly because they operate in low-light situations with smaller hand-held cameras or remotely operated surveillance cameras.

Also, if you have film or slides in the live show, they probably will not be bright enough to be recorded on the tape. Slides and film are best added electronically during postproduction through a variety of effects processes. Remember, you are lighting for a broader view of the show and what the eye sees is not what the camera sees with its very limited contrast ratio. Lighting on scenic pieces generally must be increased in intensity if they are to read properly. LED screens have pretty much eliminated this problem; even if the tour is using slides or film, it would be logical to switch to LED screens for the actual shoot so this issue of level is eliminated.

There are ways of compensating for light issues through the efforts of the video controller. The controller watches and controls, among other things, the iris of the video camera. However, you should not count solely on this ability to boost the level electronically to make the light intensity acceptable.

The Cutaway

An example of why cutaways are important is a 5-minute guitar solo that goes over great with the live audience but is dull when transposed to film or tape. Television viewers have very short attention spans and must be kept interested with visual images that add to the enjoyment of the music. Extensive studies have been made on how often to change images if you want to keep the television viewer's attention. The director must use cutaways such as the live audience's reaction to the solo or other band members' reactions to insert during the solo.

Monitors

You should request that a monitor be placed at your lighting console; ideally, it will come with a router that allows you to switch from one camera to another so you can see what options the director has. Ideally, you should have a separate monitor with the *line feed* so you can see what shot the director is using. Oh, and also demand an intercom that allows you to hear the director and even communicate with him. It will eliminate a lot of confusion. This will be a big help, and, depending on your console setup, you can stay one step ahead of the director by altering the balance of the cue. I suggest you make some "safe cues." Those are specials on each musician, the audience light, some broad set light, and anything else that might be able to fill in a hole when you see that the camera is looking into a black hole and you can't alter your active cue sequence.

Balancing Foreground with Background

Camera sensitivity to light has greatly improved over the years, especially with HD. The benefit for live performance filming is that light levels don't need to be so high that white light ambience wipes out all of your great stage looks and graphics. Do not be alarmed that the follow spot levels may be reduced significantly, as low as 30 to 40 fc. That is not much more than normal room lighting. The lower that the foreground level can be, the better for your background, especially for stage sets that are lit with more saturated colors or that are farther upstage. For brighter stage sets and backgrounds with bright LED screens, an intensity level will be determined to balance the performer with those types of special surroundings.

Key Light

Key light is generally thought of as a film term. The key light is the primary source of illumination from the direction that the camera views the scene most of the time. It is this source that very often has the lowest foot-candle (note that most film people now use *lux* to define light output) reading on the concert set. In television, it will often be the brightest.

In concert lighting, the backlight is usually the brightest. Because follow spots are usually the concert designer's only front light and because they cannot be trained on all of the musicians all of the time, it is unlikely that the camera will have enough illumination for other shots, or cutaways, on the drummer, keyboard player, or other individual band members. The lighting problem on cutaways can be solved in three ways. First, add follow spots. Not always an artistically justified solution, I know, but it will do when no additional fixed lighting can be added. Second, adding an even bank of front white light producing 125 fc will help. The video controller will be thrilled, but it takes away from the audience's interest in what is happening on stage. Third, try placing white light specials from the front on the drummer and keyboards and backup singers that can be dimmed up only as required. This will cause the least change in the live look while satisfying the video controller and the director's needs.

THE CLOSE-UP

At least one or two cameras will be fixed on the long shots to capture full stage looks, which are referred to as *safety shots*. If the director has the technical ability, he or she will keep a running record of this shot so that if at any time a close-up or the start of a guitar solo is missed they can cut to the wide shot and cross-fade into the shot they wanted when it is available. Given that most concerts are shot with a minimum of 5 cameras but more generally 8 to 10, you will quickly see the disparity of cameras on or near the stage. The chances are that some of your most proud moments of concert lighting drama will end up on the edit bay floor in the final DVD product because the director wants close-ups of the lead singer's looks of angst or a close-up of the guitarist's fingers.

Close-up shots will feature predominately in any long-form concert film, and a constant level of white balanced light for the performers is paramount for the overall presentation of the film. Follow spots and conventional luminaires or a moving luminaire with a fixed position for key light will normally suffice for long-form concert film shoots, just as long as the

angle is not too extreme. Saturated color on a performer for too long causes the viewer to get tired or can lead to eye strain. So, you may need to back off of the primary red on Alice Cooper some of the time.

CREATING BACKGROUND

There has to be some reference created for the film-maker to demonstrate distances that will give the viewer a sense of being there, too. Is the concert being filmed in a venue that is as vast as a stadium or in an intimate club? Between the stage and reverse shots into the audience, camera angles cover 360 degrees of wherever you are. Creating backgrounds with light or adding more scenery to light in addition to your normal road concert lighting rig may be a consideration for avoiding black and empty backgrounds, particularly to the off-stage right and left. Concert designers, both set and lighting, normally conceive the show as being viewed from a nice, neat 90 degrees from the front.

As an example, for an Elton John concert in Verona, Italy, at the 2000-year old coliseum, the stage set was dwarfed in comparison to the immense place. The seating was not sold behind the stage, so we built 4 60-foot-long trusses and laid them down on the seating, following the naturally steep incline of the stadium, and lit the neutral-colored bench seating as if it were a huge cyc. At each exit in the coliseum, we mounted a quartz light (45 of them total) above the portal, and the ruined architectural remains, high above the top rows, was dramatically uplit. No matter where the camera pointed—panning the audience and the historical monument or pulling back to a full stage shot—all bases were covered.

AUDIENCE LIGHTING

Audience lighting for taping or filming a live show is a must. Without it, the show might as well be shot on a sound stage where camera placement and lighting are optimal. What makes the performance *live* is the audience's reaction to the band. The producer or director needs that reaction on tape. A talk with the director will yield definite ideas on how to handle audience lighting. Generally, the discussion will focus on four questions:

1. Should the light be colored or white?
2. Is the audience light to be on all the time or just between songs?
3. Does the director want front, side, backlight, or a combination on the audience?
4. How much time and money can be spent to mount additional lighting for the audience?

I do not like to light an audience from the onstage angle as it puts constant light in their eyes and makes it difficult for them to concentrate on the stage performance; however, there are the wow moments or sing-along moments that any concert designer will take advantage of with "audience blasters." Backlight is great for showing the size of the audience on a wide shot, and sidelight will pick up enough faces to satisfy most directors without annoying the whole crowd. You see many audiences lit with colored pools of front light, especially in the back of the auditorium or arena. It looks great on camera, but it can be very distracting if you are the one sitting looking into those luminaires.

Just accept the fact that whatever you do is a no-win situation and make the best of it. The audience will hate being distracted no matter what you do, so you might as well give the video or film what it needs with as much consideration to the audience as possible. I recommend sidelight as being the best alternative, and it should only be brought on between songs and during a few fast numbers. Remember that the reactions can be cut in anywhere so the specific song being performed at that moment makes no difference as long as there is no reference to the stage in the shot.

ACCOMMODATION

Whatever the lighting director does to accommodate video or film on a live concert will have adverse effects on the artist and the road crew's normal operation of the tour, so try to understand the crew's problems and the pressures they are getting from the artist. If you go into this with an open mind and a willingness to cooperate, much can be accomplished

and an exciting show can be recorded that satisfies everyone.

LONG-FORM CONCERT VIDEOS

So you screamed you head off for 2-1/2 hours for your all-time favorite performer. You sang along with every song, probably out of key, but you didn't care. The weird and wonderful part about concerts is that when they are over, all you have is your memory of the experience. But, did you notice all those film or video cameras around, maybe a jib boom on a track running back and forth in front of the stage? Your experience may have been captured in a long-form concert DVD, including the "B" roll of the events before and after the concert. Every artist and manager wants to make a DVD of the concert, not only for archival reasons but also for commercial reasons. It is lucrative and could make it to cable television or a major network broadcast.

For most concert-goers, there is no substitute for the thrill of being a part of a live show. But, if the long-form concert video is as close as we can get, the concert lighting designer has a chance to move into another facet of the business without really trying. All it takes is to go on the road with a band and inevitably a camera crew is going to show up. The problem is, how can a concert lighting designer be prepared for involvement with a film crew and director?

The day of the concert, film people that you have never met before are going to come in and invade your world. Typically, the film company will bring a qualified film (or video HD) lighting designer as tangible insurance that all things in the basic film lighting areas are covered. Recognize that the visiting lighting designer is now the boss; you have been demoted. It is a bitter pill for some lighting director to swallow, particularly if the visiting lighting director is condescending or shows little patience. It makes the day far more difficult, but it happens. Accommodating the situation with the film director of photography (DP) or gaffer, whether you like the person or not, will be far easier if you understand some basic Film 101. Learn some film parlance.[1] There can be few experiences that will match the thrill of working on a film or video/DVD, more so because you will also enjoy some ownership of the finished product.

LIVE VIDEO

Changes made to the background lighting and color changes are usually not requested. The director normally wants to keep the live feel. Many times that may be created by the haze showing the air light on the wide shots, which gives this live look in the minds of directors I have worked with. However, the musicians need to be treated with some white light that is balanced for the camera levels. Balancing the intensity of the follow spots may require adding a *neutral-density* filter that reduces the output without altering the color. In some cases, the film lighting designer may demand that the follow spots be balanced for 3200 Kelvin, which is the standard for indoor video and most films. This can be accomplished if the gaffer or lighting designer has brought along a color meter that shows what the color temperature is. They should also have brought the appropriate correction filters with them, and you certainly should not be expected to have them.

If your show is using LED screens, then balance may be an issue here, also. Because most LED screens were designed for use outdoors, their color temperature range and lumen output are high even for a rock concert. Many are run at as little as 30% intensity. If the concert is being shot on video, the video controller will probably have a say as to what the balance should be between the follow spots, the stage lighting, and the screens.

[1]Blain Brown, *Motion Picture and Video Lighting*, Sec`ond ed. (Burlington, MA: Focal Press, 2008).

IV

THE AFTER-WORD

27

POSTSCRIPT: LOOKING BACK

The field of concert lighting has moved well beyond the narrow label of rock & roll lighting. In the past decade we have seen methods originally developed for tour lighting adopted by or adapted to dance, theatre, Broadway, theatre and dance bus and truck companies, television, theme parks, architecture, and, yes, even opera. Film is the latest to use trussing, multicable, and moving luminaires. The use of the PAR-64 luminaire and freer use of bold color media attest to the acceptance of some of these approaches.

Early concert-style rock & roll performances on television were accepted by the viewing public so well that the television people took notice, even if they did not jump in with both feet. I introduced the use of color projected on people rather than limited to the cyclorama on the *Don Kirshner Rock Concert* series in 1973. Now the PAR-64 finds wide use on location video. It is becoming rare to find a television studio where a PAR-64 or two is not in use. Now LEDs are showing up imbedded in sets. Moving luminaires are a staple on game shows.

I am not advocating the ouster of the venerable Fresnel, but I am an advocate for continued change. Change happens in two ways: through new inventions and, probably more realistic for theatre and its sister media, through the borrowing of techniques and equipment. Lighting has never been a heavily financed area of research in the theatrical arts; therefore, we must take what we can from other sources. One reason for the surge in new equipment is that the manufacturers have all these new venues that are buying their concert products.

Concert designers and their equipment suppliers have taken found space and created "theatre." When people are able to reach beyond the rock & roll label, they can see what real advances have been made, both in design potential and in the high-tech electronic explosion spearheaded by the intelligent, moving luminaire.

The prejudice against concert lighting has vanished. All entertainment media users are smart enough to know a good thing when they see it and when it is economically viable to use. Take the good elements and discard the bad; improve techniques and adapt them to other areas. This is the essential character of theatre, the great adapter.

This book reveals only a small portion of what concert tour lighting has to offer. The Bibliography provides the names of magazines and books about concert lighting. For those who need to broaden their basic knowledge of lighting, this list of resources is essential to making use of the full potential of the material discussed in this book. There are even schools that teach this high technology in graduate theatre programs or programs geared to the media. I hope you will look into these to add to your store of knowledge.

WHERE IS CONCERT LIGHTING TECHNOLOGY HEADED?

Concert, touring, and all theatrical lighting will continue to embrace digital technology and "green" energy efficiency. More manufacturers are developing

Follow spot High-power, narrow-beam light suited for long throws (typically 100 to 300 feet), generally with an iris, shutters, and color changer; it is designed for hand operation to follow the movements of performers.

Found space Any space used for theatre, concerts, or television productions not specifically designed as a performance area.

***f*-Stop** A measure of the light-transmissive ability of a camera or projection lens.

Gobo A pattern or breakup placed in front of a hard-edged light designed to cast a specific design or modeling onto a surface.

Grand tour A tour taken by a star, usually with only a piano, to perform excerpts of famous opera and classical works.

Groupies People, often girls, who are so devoted to a popular recording artist that they collect memorabilia, join fan clubs, and go out of their way to be everywhere the artist is.

Hanging point The point on the truss from which the attachments are made for rigging.

HD High-definition.

High-density dimming The use of advanced engineering and microprocessors to miniaturize the electronic dimmer.

HMI Mercury (Hg) medium short-arc iodide lamp (mercury/argon additives) in tubular, sometimes double-ended form at 5600 Kelvin or higher. In entertainment this is preferred definition. Manufacturers technical term is Hydrargyrum Medium Arc Iodide.

Hod A number of individually jacketed cables, usually three #12 wires, taped together for easy transport and layout on a lighting pipe.

HTI Metal halide short-arc lamp at 5600 Kelvin. Color temperatures are available up to 7200 Kelvin (daylight).

IATSE International Alliance of Theatrical Stage Employees, the international bargaining unit for stagehands, including property masters, audio technicians, carpenters, riggers, electricians, wardrobe mistresses, and other backstage employees.

Iso An abbreviation of the word *isolated*; in television it denotes the switching of a camera onto a tape reel other than the master tape.

Kelvin temperature The unit of temperature used to designate the color temperature of a visible light source.

Key light Motivating source of illumination that establishes the character and mood of the picture.

Key word Method of calling cues that does not rely on visual or mechanical signaling devices; an expression such as "go or out" prompts someone to react in a planned manner.

Keystone correction Compensating for off-angle projection of the image on a surface.

LAN Local area network.

Layering The use of color to create depth and separation by using different shades or saturations of a single color.

Letter of agreement A legal agreement generally written in plain language and in business-letter form rather than legal form.

Light show Mixture of theatrical lighting, projections, film, and blacklight images used to create an environment for the audience (popular in the 1960s).

Looks Planned patterns of light and often color combinations that are programmed and used one or more times in a concert.

Lumen per watt The number of lumens produced by a light source for each watt of electrical power supplied to the filament.

Lumens A unit of (light) flux.

Luminance The true measured brightness of a surface.

Luminaire The modern term for all theatrically designed lighting instruments.

Matrix or **pin matrix** A device used to group channels to one or more master controllers, often employing a small pin or patch placed on the console.

Media server Computer dedicated to the task of playing and organizing video clips, stills, or 3D images during a performance.

MMA MIDI Manufacturers Association.

Modeling Three-dimensional rendering of lighting system and components of the stage set relative to stage dimension and height exactly as they would be in real time.

MOR Middle of the road, designating music that is marketed between easy listening and rock & roll, such as Barry Manilow, Air Supply, and John Denver.

Multicable More than one complete electrical circuit housed in the same flexible protective jacketing.

Offline editing Preliminary editing of videotapes, using copies of the original tapes; usually performed on a low-cost 1/2-inch editing system that allows the director to try various edits before the final edit without expensive edit bay and editor costs (often called a *work tape*).

Online editing Final editing of videotapes using the original master tapes to produce a finished program.

Package dimming The grouping of typically 6 or 12 dimmers in a single, integrated enclosure, often used for portable work.

PAR-64 Parabolic aluminumized reflector luminaire used with a quartz lamp sealed behind one of a group of lens configurations. The "64" designates the diameter of the lens in eighths of an inch.

Patching Interconnecting dimmer output to loads via a patch panel, a system that usually uses a plug and jack on the line side of the load only.

Per diem Daily allowance paid to the crew and cast by their employer while they are away from home; it is used to cover nonreimbursed expenses such as food and laundry.

Pickup point The point of attachment to the building's load-bearing structure used for rigging.

Pigtails Short 3- to 5-foot cables, usually #2/0 or #4/0 welding cable, used to connect to the house power panel.

Pink contract An agreement issued by IATSE to qualified members who want to travel outside of their local jurisdiction.

Pin matrix *See* Matrix.

Pixel mapping Video scaling technique used in display devices. A monitor that has been set to 1:1 pixel mapping will try to display an input source without scaling it, such that each pixel received is mapped to a single pixel on the monitor. This will often result in a black border around the video unless the input resolution is higher or the same as the monitor's native resolution.

Port, Ethernet Most common connection to LAN networks for high-speed computers. Looks like a slightly larger phone jack.

Prep The time devoted to organizing and packaging materials such as lighting or sound prior to the rehearsal or tour, usually in a location other than the first venue.

Programming The real-time mixing of multiple video camera shots or taped feeds onto one master tape or into a live line transmission.

Proscenium An arch placed so as to separate the audience from the acting or performance space, often called a "picture frame."

Psychovisual Relating to the psychological use of color to effect emotional response in the viewer.

Punch light A luminaire with a highly concentrated beam that allows long throws.

RAID (redundant array of independent disks) A disk subsystem that is used to increase performance or provide fault tolerance or both. RAID uses two or more ordinary hard disks and a RAID disk controller. RAID has also been implemented via software only.

Ray light Reflector with a separate 120-volt lamp, usually of 600 watts, which fits into a PAR-64 housing. This unit creates a very narrow beam of light, the same as an ACL but without the problem of matching 12-volt to 120-volt dimming.

RGBHV Red, green, blue, high voltage.

Rigging The process of installing lines or motors to support trusses, scenery, etc., in their required positions.

RJ45 Another name for Ethernet connector.

RMS Root mean square, a method of averaging a sine wave to arrive at a mean average.

Road company The authorized production of a play or musical that is produced in one city and then performed in multiple cities for limited engagements around the country.

Roadies An early term used for people who traveled with popular bands and took care of the musical equipment.

Saturation The purity of a color; the extent to which a color has been diluted.

SCR (silicon-controlled rectifier) dimmer Solid-state electronic device used to control lamp brightness by cutting off part of the cycle or the alternating current supply in a specific type of dimmer.

Scrim A gauze-like curtain that, when illuminated from the front, appears opaque; if light is brought up from behind the scrim, it becomes transparent.

SD Standard definition (standard for digital video disk recorders).

Set The order of songs to be performed, generally used for musical performances where no dialog or scripted words are recited.

Sidelight The source used to rim faces and model profile shots, often at or near head height.

Soft patch An electronic means of assigning dimmers to control channels without physically wiring jumpers on a terminal board or patch panel.

Span set Circle made of nylon web strands encased in a nylon covering that is wrapped around areas of load-bearing structures to prevent chafing.

TCP/IP Transmission Control Protocol/ Internet Protocol.

Theatrical smoke Smoke created either through a chemical reaction or combustion (often caused by heating an oil-based liquid until it vaporizes).

Torm Slang word for "tormentors," which are the side curtains that hang next to the proscenium opening; these curtains can be adjusted to alter the width of the performance area.

Trade usage Custom or widely used practice common to a business, usually unwritten but believed to be generally understood.

Truss Metal structure designed to support a horizontal load over an extended span. In theatre and concert work, it is the term applied to the structure that normally supports the lighting over a 40- to 60-foot clear span.

Tungsten-halogen lamp (quartz) Type of lamp design achieving an almost constant output and color temperature with a higher lumen output through the working life of the lamp. The halogen vapor (or bromine or iodine) facilitates a chemical recycling action, preventing the blackening of the bulb wall with filament particles.

Ward, E., Stokes, G., Tucker, K., 1986. Rock of Ages: The Rolling Stone Illustrated History of Rock and Roll. Rolling Stone Press, New York.

Wilkins, T., 2007. Access All Areas: A Real World Guide to Gigging and Touring. Focal Press, Boston, MA.

Woodbridge, P., 2000. Designer Drafting for the Entertainment World. Focal Press, Boston, MA.

Woodward, A., 2008. Stage Automation. Entertainment Technology Press, Cambridge, UK.

Magazines and Journals

Church production, sound and lighting for houses of worship. <www.churchproduction.com>.

Event solutions, event and tradeshow providers. <www.eventsolutions.com>.

Light & sound America, concert lighting, sound, projection. <www.fulcoinc.com/lsa.htm>.

Lighting & Sound International. PLASA's flagship publication for entertainment industry in England and Europe. <www.lsionline.co.uk>.

Live design, theatre, concert scenic, costume, lighting, sound. <www.livedesignonline.com>

Pollstar, touring talent. <http://pollstar.com> Pollstar also publishes yearly guides: *Concert Venue Directory*, *Concert Support Services*, and *Artist Management*.

Professional lighting & production, lighting, sound, production. <www.professional-lighting.com>.

Projection, lights and sound news (PLSN), concert, television, touring technology; sound; light; staging; projection. <www.plsn.com>.

Protocol, ESTA journal. <www.esta.org>

Sightlines, USITT publication. <www.usitt.org>.

Stage directions, theatre acting, technician, and teatcher. <www.stage-directions.com>.

Technologies for worship, houses of worship production. <www.tfwm.com>.

Theatre Design & Technology. USITT publication. <www.usitt.org>.

Non-Traditional Schools for Lighting

Full Sail University. <www.fullSail.edu>.

North Carolina School of the Arts. <www.ncarts.edu>.

Rigging Seminars, Inc. <www.Riggingseminars.com>.

The theatre academy. Los Angeles City College. <www.lacitycollege.edu>.

Annual Conferences and Shows by Associations

Catersource Event Solutions. <www.event-solutions.com>.

Entertainment Services and Technology Association. <www.esta.org>.

InfoComm. <www.infocomm.org>.

Live Design International (LDI). <www.ldi.com>.

Professional Lighting and Sound Association (PLASA). <www.plasashow.com>.

Projection, lights, and staging news. <www.plsn.com>.

Themed Entertainment Association (TEA). <www.teaconnect.org>.

United States Institute for Theatre Technology. <www.usitt.org>.

Index